Eylaf Bader Eddin

Translating the Language of the Syrian Revolution (2011/12)

Studies on Modern Orient

Volume 43

Eylaf Bader Eddin

Translating the Language of the Syrian Revolution (2011/12)

DE GRUYTER

ISBN 978-3-11-221402-2
e-ISBN (PDF) 978-3-11-076769-8
e-ISBN (EPUB) 978-3-11-076774-2
DOI https://doi.org/10.1515/9783110767698

Library of Congress Control Number: 2023944629

Bibliographic information published by the Deutsche Nationalbibliothek
The Deutsche Nationalbibliothek lists this publication in the Deutsche Nationalbibliografie; detailed bibliographic data are available on the internet at http://dnb.dnb.de.

This volume is text- and page-identical with the hardback published in 2024.
This book is published open access at www.degruyter.com.

Cover image: Seagulls Circle the Heavens' Gates, 2017, ink on paper, 50X65 CM.
From project Assemblage, Dark Nights onto Rolling Waves by artist Mohamad OMRAN.
In collaboration with the author Odai Al Zoubi
This project was accomplished with the support of Ettijahat-Independent Culture and the Goethe Institut
Typesetting: Integra Software Services Pvt. Ltd.
Printing and binding: CPI books GmbH, Leck

www.degruyter.com

Acknowledgements

I dedicate this research to the Syrian people who believe(d) in the dream of a Syria free from the Assads.

First, I am very grateful to my supervisors: Prof. Dr. Richard Jacquemond and Prof. Dr. Friederike Pannewick. I would have been unable to complete my research without their constant support.

I am very grateful for the comments, feedback, and reviews from my colleagues and friends, in particular Simon Dubois, Cecile Boëx, Christian Junge, Georges Khalil, Anne-Marie McManus, Rachid Ouaissa, Laura Ruiz de Elvira, Lionel Ruffel, Anna Poujeau, and Rahaf al-Doughli.

I am indebted to the Friederich-Ebert-Stiftung (FES) for granting me a three-year award that gave me the opportunity to dedicate all of my time to turning my resulting doctoral thesis into this book. I would particularly like to thank Kathrein Hölscher, Beate Eckstein, and Yvonne Plenckers, the academic advisors at FES who listened to my concerns and helped me to overcome the professional obstacles I faced during the writing process.

Fieldwork, a travel assistantship, and part of the publication costs were supported by ZEIT-Stiftung Ebelin und Gerd Bucerius. Without its generous support, I would not have been able to investigate, conduct interviews, or travel to academic events in 2014–15. I am very grateful to Jane Bartels and Dr. Anna Hofmann. Other funding was also received from the Marburg University Department of Arabic Studies and The Center of Near and Middle Eastern Studies, the Carl and Charlotte Schott Stiftung, and IREMAM (UMR 7310, CNRS & Aix-Marseille Université) at Aix-en-Provence, France.

I am very thankful for my colleagues and friends at the universities of Aix-Marseille and Marburg, Christian Junge, Alena Strohmaier, Malte Hagener, and Elyze Zomer, for the friendly support they gave me while I was writing.

Some of the arguments set out in this book were first published in Arabic by the Etijahat and Mamduh Idwan publishing house, after receiving the Sadiq Jalal al-Azm Award. I owe a great deal to their generosity. The final phase of writing this book was during my contract with SYRASP, ERC grant, No. 851393.

Last but not least, I warmly thank all of the Syrian activists and non-activists, academics, intellectuals, and writers who agreed to be interviewed and have helped to enrich the content of this research, especially Syrian Revolution Archive, Tamer Turkmane, and Creative Memory of the Syrian Revolution, Sana Yazigi and the team.

https://doi.org/10.1515/9783110767698-202

Acknowledgements

I dedicate this [illegible] of the Syrian people who believed in the dream of [illegible] Syria [illegible]

[illegible] to my supervisors, Prof. Dr. Richard Jacquemond and [illegible] have been unable to complete [illegible] their [illegible].

I am very grateful for [illegible] feedback, and [illegible] my colleagues and friends, in particular [illegible] [illegible] [illegible] Latifa [illegible] and [illegible]

[illegible]

[illegible] process.

[illegible] as a part of the [illegible] [illegible] support. I would [illegible] conduct [illegible] funding [illegible] Arabic [illegible] of [illegible] [illegible] HMR [illegible] [illegible] en [illegible]

[illegible] colleagues and friends [illegible] [illegible] Strohmaier [illegible] Elyas Zahra, [illegible] friendly [illegible] they gave me while [illegible]

Some of the arguments [illegible] this book were first published in Arabic [illegible] the [illegible] publishing house, after receiving the [illegible] Award. I owe a great deal to their generosity. The final [illegible] of this book was during my contract with SYRASE [illegible] grant No. [illegible]

Last but not least, I warmly thank all of the Syrian activists and [illegible] [illegible] and writers who agreed to be [illegible] helped to enrich the content of this research, especially [illegible] and Creative Memory of the Syrian Revolution [illegible] the team.

A note on transliterations and translations

I have followed the *International Journal for Middle East Studies* (*IJMES*) guidelines for transliterating Arabic. All short Arabic titles, words, quotes, expressions, and so on are directly transliterated and followed by the translation in brackets. However, when using Arabic words or phrases longer than three words, the Arabic script is used instead, followed by a translation. When I refer to the same source published in Arabic and English, the transliterated authors' names refer to the Arabic source and the non-transliterated names refer to the English translation. In addition, all names, excepting authors, are written according to the English spelling preferences as per their personal social media accounts or as otherwise known, such as Hafez al-Assad, Bashar al-Assad, Samih Choukair, Wael Kfoury, and Najwa Karam, etc. If there is no English spelling of a person's name available online, I have transliterated it. All translations are my own, except when citing or quoting another source. In order to retain the pronunciation and the musicality of songs and some slogans, they are transliterated as they are pronounced in their songs/slogans in Syrian colloquial Arabic rather than converted to Modern Standard Arabic.

https://doi.org/10.1515/9783110767698-203

Contents

Appendices

Introduction

لا أجهزة تنصت في الوطن العربي في أي مكان، لأنه في الأصل لا أحد يتكلم.
محمد الماغوط من "سيّاف الزهور"

> The Arab world is astonishingly devoid of bugs [covert listening devices] altogether, simply because nobody speaks there.
>
> Muhammad al-Maghut, *Sayyāf al-Zuhūr* [The Flower Headsman]

In September 2011, after working with activists designing and writing slogans, banners, and other revolutionary materials for use the following day, I wondered what the impact of all of the work we had produced would be if it were translated into English or another language. This was the unconscious beginning of my interest in the differences in the revolutionary language once it was translated into English, and contributed to my decision in 2013 to write my master's thesis about the difficulty of translating certain slogans of the Syrian revolution. I first realized this in 2012 when I was interpreting an interview for a Norwegian journalist in Beirut. The activist said enthusiastically in Arabic, with his body language showing great pride and courage, "We chanted 'Oh Hafez, curse your soul.'" After I interpreted this for the journalist, I waited for an admiring reaction to such a profound slogan that destroyed the symbolism of the Assads. I gave her time to react. She did nothing. I waited longer. No reaction. I was disappointed when she was surprised to learn that the protestors were cursing Hafez, rather than Bashar, and explained to her the context and importance of this slogan. This started to get her attention. I explained the dimensions of the slogan in greater depth, and she then showed more interest in and understanding of it. At that moment, I realized that the task of translation is not simply to convey words, sentences, and expressions from one language into another, but that it is a plural act that conveys different layers of meaning, interpretations, attitudes, narratives, ideologies, and discourses. Moreover, it is a negotiation between the source and target languages that, through a translator, determines what is shown to and what is hidden from the target language text. Such an act converts the act of translation from a transfer of equivalent words between two languages into a cultural dialogue between two cultures by unpacking the thick content of this language through contextualization.

While the Syrian revolution has obviously produced significant social and political changes, the far-reaching consequences of its cultural and discursive transformations are yet to be seen. For activists, academics, and journalists, however, it was initially a revolution of language, destroying the symbolic oppression and torpidity of the old regime and creating a new language that enabled them to defy, inform, narrate, and translate ongoing events and transformations. This language was

https://doi.org/10.1515/9783110767698-001

initially created to defy, subvert, and challenge the regime and its discourse. The language of the Syrian revolution embodied not only the revolutionary context of the protests, but also the long history of the regime's symbolic violence. This understanding of the social and political changes in the language prompts its scrutiny from a linguistic and philological perspective, as well as an examination of how the speech act and performance became united via the discourses of the regime and the revolution, and later to an English-speaking audience through translation. For this reason, I use the term "revolution" in this book for two main reasons: first, because there was a revolutionary shift in the use of language, and second because this is how activists on the ground wanted to describe it. Terms other than revolution might suggest adopting the perspectives of other narratives that cannot be hidden; it is impossible to be neutral when labeling a political phenomenon.

In various examples, I use the word "Syrians" to refer to Syrians in general, by which I mean all those under the Assads' political regime. In doing so, I by no means wish to generalize that all Syrians surrendered to this domination, but rather that they lived through it. Some of them were obliged to be silent and others believed in the Assads' symbols. There were also those who opposed this domination and paid for it with several years of their lives in prison. It is important to emphasize that Syrians cannot be categorized into a simple binary, and in other contexts I specifically refer to Assad loyalists or pro-revolution supporters and activists in order to avoid generalization.

The language of the revolution is thus approached from a philological and linguistic perspective, and is also considered as cultural capital in sociological terms. This approach highlights language as primarily a political product: language used *in* a revolution but also language *as* a revolution and as a metaphorical weapon and method deployed by both sides. This raises many questions with regard to language. How can language be a historical, political, cultural, economic, and social indicator of transformations in a society through its manifestations such as slogans and songs? What is the relationship between symbolic violence, different types of symbolic capital and their consumption in different settings, and dominant and dominated discourses in language and practices?[1]

1 Pierre Bourdieu, *Language and Symbolic Power*, trans. Gino Raymond and Mathew Adamson (Cambridge: Polity Press, 1991), 14; Pierre Bourdieu, *The Logic of Practice*, trans. Richard Nice (California: Standard University Press, 1990); Pierre Bourdieu, *Distinction: A Social Critique of the Judgement of Taste*, trans. Richard Nice (Cambridge: Harvard University Press, 1984). Symbolic capital is a term used by Pierre Bourdieu, essentially referring "to accumulated prestige and honor" that is used as a form of both power and violence. Symbolic capital cannot be acquired without other types of capital, mainly economic and cultural. In this book, the concept of symbolic capital is used as an open umbrella term that includes other types of capital. The openness of the term also em-

This revolutionized language has been transferred from Syria to the world by highly competent narrators (native and non-native academics, journalists, activists, etc.), as seen in news items, blogs, books, and translations on Syria. This study is divided into two main sections: language, and translation/translating. The first section overlaps the second, although I show it is a separate act through thick descriptions of the language of the revolution versus that of the regime and later the translation of this language into English. In the first section, I analyze the language of the regime, which provides the necessary context for a subsequent parallel analysis of the language of the revolution. In this section, I demonstrate the economics of language and discourse in Syria through the dynamic interaction between discursive products of the regime and the revolution. This analysis of the language shows the multilayered context of the language of the regime and that of the revolution, using thick translation as a tool of analysis.[2]

In contrast to other Arab revolutions, the Syrian revolution was essentially neglected and abandoned by the English-speaking world. While the Tunisian and Egyptian protests were archetypal revolutions for the English-speaking sphere from the early days of protests, things were very different for abandoned revolutions such as those in Syria and Yemen. This can be explained by a recognition of Syria as located at the margin, lacking journalists, translators, and agents to connect it with the center or with actors from the center.[3] This is where the second section of the research comes in, which aims to demonstrate how the revolutionary language was translated into English. I begin by exploring the context of translations from Arabic into English and the situation of the English translation market, based on

phasizes that forms of cultural, economic and symbolic capital are interchangeable, as the research will illustrate.

2 Clifford Geertz, *The Interpretation of Cultures: Selected Essays* (New York: Basic Books, 1973), 27; Kwame Anthony Appiah, "Thick Translation," *Callaloo* 16, no. 4 (1993): 808–19, https://www.jstor.org/stable/2932211. "Thick translation" is a term originally derived from Clifford Geertz's "thick description," meaning a detailed, contextualized account that focuses on analysis of meaning rather than simple observation. It is a method that is used for "setting down the meaning particular social actions have for the actors whose actions they are, and stating, as explicitly as we can manage, what the knowledge thus attained demonstrates about the society in which it is found and, beyond that, about social life as such." Based on this, thick translation was developed by Kwame Appiah, who used this term to mean a detailed translation that makes use of annotation and glosses to reflect the sophisticated context of the source language text.

3 This does not mean that Egypt and Tunisia are considered centers. The openness of Egypt and Tunisia to foreigners helped to give a voice to their revolutions, in contrast to Syria which was limited to the regime media. An example of this openness can be seen at the American University in Cairo, where Samia Mehrez taught a course on the Egyptian Revolution during the protests. This collective event, which resulted in a book, was attended by different foreign and local agents. Such initiatives never happened in Syria or Yemen.

the numbers of books translated from Arabic into English, and then consider two levels of translation: first, translation as an act of importation into the dominant discourse, represented by three books in which the revolutionary language is translated from the margin to the center; and second, translation from the margin to the center, as represented by activist translations from Arabic into English.

Political Activism and Academic Research

Writing about contemporary Syria, especially after 2011, is difficult for scholars in the social sciences for many reasons. The voices of most Syrians were already muted under a brutal regime, but after March 2011 it became almost impossible for Syrians to speak to the international community from inside Syria. Most foreign academics and researchers left the country because of the personal risks associated with the escalation of regime violence. In addition to this, and most significantly, the regime sought to silence Syrians' voices entirely by barring all foreign and international media outlets from covering the protests as of the second month of protests in 2011.[4]

In Egypt, the openness of the revolutionary field was essential to more openly translating the Egyptian revolution from within, as many foreign and native actors were on the ground reporting or dealing with the protests. In the case of Syria, primary sources were also mostly inaccessible; frequently, the only sources available for research consisted of online media and people who had fled Syria to seek refuge abroad. While scientific research remained feasible, it became far more difficult and raised questions of authenticity in relation to digital sources as well as the ethical, methodological, and contextual issues posed by research actors who had fled the country in fear of their lives. In addition to this, conducting fieldwork from outside prevents researchers from observing a particular phenomenon in its original setting, restricting scholars to information from witnesses and testimonies from people who have left the research context and relinquished the privilege of observing the setting themselves, through their own eyes. Finally, in most cases academic scholarship on Syria was unable to examine the protests from multiple perspectives, since acceptance of an individual (researcher) by one side precluded

4 Khālid Mamdūḥ al-ʿAzzī, "al-Iʿlām al-Sūrī al-Rasmī wa-Thawrat Rabīʿ Sūriyya" [The Syrian Official Press and the Syrian Spring Revolution], *Zamān al-Waṣil*, June 11, 2011, accessed July 18, 2023, https://www.zamanalwsl.net/news/article/19906/; ʿAbdullāh Turkumānī, "Isṭūrat al-Mūʾāmara fī al-Iʿlām al-Sūrī" [The Legend of Conspiracy in the Syrian Media], al-Markaz al-Sūrī lil-Iʿlām wa-Ḥuryyat al-Taʿbīr [Syrian Center for Media and Freedom of Expression], July 7, 2011, accessed July 18, 2023, https://bit.ly/3dxil17.

their acceptance by the other. This obliged researchers to essentially depend on information gathered from one side, with the scant information available from the other side playing only a marginal role. To put it differently, there is generally a missing component in remote research on Syria: the regime side. Researchers have focused on the cultural resistance of the revolutionists and neglected regime attempts to subvert this resistance and its symbolic products, or quite simply, the symbolic violence of the regime has not been seen a reason for this resistance. This interconnectedness between the regime and the revolutionary side presents difficulties for researchers conducting fieldwork or interviews with representatives of multiple factions.

As such, with the exception of a handful of scholarly works, contemporary Syria has not been thoroughly researched from within over the last decade—and only rarely before that. This is why, after I arrived in Europe in 2012, I decided to utilize my past activist experience in Syria in my postgraduate studies. The difficulty of accessing data and fieldwork creates an opportunity for local researchers who come from the same cultural and social background as the research topic and are likely to have partially or fully experienced the phenomena under analysis. The combination of academic research, political activism, and political commitment is not a new practice. We can, for example, look to the experience of Jewish scholars who developed new academic approaches based on a combination of their personal experiences and observations, theoretical understanding, and methodological practices, creating a literature about the Holocaust of the European Jews that became the most important scholarship for understanding those persecuted and threatened with death under the Nazi regime. This call for allowing political and personal experience to help create academic knowledge is not meant as an emotional appeal, but rather recognizes the need to let marginalized people, activists, politicians, and subalterns themselves theorize using their own experience and find methodological and theoretical approaches to deal with experiences of injustice and oppression, traumatic incidents, symbolic violent domination, and euphoric political events through empowering their agency in producing knowledge. Such an approach would not exclude experts in social sciences or require them to step aside, but would instead constitute a collaboration between academics, political academics, and activists, allowing them to share and help one another to produce knowledge on a given topic.

This study is no exception to the above examples and is written by an academic whose political activism on the ground ended in November 2012. Being part of the research setting provided me with many privileges that facilitated the progress of my study. First, I was seen as a member of the same community, which helped me to be seen as a legitimate witness for the research group, setting, and phenomena, in addition to "sharpen[ing my] abilities for critical reflection" as a member of the community.[5] It also helped build trust among interviewees from the same community, though it simultaneously became more difficult to interview people who shared the same background but had different ideological attitudes.[6] My participation in political activism in Syria during the first year of the revolution helped me a great deal to connect with academics, journalists, activists, and all types of actors in the Syrian revolution. My involvement in the revolution facilitated reaching people from the same community and even helped build a high level of trust, interest, and confidence in what I was researching and studying. This same involvement made it very difficult to connect with the regime community and supporters. As my study shows, I was able to reach some pro-regime activists, but when my safety was threatened, I stopped interviewing pro-regime individuals. This decision was not made to exclude an important part of Syrian society, but to protect my personal safety, which would be expected from any non-activist researcher.

Second, being part of the research topic provides the researcher with an understanding of the many linguistic, cultural, and empirical dimensions that take longer for outside researchers to grasp. This is illustrated in the repertoire that a native researcher might have for interpreting events, which enables more in-depth cultural interpretation for fieldwork.[7] In addition to this, a researcher who belongs to the same cultural and social group is able to conduct the research "in a more sensitive and responsive manner" than an outside researcher who will take a long time to acquire the necessary linguistic and cultural skills.[8] Furthermore, without having personally lived through and experienced the symbolic violence of the Assad regimes, it would be much more difficult to write about or analyze it. In order to ensure the accuracy of my memories and experiences, I confirmed with other individuals and archive materials that I was not creating fictional stories or inaccurate memories. Examples of this include the use of Assad's sayings in exams, using "*Minḥibbak*" [we love you [Bashar]] as a password for free internet access, and so on. It took me a long time to be able to process and analyze my personal experiences; while this may have created distance between myself and the

5 Amanda Coffey, "Ethnography and Self: Reflections and Representations," in *Qualitative Research in Action*, ed. Tim May (London: Sage, 2002), 314–31.

6 Pranee Liamputtong, *Researching the Vulnerable: A Guide to Sensitive Research Methods* (London: Sage, 2007); Emily Finch, "Issues of Confidentiality in Research into Criminal Activity: The Legal and Ethical Dilemma," *Mountbatten Journal of Legal Studies* 1, no. 2 (2001): 34–50.

7 Deborah Court and Randa Abbas, "Whose Interview Is It, Anyway? Methodological and Ethical Challenges of Insider–Outsider Research, Multiple Languages, and Dual-Researcher Cooperation," *Qualitative Inquiry* 19, no. 6 (2013), 480–88, http://dx.doi.org/10.1177/1077800413482102.

8 Russell Bishop, "Freeing Ourselves from Neocolonial Domination in Research: A Kaupapa Māori Approach to Creating Knowledge," in *The Landscape of Qualitative Research*, 3rd edition, eds. Norman K. Denzin and Yvonna S. Lincoln (California: Sage, 2008), 148.

research context, it also provided me with a greater social, political, and cultural understanding of my personal experiences.

However, the concepts of insider and outsider researchers do not necessarily mean solely people who are from inside or outside of a geographic place,[9] but can also apply to researchers from the same research location if they come from a different social class, religion, or political background that could make them be seen as insiders or outsiders by interviewees. While a pro-Assad researcher, for example, is an insider geographically, they will find it more difficult to research this topic than a geographic outsider. Similarly, my specific situation of being against the Assad regime would likely prevent me from doing pro-regime fieldwork.

The positionality of my research was also affected by the period of time that I spent in transition, thinking about and distancing myself from the events and my experiences in Syria and equipping myself academically to present the current research. The eight-year period that separates the end of my revolutionary activism in November 2012 from my doctoral thesis in November 2020 helped me to reconstruct and translate what I witnessed and experienced in this work. This personal and academic development allowed me to interview pro-regime activists and obliged me to admit the importance of the language of the regime and its influence on the revolutionary one, as shown in this research. Such a statement would have been impossible for me to make in 2011.

Being part of the Syrian revolution helped me to analyze the revolutionary language, and its impact can be seen in the use of certain terms and expressions in this research. This is something that I do not hide, and that can be read through my words. Coining or using a term for something that happened during any period of history cannot be a neutral act. Additionally, choosing between competing terms indicates a judgement regarding the historical event itself, as I show in Chapter 4 in reference to the Israel-Syrian conflict in 1973. As a faithful translator of a text from Arabic into English, I translate what happened in Syria in 2011 as a "revolution," just as those protesting in the streets did. Second, calling Syria 2011 a revolution is part of the worldliness of the text I am dealing with. Calling for the materiality of the text in the Saidian sense means showing its worldly connections, including political ones, which necessitates calling Syria 2011 a revolution based on its connections in the Syrian context, rather than the English-language one.[10]

In addition to my position in the Syrian activism space, this book is based on extensive fieldwork stays and visits, which were conducted in 2015. The main task

---

9 Robert K. Merton, "Insiders and Outsiders: A Chapter In The Sociology Of Knowledge," *American Journal of Sociology*, 78, no. 1 (July 1972): 9–47.

10 Edward W. Said, "The Text, the World, the Critic," *The Bulletin of the Midwest Modern Language Association* 8, no. 2 (1975): 1–23.

of my fieldwork in 2014–15 was to verify the data I had collected in 2011–12 in Syria and gather additional narratives, witnesses, and archives from other Syrian activists. My fieldwork interviews were thus conducted in countries that were considered to be gathering places for activists and Syrians in general, including Turkey (in Gaziantep, Istanbul, Ankara, and Urfa), France (in Paris, Nantes, Troyes, Marseille, and Bordeaux), and Germany (in Berlin, Frankfurt, Bonn, Dusseldorf, Dresden, and Leipzig).

Since interviews seeking people's perspectives and narratives cannot be fully structured, because "people always say something to split the structure," I used a semi-structured interview format.[11] In semi-structured interviews, "the thematic direction is given much more preference and the interviews may be focused much more directly on certain topics."[12] Interviewees had the freedom to speak, but I was always in control of the thematic direction in order to keep the interview on course. In this sense, I sought to conduct interviews "with the purpose of obtaining descriptions of the life world of the interviewee in order to interpret the meaning of the described phenomena."[13] Overall, I conducted fifty-five semi-structured interviews, most of which took place in Turkey. In addition, I used digital platforms such as Skype, Facebook, and WhatsApp to conduct remote interviews with intellectuals, academics, journalists, and pro-Assad supporters who lived in places I was not able to reach. I made a specific effort to interview pro-Assad artists and intellectuals, but only a few were willing to speak with me. Some refused when I told them the topic of my research or after they discovered its political perspective. For safety reasons, I decided not to conduct any further interviews with Assad supporters. In this book, I do not simply seek to contextualize the language of the regime and the revolution within the research timeframe of 2011–12, showing the pluralism of meanings through translation, as examined in the second and third chapters. Nor do I aim to provide technical solutions through analysis of sample translations or offering alternative translations, or to simply sketch out the translation of this language from the margin (Syria) to the center (the English-speaking audience). Instead, I aspire to an in-depth discussion of different arguments and questions related to translation, activism, and language itself. The main question is how the revolutionary language in Syria was transferred into the English-speaking market. Along with tracing the act of translation, I raise the question of the importance of the language and its performativity in discursively explaining the profound histo-

---

11 Ian Parker, *Qualitative Psychology: Introducing Radical Research* (Buckingham: Open University Press, 2005), 53.

12 Uwe Flick, *An Introduction to Qualitative Research*, 4th ed. (London: Sage, 2009), 211.

13 Steinar Kvale and Svend Brinkmann, *InterViews: Learning the Craft of Qualitative Research Interviewing* (Los Angeles: Sage, 2008), 3.

ries of Syrians, including the symbolic domination of the regime and its subversion by regime opponents who opposed the regime more explicitly in 2011. The positionality of translation, through its actors, settings, and locations, helps us to understand the modes of reception of the translation. Moreover, it sheds light on the translation's interaction with its mechanisms and the transformation processes that differ when the translation production chain is changed. How would a change of translator, funder, or place of translation affect the translated work? Would it create a different mode of reception? What is the importance of English agents choosing translations of Syrian themes on the revolution; does it reflect how Syrian activists themselves want to be voiced? Is there a difference if the voicing act is rendered by agents from the margin or the center? How is importing a translation into the market different in the absence of national or government funding? Within the framework of this discussion, the study tries to explore how translations of the language of the Syrian revolution are reshaped after leaving their originating discourse and entering the English one. How do these transformations of Syria from Arabic into English retain the thick histories and multilayered meanings in another medium that depends on narratives other than the Arabic one(s)?

This travel of discourse and language from the margin to the center is not even, or systematically fixed. It is affected not only by the place of translation but also by the agents, publishers, funders, and market economics, all of which shape translated books, as the questions raised suggest. Moreover, when translations of the revolutionary language from this periphery discourse (Arabic) reach the center discourse (English), they not only offer a narrative or story that is not identical—as is the case for all translations because it is an interpretation—but instead provide one that reduces and simplifies it, presenting a shallow depiction of the revolutionary language in English. This is why Syrian activists since the beginning of the protests in 2011 have been surprised to see the English terms used for their protests—unrest, conflict, clashes, rebellion and, eventually, civil war—in contrast to other Arab countries that witnessed similar protests. This difference between terms for Syria 2011 originating inside and outside of Syria and in English shows not only different appropriations of the Syrian event but also represents discourse(s) and narrative(s) different to the one(s) started in Syria but inherently familiar to the English reader.

In order to answer the above questions, the first two chapters aim to determine the extent to which the revolutionary language was a continuation of or a break from that of the regime. This detailed examination confirms the deep roots of the revolutionary language that emerged in 2011 and shows that the latter was related to past contexts, buried in the language of the regime. It also sheds light on the complexity of creating meanings and interpretations of each discursive product. Moreover, tracing symbolic products in the Syrian symbolic market explains the dynamism and interaction of symbolic products and their symbolic value in their symbolic market, which adds more symbolic layers to the products themselves.

Domination and Culture in Syria

The language of the Syrian revolution was not born in a void, cut off from all context and only following the other Arab protests in 2011. Instead, I argue that it was born from several decades of Assad's domination of Syria since 1970. I am not the first to have tried to cover new facets of this domination. All of the scholarship and literature agrees on the fact that symbols, discourse, and language, including audio and visual productions, were scarcely researched in the field of Syrian studies or were considered to be of inferior importance to political and economic studies, to the extent that "most of the scholarship tends to ignore the cult and its implications for political life together" as Lisa Wedeen says in her first book on Syria.[14] While Wedeen argues that Assad's cult is "a strategy of domination based on compliance rather legitimacy," and Salwa Ismail similarly analyzes the performativity of symbolic violence by the regime, I try to build on their arguments on structuring the frames of this violence.[15] Wedeen describes it as ambiguous and analyzes it through its outcome, while Ismail is more focused on its functionality in Syrian everyday life through narratives. I seek to offer a more conceptual structure for the products of the regime language represented by slogans, banners, songs, and everyday terms. In contrast with previous scholarship, this study is based on extensive sources that have never previously been presented or analyzed, collected by personal observation and participation in consuming this language in Syria. According to Wedeen, the images of Assad presented by the regime served to strengthen Assad's power to the extent that "his regime can compel people to say the ridiculous and to avow the absurd."[16] While it is correct that the regime acquired power and gained authority from compelling people to repeat ridiculous claims and other nonsense, it was not only about power. The regime could have power without obliging Syrians to engage in these activities, as was the case at the start of Assad's domination after the massacre in Hama, when he did not need to circulate his symbols and cult—which was not yet a cult at that time—in order to impose his domination and control of all Syrians.

14 Lisa Wedeen, *Ambiguities of Domination: Politics, Rhetoric, and Symbols in Contemporary Syria*, 2nd ed. (Chicago: University of Chicago Press, 2015), 4.
15 Wedeen, *Ambiguities of Domination*, 6; Salwa Ismail, *The Rule of Violence: Subjectivity, Memory and Government in Syria* (Cambridge: Cambridge University Press, 2018).
16 Wedeen, *Ambiguities of Domination*, 12.

Taking a very similar approach to Wedeen, Ismail sees violence as a key strategy for the Assad regime to rule Syria, using it "not only to harm the body, the mind and affect but, also, to reconstruct, shape, discipline and normalize the subject of government."[17] This view of violence in Ismail's book depends on prison and arrest as a key term for diffusing violence in the Syrian social context. Similarly to Ismail and to some extent connecting prison to everyday life, miriam cooke's book about the production of art in Syria also suggests a similar practice to regime domination by providing specific spaces for criticism through what she calls "commissioned criticism." Commissioned criticism is an effective method of *tanaffus,* or deep breathing, that provides a safety valve to release built-up pressure, akin to how a tire might explode as a result of too much air pressure without some form of release.[18] The concept of *tanaffus* is linked by cooke to the "*sahat al-tanaffus*, literally, the breathing yard" within prisons.[19] As I interpret her work, she sees Syrians metaphorically as prisoners who need to breathe deeply or they will explode. Syria, as the wider prison, has "commissioned criticism" that works like a breathing yard for the Syrian public to keep them from exploding. At the same time, however, these yards are horrible places for Syrians because in some prisons "the previous night's victims were hanged there."[20]

Prison is a keystone for these various studies of domination, violence, and censorship, all of which assert the relationship between all of the different methods of violence and domination. As a Syrian, I see political detention as the highest form of violence that a Syrian might be exposed to, although prior to 2011 many Syrians had never passed through the detention experience. For this reason, and in order to follow the argument of the relationship between violence, domination, and prison, I have collected the official numbers of political prisoners in Syria from reports produced by human rights organizations. When examining data from these reports, shown in the figure in Appendix 1, it is important to recognize that the numbers in these reports frequently conflict and are inaccurate because of the Syrian political situation. The regime actively prevents news from getting out, so the information is secretly collected from families of the arrested or from people who have recently been released. No organization has access to accurate numbers, and there have been repeated denials by the regime of the existence of a political prison, including

---

17 Ismail, *Rule of Violence*, 9;

18 miriam cooke, *Dissident Syria: Making the Oppositional Arts Official* (Durham: Duke University Press, 2007), 71.

19 Ismail, *Rule of Violence*, 133.

20 Ismail, *Rule of Violence*, 133.

by Bashar al-Assad in one of his interviews.[21] Some annual reports add new events that had not been included in previous reports or information about incidents that occurred or which they were notified about after issuing the report. According to human rights lawyer Anwar al-Bunnī, all of these reported arrest statistics are inaccurate and the real figures are likely much higher:

> It is very difficult to estimate how many Syrians have been arrested by regime authorities and are in its detention centers and prisons, but we can give approximate numbers throughout the historical periods. From 1980 to 1983, there were more than 20,000 prisoners. Then the number of arrested people decreased to where there were tens of people arrested every year, which lasted up to 2001. The arrest campaigns returned after 2001 to reach civil society and Syrians who had returned from Iraq. There are still about 5,000 prisoners from before 2011 whose fates remain unknown.[22]

The graph of political prisoners in Syria (see Appendix 1), drawn from the statistics of human rights organizations as noted above, shows that reports of physical violence (detention) gradually reduced between the 1980s and 2010. The number increased suddenly in March 2011, which illustrates that in the three decades prior to 2011, the regime controlled the country by means other than the physical violence of detention. Detention represents the highest degree of punishment used by the Syrian regime, because prisoners are deprived of their civil rights if they are released—including withdrawing their passport, and rendering them ineligible to work for the government, or benefit from services offered by the state—and they are exposed to daily torture, mental and physical disorders, and diseases. Physical violence represented by prison, according to the graph, was less practiced. To put it differently, the data indicates that in the early 1990s, Syrians were not dominated or controlled by detention centers and prisons as much as by the pervasive violence that was constantly being created, taking hold of the Syrian conscious and unconscious imagination, beginning with the distribution of products by the regime and later reinforced by individual Syrians who created a fictional palace of fear, imagining the violent reaction of the regime to any Syrian who dared challenge it.

The question that I asked myself and all of my interviewees who lived in Syria was, if the most frightening action with which the regime threatened Syrians was being taken to one of these detention centers, why were Syrians compliant with the regime even if they had not experienced incarceration during their life? It is tricky to answer this question simply, but it is rare in the current period for different Syrians from different classes and different regions to have no relatives

21 Interview with Bashar Assad, Rīm Maʿlūfī. Youtube. Syrian Presidency, June 9, 2022, accessed September 28, 2022, https://www.youtube.com/watch?v=0V4yj7YVRk8 (no longer available).
22 Anwar al-Bunnī (Syrian human rights lawyer), interview with the author, April 23, 2017.

who have not passed through a political prison. In addition to this, as Ismail states, the narratives and echoes of detention were present all the time and affected the mind. I argue that the violence of imprisonment was an ever-present ready method to use when *tanaffus* was not enough to ensure tight domination over Syrians, or when a Syrian departed from the compliance expected by the regime. An essential component for dominating Syrians was creating an appropriate social context in which individuals could learn from and adopt their behaviors so as to profit from them and show obedience to the regime. "As if" is Wedeen's model for Syrians that made them dominated by the regime such that they acted "as if" they believed in what they are acting. This system is constructed on the basis that "people are not required to believe the cult's fiction, and they do not, but they are requried to act as they did."[23] Similarly to the "acting as if" described by Wedeen, it is my view that many Syrians were certain that using cult symbols, visuality, music, slogans, and different symbolic products were more of a rewarding tool that they competed to possess in order to change their social class, or to profit symbolically from the position they might acquire from consuming such behaviors or improvising new ones to praise the regime. This does not preclude the existence of many regime loyalists who believed in the Baath ideology. Even if someone did not believe in what was said, they could be certain that it was likely to suit numerous interests, and if this was not the case, then at least the impending punishment would be prevented. In other words, there was a social context that encouraged Syrians to compete to acquire, use, and consume what was available in the public space, and improving new techniques and products to the regime prepared a fertile atmosphere for a social context in which individuals could perform and gain at the same time.

This social context, or as Pierre Bourdieu terms it differently, the field, game, and market refers to "the structured space of positions in which the positions and their interrelations are determined by the distribution of different kinds of resources or capitals."[24] It is the place where all of these narratives and testimonies hover around and also the location where individuals compete to create, learn, and acquire new skills depending on their social class and the symbolic capital they are born with in order to survive and profit from what exists and exchange different types of products with others. While individuals may internalize and externalize the practices and symbolic products of a society in different ways, they remain within the main schemes of the *habitus*, defined by Bourdieu as "the set of dispositions which incline agents to act and react in certain ways."[25] Moreover, according

23 Wedeen, *Ambiguities of Domination*, 30.
24 Wedeen, *Ambiguities of Domination*, 13.
25 Bourdieu, *Language and Symbolic Power*, 12.

to Bourdieu, there "is a dialectic of the internalization of externality and the externalization of internality."[26] In this context, individuals are exposed to discursive patterns by the regime and the class of the regime including loyalists and pro-regime supporters, a coercive act of externalization to be internalized by the Syrian individual in society. This relationship between how individuals externalize and internalize practices is known as "euphemization," which is a method by which an individual self-censors in a way that suits the surroundings and the class to which they belong.[27] However, this euphemization is a result of the ideological and discourse production of the dominant class of the regime, coercing the individuals surrounding them to behave in a similar way. It enhances the ideological functionality of the field and the production of ideology and discourse as a main source of symbolic power in the first instance, but also subsequently as a symbolic violence as a result. Analyzing the discourse and language of the Assads leads us to think about the symbolic power in discourse and its violence, because it illustrates how to "impose an apprehension of the established order as natural (orthodoxy) through the disguised (and thus misrecognized) imposition of systems of classification and of mental structures that are objectively adjusted to social structures," as Bourdieu analyzes the sources and effects of symbolic power.[28] It is clear that many aspects contributed to creating the symbolic power and domination of the Assads through the final product of it as an ideological discourse in the Syrian market through different products. In the sections below, I discuss these aspects and then go on to analyze the cultural products that can be found in the Syrian market, in order to understand the mechanisms of domination and how Assad gradually changed from the human being who led Syria into the godlike figure that dominated Syria. These three main aspects were the propagandists, Baath organizations, and finally publishing houses, which controlled what to consume in the market, in which spaces, and what to read in the book market.

The Propagandists

The cult of Assad as the country's sole absolute leader was not a distinctive characteristic witnessed only in Syria in the last half century, but was also an essential element for authoritarian regimes in places such as Iraq, Libya, Egypt, Tunisia, and

---

26 Pierre Bourdieu, *Outline of a Theory of Practice*, trans. Richard Nice (Cambridge: Cambridge University Press, 1977), 72.

27 Bourdieu, *Language and Symbolic Power*, 84.

28 Bourdieu, *Language and Symbolic Power*, 169.

other Arab countries.[29] This cult appears to have been imported along the lines of the Stalin model, as "the Syrian government hired Soviet specialists to help orchestrate spectacles and train participants."[30] This section focuses on a group of individuals who had a leading role in creating regime propaganda and discourse. These figures created the conditions necessary to establish the groundwork for filling the Syrian symbolic market with products relating solely to Assad and his family, both visually (e.g., images, murals, banners, etc.) and musically (e.g., national songs, various anthems, etc.). Determining exactly who helped Hafez al-Assad create his cult in Syria is a tricky process. I began my study with books that mentioned the roles of specific individuals, researching each one to determine whether they were active in structuring the regime machinery of symbolic products. The starting point was Patrick Seale's book about Assad, which was written to a great extent under the eye of Hafez al-Assad himself, and which mentions a few names of those who were responsible for composing Assad's propaganda. I then continued to examine people who occupied the same positions in the Syrian regime, although uncertain whether they played a role in the making of Assad's propaganda and symbolic products. In addition, I conducted seventy-nine interviews with journalists, activists, writers, and other individuals who experienced the 1980s in Syria. I first compiled, then narrowed down a list of those potentially involved, eventually settling on four individuals: George Ṣaḍḍiqnī, Aḥmad Iskandar Aḥmad, Muḥammad Salmān, and ʿAli ʿUqla ʿIrsān. All served as Ministers of Information, except for ʿIrsān, who was the president of the Arab Writers Union, which was responsible for publications in Syria. It is most likely these figures who had a major impact on Assad's appearance of omniscience. Syrians claim that Aḥmad Iskandar Aḥmad was the mastermind and, according to Seale, he was "the inventor of the cult [of Assad]."[31]

A lack of resources meant that I was dependent on political historical events, interviews with witnesses, marches, books, speeches, and videos to understand the factors that dominated Syrians discursively. The outcome of these discursive apparatuses (propagandists, popular organizations, and publishing houses) is clear evidence that the Assad regime did not create these products by chance, but produced them systematically.

---

29 Joseph Sassoon, *Anatomy of Authoritarianism in the Arab Republics* (Cambridge: Cambridge University Press 2016) 185–220.

30 Wedeen, *Ambiguities of Domination*, 27.

31 Patrick Seale, *Asad: The Struggle for the Middle East* (Berkeley: University of California, 1990), 339.

George Ṣaḍḍiqnī (1973–4)

George Ṣaḍḍiqnī (1931–2010) was the Minister of Media and Information for one year during the most difficult period of Assad's political life. He was among the Baathists pardoned by Assad and allowed to return to Syria and participate politically under his rule.[32] Wedeen's investigations and the witnesses she interviewed during her fieldwork in Syria in the 1990s confirm that Ṣaḍḍiqnī objected to putting Assad's image on both special and everyday Syrian items, such as notebooks, walls, and so on.[33] Ṣaḍḍiqnī thought that having the image of Assad everywhere in the Syrian public sphere might raise religious tensions, which is why he refused to help or be part of creating the Assad cult, and might also explain why he served only one year as Minister of Media and Information. He had the honor of being Assad's spokesman and the person who transferred Assad's speeches to the media during that time, which was particularly important during the War of October. At that time, Syrians called him the Minister of *Tishreen* [October], in reference to the war against Israel.[34] It is interesting to note that in a 2008 interview, Ṣaḍḍiqnī mentioned nothing about himself as Minister of *Tishreen*, given how important the War of October was for the regime, which refers to it as "the Liberation War against Israeli aggression."[35] In this interview, Ṣaḍḍiqnī referred to himself as a linguist and a member of the Arab Academy of Damascus, whose sole concern was the decline of classical Arabic and its replacement by dialects, and avoided discussing his political life, despite the fact that during his ministry "Syria celebrated what was officially styled the victory of the October War of 1973."[36] His objections to supporting the cult of Assad and the fact that he did not mention his former position as Minister of Media and Information during the 2008 interview, combined with his unwillingness to discuss his political life and his short term in office, show that Hafez al-Assad was not satisfied with his performance in that position and explains his replacement by Aḥmad Iskandar Aḥmad.

---

32 Seale, *Asad*, 171.

33 Wedeen, *Ambiguities of Domination*, 33.

34 Kāmel Saqir, "Sūryyah:Tadahhwur al-Ḥāla al-Ṣiḥīīya li-Wazīr ʾIʿlām Ḥarb Oktobar wa-Aḥad Munaẓīrī al-Baath George Saddiqni" [Syria: The Deteriorating Health of the October Minister of Information and Baath Intellectual George Saddiqni], *al-Quds*, May 26, 2010, accessed June 1, 2020, https://bit.ly/3eCCcgf (no longer available).

35 George Saddiqni, "Muqābalah maʿ George Saddiqni" [An Interview with George Saddiqni], interview by Maysāʾ al-Jardī, *Althawra*, January 28, 2008, accessed June 1, 2020, https://bit.ly/3cj7OWJ (no longer available).

36 Wedeen, *Ambiguities of Domination*, 34.

Aḥmad Iskandar Aḥmad (1974–83)

Aḥmad Iskandar Aḥmad (1944–83) was the most controversial of all Syria's ministers of Media and Information because of rumors about him and his youth when he was appointed as a minister in 1974. Opponents of the regime used these rumors and exaggerations to depict him as a notorious figure and sinister mastermind of the regime. Through his cunning and wisdom, according to many Syrian articles, he made Assad the god of Syria. Even today, many Syrians see him as the Syrian Joseph Goebbels for promoting Assad over Baathism, overlooking the fact that under Assad the Baath ideology was only a tool to stabilize the rule of the Assad family rather than to promote the doctrine of Baathism. Aḥmad's legend and reputation made regime supporters proud to have such intelligent and talented people in the government. Of course, at the same time, pro-regime individuals did not dare say anything against his presence. Others argue that he was not so important a person, though Seale's declaration that "the inventor of the cult was Aḥmad Iskandar Aḥmad" made him famous.[37] Due to the debate about this figure, I attempted to shed light on the ambiguities, exaggerations, and rumors surrounding him with the small number of documents regarding this and via interviews with those who met him in person and worked with him. Many journalists focus on Seale's assertion that Aḥmad was the inventor of Assad's cult, despite the fact that he is mentioned seven times in the same book, simultaneously associated with different locations and situations, all of which were ignored by journalists who focused their imaginations on him as the creator of the cult. He was not only a propagandist but also a politician, though anything other than his propaganda work is largely unknown. Other than his notorious appearance in Seale's book and vague mentions of him in various media articles, he is not discussed in any studies, articles, or books. I therefore sought interviews with people who knew him personally or had worked with him when he was a minister. My investigations revealed two competing narratives: the first defending him as a patriot who tried to use his close relationship with Hafez al-Assad to benefit Syria, and the second presenting him as an ordinary person who profited from his authority and became the devil used by Assad to execute the cult formation plans. I was able to reach his daughter Lamā Aḥmad Iskandar Aḥmad in Germany, where I conducted an interview with her. She told me that:

> My father was born in Homs in 1944. He got the best grades in his high school, which qualified him for a scholarship to Ain Shams University in Egypt. He was dismissed from his BA studies because he was Baathist. Later, he was able to [get] his degree from Egypt specializ-

37 Seale, *Asad*, 339.

> ing in Library Science and another one in English Literature from Damascus University. This equipped him with the best competence to be a journalist. When he wrote an article in the al-Thawra newspaper, the newspaper sold out. Because of his daring critique of the Syrian authority, Assad liked him and became very close to him. He was appointed Minister of Information when he was twenty-nine, the youngest minister ever in Syria. He was outspoken and had a special charisma. The fact of the matter is that he was very close to Dictator Hafez al-Assad. Relative to this, my father participated in creating Assad's cult for the period before the 1980s, not after. He passed away due to cerebrospinal fluid that later developed into brain cancer. He was Assad's right-hand man, to the extent that he had more authority than just being the Minister of Information. I am sure that Assad profited from my father's skills in language, at least in speechwriting.[38]

Seale mentions Aḥmad in numerous stories that confirm his daughter's account. On the other hand, according to historian and novelist Khairy Alzahaby (1946–2022), Aḥmad was a short story writer who was overestimated by journalists who wrote about him, and his writing level and style were very weak. In Alzahaby's opinion, there was nothing exceptional in his character, but what made him famous was the statement that Seale had made about him.[39] Tawfīq al-Ḥallāq, a journalist active when Aḥmad was a minister, told me that he met Aḥmad twice and that he protected al-Ḥallāq's program "Al-Sālib wa al-Mujab" [The Negative and the Positive],[40] which would have been censored and cancelled without his help, especially as it directly criticized the regime. Al-Ḥallāq argued that Aḥmad tried to find a balance between Assad's strict policies and the limited freedom that was offered at that time. Drawing on these unclear pictures of Aḥmad and the lack of any documents, there are three possible essential conclusions about his role in developing the cult of Assad. The first is that Aḥmad Iskandar Aḥmad played no role in Assad's cult because he was dying by 1983, and the cult started after the tensions with the Muslim Brotherhood ended with the massacre in 1983, since it would not have been possible for Assad to create a cult for himself in Syria amid such political tensions; the second is that Aḥmad had prepared everything for Assad in advance but did not have time to see it come to fruition; and the third possibility is that Assad came up with the plan and relied on Aḥmad to execute it. After his visit to North Korea in 1974, Assad developed a theory of how to make it happen, believing that Aḥmad was the best person to put such a plan into motion and make it successful.[41] This comes across in Seale's discussion of how Assad's wife dreamed that Assad would

38 Lama Aḥmad Iskandar Aḥmad (daughter of Aḥmad Iskandar Aḥmad and a former diplomat who defected after the Syrian revolution), interview with the author, May 16, 2017.

39 Khairy Alzahaby (Syrian historian and novelist), interview with the author, May 19, 2017.

40 Tawfīq al-Hallāq (Syrian journalist), interview with the author, May 25, 2017.

41 Dae-Sook Suh, *Korean Communism, 1945–80: A Reference Guide to the Political System* (Honolulu: University of Hawaii, 1981), 197.

be the most important Arab leader.[42] Trying to elucidate the role of figures such as Aḥmad in creating Assad's cult does not necessarily construct specific structures for censorship in Syria as these might remain ambiguous, as Wedeen describes in her book. At the same time, his protection of al-Hallāq's program appears to reflect the strategy that the regime followed in order to keep the valve open to prevent people from exploding.

Muḥammad Salmān (1987–2000)

Muḥammad Salmān was one of the most important leaders of the Baath Party in Syria. He spent some thirteen years as Minister of Information, an indication that Hafez al-Assad was satisfied with his performance. One of his achievements was to broker "a reconciliation between the regime and some of the oppositional thinkers and writers [like] Zakaria Tamer."[43] In a 2010 interview, he claimed that censorship in Syria "has become open recently except in the topics that offend the structures of Syrians."[44] This suggests that there was a development and a shift in the distribution of symbolic products. The 1990s witnessed a change to the system of censorship because of the unprecedented level of corruption in the regime. The regime wanted to divert people's attention from this and, at the same time, to make use of *tanaffus* in order for people to be able to continue under repression.

During Salmān's tenure as Minister of Information, the focus on the twin goals of diversion and *tanaffus* were the reason for Syrian dramas changing significantly, seemingly testing the bounds of freedom allowed at that time, with shows touching on very sensitive topics. Television series such as *Buqʿat Ḍaw* [Spotlight] and Marāyā [Mirrors], for example, became very famous and daring in their criticism of the regime and its corruption. This explains why Marāyā ended after 2011 and why Buqʿat Ḍaw followed the lead of the regime propaganda in describing the Syrian protests—because their messages or the limited criticism allowed by the regime could not be useful within the context of the Arab revolutions: since the protests had already exploded, there was no need for a safety valve. In theater, Humām Ḥūt, a Syrian actor and playwright, was very famous for producing several pieces of theater criticizing the corruption but not the regime, and has said that: "Bashar al-Assad asked me, personally, in the first year of his reign to leave politics [alone]

---

42 Seale, *Asad*, 165.

43 Muḥammad Salmān, "*Muqābala maʿ Muḥammad Salmān*" [An Interview with Muhammad Salmān] by *Al-Quds Al-Araby*, December 4, 2010, accessed July 18, 2023, https://bit.ly/2wVifB8.

44 Salmān, "*Muqābala maʿ Muḥammad Salmān*."

and focus more on the topics of resistance and challenging the Zionist entity and imperialism, and to promote the slogans of the regime for his audience."[45]

Salmān derived his authority in the Ministry of Information from his relationship to the regime, since his daughter was married to Ali Dūbā, who was the head of Military Intelligence in Syria and an advisor to Hafez al-Assad. According to testimonies from interviews, Salmān was famous for saying "the president is a good man, but the people in power around him are bad."[46] This opened the way for Assad to be forgiven for any corruption that happened in Syria, on the grounds that the people who were around the leader kept him from knowing the reality of the situation.

ʿAli ʿUqla ʿIrsān

In addition to the cult, and control over drama, the book market in Syria was an essential element in the glorification of Assad. Although ʿult ʿUqla ʿIrsān (b. 1941) did not serve as Minister of Information, he was very important in equipping the repressive machine of the regime by presiding over the Arab Writers Union in Syria for thirty years. From a brief look at the status of the Union, it is clear that the most important Syrian writers and thinkers were banished from it, including Saʿdallāh Wannūs, Adonis, Mamdūḥ ʿIdwān, Ḥanā Minā (despite being one of the founders), Salīm Barākat, Ibrāhim Maḥmūd, and others. After the fall of the Iraqi regime, the entire Iraqi Writers Union was banned from participating.[47] In its founding statement, the Arab Writers Union implicitly claims to be the spokesperson of Arabs and promotes Arabic leadership in culture without mentioning Syrian authors. While the Union was formed in Syria and its members are mainly Syrian, with a few authors from other Arab countries, calling itself the "Arab Union" was an indication of its self-positioning as the Arab leader, as was its mission "to enhance the resistance and steadfastness of spirit of the Arab citizen."[48] By addressing not Syrians but Arabs in the statement, it declares itself to be the only union that defends the Baath Party claim of Arabism.

---

45 Humām Ḥūt, "*Muqābala maʿ Humām Ḥūt fī Barnāmij Jū Shū*" [An Interview with Humām Ḥūt on the Joe Show program], *Joe Show*, YouTube video, uploaded July 30, 2016, accessed July 18, 2023, https://www.youtube.com/watch?v=nqZ6_gaceTQ.
46 Salmān, interview.
47 Aḥmad Ḥaydar, "*Al-Sayīd ʿAlī ʿUqla ʿIrsān (Kifāyah)*" [Mr. ʿAlī ʿUqla ʿIrsān (enough)], *Alhewar*, September 3, 2005, accessed July 18, 2023, http://www.ahewar.org/debat/show.art.asp?aid=44617.
48 Arab Writers Union, "*ʿAn al-Ittiḥād*" [About the Union], accessed June 1, 2020, https://bit.ly/2zRZSi7 (no longer available).

ʿIrsān's importance lies in his control over all printed works in Syria. The main publisher was the Arab Writers Union, represented by ʿIrsān himself, who had to approve all publications before printing. Khairy Alzahaby says that "if a book managed to be published [without his approval], ʿIrsān had other ways to fight it by increasing its price in the bookshops so nobody bought it."[49] The Arab Writers Union was only one of the organizations with which ʿIrsān was involved. He was also the founder or co-founder of many cultural organizations, ensuring that all cultural and artistic life was either affected by or under his control and could be paralyzed by him. Some of the positions he held were "president of the Artists Syndicate in 1970, director of theaters and music from 1969–76, a member of the leadership of the Baath Vanguard Organization and Youth Revolution Federation, and assistant to the Cultural and National Guidance Minister in 1976."[50] After Hafez al-Assad came to power, he eradicated all syndicates and labor unions, and then relaunched organizations that were compatible with the regime.[51] ʿIrsān appears to have been a vital person in recreating and reestablishing Baath organizations and unions, as a founding member of the Actors Guild, Revolutionary Youth Union [Ittiḥād Shabibat al-Thawra], Vanguard Union [Munaẓamat Ṭalāʾiʿ al-Thawra], and others.[52] However, this seizure of the book market was not final and unclear. It remained possible to bypass censorship and publish books by publishers outside of Syria, or in very rare cases books were approved by the censor. One example of this is Faraj Bayrekdar's poetry volume "Marāyā al-Ghīyāb [The Mirror of Absence] which was published in Syria but never released because the secret service found out about it in the last stage.[53] This is another example of how if a book passes censorship, it is restricted in different ways. Lebanon, however, always remained as a possible country in which to publish, especially after the regime army left its territory.

---

49 Alzahaby, interview.

50 Kāmil Salmān al-Jabūrī, *Muʿjam al-Udabāʾ min- al_ʿAṣr al-Jāhilī ḥatā 2003* [The Dictionary of Writers from the Pre-Islamic Era to 2002] (Beirut: Dar al-Kotob, 2003), 299.

51 Khulūd al-Zughayīr, *"Sūryā fī-Muʿtaqal al-Baʿth|Assad Qiṣat Waṭan* (2)" [Syria in the Prison of Assad/Baath, A Story of a Home (2)], *Souria Houria*, April 28, 2012, accessed July 18, 2023, https://bit.ly/36LIcQz.

52 Al-Bābtīn Foundation, *"Ḥayāt ʿAli ʿUqla ʿIrsān"* [The Life of ʿAli Uqla ʿIrsān], *Abdulaziz Saud Al-Babtain Award for Creative Poetry*, May 13, 2017, accessed June 1, 2020, https://bit.ly/2MegJ1c (no longer available).

53 Interview with the author, Faraj Bayrekdar, May 15, 2022.

Popular Organizations

The four figures I have discussed—George Ṣaḍḍiqnī, Aḥmad Iskandar Aḥmad, Muḥammad Salmān, and ʿAli ʿUqla ʿIrsān—played a vital role as founders of Assad's discourse or took upon their shoulders the task of assisting the regime to seize all of the possible free spaces in Syria. This seizure allowed the Syrian market, and individuals like those mentioned above, to create and establish new practices that could be seen not only at the level of discursive products but also in terms of the discursive domination of Syrians. The individuals discussed were notable for a number of reasons as well as being responsible for laying the cornerstones for the spread of Assad's image throughout all visual spaces and the foundation of pro-regime organizations that impacted Syrians at every stage of their life course.

Thinking again about habitus as "a set of dispositions which incline agents to act and react in certain ways, the dispositions generate practices, perceptions and attitudes," the popular organizations helped to shape and impose a certain set of dispositions on Syrians.[54] To put it differently, to be able to control the practices, perceptions, and attitudes of individuals, it is highly important to be able to create a specific atmosphere that can affect their dispositions, since the latter are "acquired through [the] gradual process of inculcation in which early childhood experiences are particularly important."[55] This is why, after Assad came to power, he first changed the education system, creating Baathist institutions and organizations to train the younger generations, and threatening the older generations—who missed the particular educational opportunities that would have shaped their dispositions—with arrest, torture, and execution. The real violence of the regime was thus experienced by the older generations, while the younger generations were controlled by fear and trained dispositions. These dispositions and the way they worked with other individuals on the ground were interwoven due to the habitus schemas of each Syrian.

Because of the importance of controlling education, three popular organizations were founded that provided education and treated Syrians as individuals, rather than solely collectively as school or university students. For example, they recruited individuals to inquire about their colleagues' absences from activities or why they were not active, by phone calls or personal visits. The civil society organizations shaped Syrians' language and strengthened Assad's control by imposing compliance in a new form. These organizations were (misleadingly) called *Shaʿbbī* [popular], indicating that they were in demand by the people, rather than the gov-

54 Bourdieu, *Language and Symbolic Power*, 12.
55 Bourdieu, *Language and Symbolic Power*, 12.

ernment or the regime. These "popular" organizations were the Baath Vanguard Organization, Youth Revolution Federation, Baath Party, and Student Union.

Before the emergence of Assad's cult in Syria, these organizations had had an impact on Syrians' education since they were children, having been founded under the Baath Party to educate and acculturate generations of Syrians from 1974 onward.[56] The objectives of the Baath Vanguard Organization, according to its official website, are "educating children according to Vanguard doctrine with out-of-school activities like external camps and training courses to build and improve their skills."[57] When children turn thirteen, they go directly to another Baathist organization called the Youth Revolution Federation, which was founded in "1968 by Assad presidential decree."[58] The Youth Revolution Federation "receives its members from the age of thirteen to thirty-five."[59] In one of Assad's speeches in 1985, he tried to illustrate the message of members of the Youth Revolution Federation. In addition to protecting the revolution, the message "does not end at a specific age but it extends to the coming years."[60] As part of the systematic education overseen by these two organizations, most Syrians join the Baath Party; it is practically mandatory in order to receive "privileges" reserved for members, including political or social positions, jobs, and even the possibility to attend a university. Numbers and statistics support the above dry statements of these organizations as being created to facilitate the easy joining of pupils and the younger generations in order to convert the majority of Syrians to Baathism. In early 1970, Baath members in Syria numbered an estimated "65,000 and it [the number of members] grew sixfold to reach 374,000 in 1981. The numbers reached a million in 1992 and 1.4 million in 2000."[61] There are two types of members of the Baath Party: *Naṣīr* [candidate-train-

56 Baath Vanguard Organization, "*Ma'lūmāt 'an al-Munaẓama*" [Information about the Organization], Baath Vanguard Organization official website, May 13, 2017, accessed June 1, 2020, https://bit.ly/2Mj9FAg (no longer available).

57 Baath Vanguard Organization, "*Ma'lūmāt 'an al-Munaẓama.*"

58 Al-Wiḥda Foundation for Press Printing Publishing & Distribution, "*Bi-Munāsabat al-Dhikra al 46 li t'sīss Munaẓamat Itiḥād Shabibat al-Thawra*" [On the Occasion of the 46th Anniversary of Founding the Revolutionary Youth Union], May 13, 2017, accessed June 1, 2020, https://goo.gl/TzfCpe (no longer available).

59 Al-Wiḥda Foundation for Press Printing Publishing & Distribution, "*Bi-Munāsabat al-Dhikra al 46 li t'sīss Munaẓamat Itiḥād Shabibat al-Thawra.*"

60 Hafez al-Assad, "*Kalimat al-Ra'īs Hafez al-Assad fī-al-Mu'tamar al-Rābi' li-Itiḥād Shabibat al-Thawra 1985*" [Assad's Speech at the 4th Conference of the Revolutionary Youth Union], 1985, President Bashar al-Assad unofficial website, accessed May 13, 2017, https://bit.ly/2NCZO9D (no longer available).

61 Kamal Dib, *Tārīkh Suriyya al-Mu'āṣir: min al-Intidāb al-Faransī ilā Ṣayf 2011* [Contemporary History of Syria: from the French Mandate to the Summer of 2011] (Beirut: Dar al-Nahar, 2011), 412.

ees/supporting members] and *ʿĀmil* [active members]. *Naṣīr* are not counted as members, because this is considered the preparatory stage before becoming an active member, which "lasts at least three years."[62] Hanna Batatu reports that in 1967, there were "about five thousand 'active' or full members and probably eight times as many candidate-trainees and supporting members."[63] Based on this and only theoretically (it is not possible to count the numbers without statistics), if we assume a rough proportion of eight times as many supporting members as active members, this would equate to 11.2 million supporting members (including active members) in 2000: more than half of the approximate population in Syria.

These organizations essentially reward their members for consolidating the cult of Assad. High school pupils are able to attend the best universities and choose the best majors for participating in parachuting or enrolling in *Ṣāʿiqa* military camps. Going to these camps ensures pupils get higher grades, which is a way to achieve their ambitions of joining more prestigious university majors. The educational system in Syria was originally based on the French model and continues to follow it. When attending university, pupils are assigned to university specializations based on their final results and grades. Let's say that a pupil needs five more points in order to study Biology. If a student has previously enrolled in *Ṣāʿiqa* camps, up to five points are made available so the student can join their preferred university specialization.[64] It is important to note that this award is only possible for those who have joined the Baath Party's popular organizations. The system of rewarding those who praise Assad worked well to protect and strengthen his authority and control people's language and discourses, affecting their practices, perceptions, and attitudes. Later, Syrians became more accustomed to praising Assad without believing, especially for the 1980s and subsequent generations, who did not remember or never learned what happened in Hama; as Lisa Wedeen says, "they act as if they believe."[65] This generation lived on the narrated regime violence of the atrocities that were committed in the 1980s, with discipline and compliance taught by the transfer of the traumatic events witnessed by the older generation. This might explain why the number of arrests by the regime decreased from 1990 to 2011, since the opposing individuals had either been arrested or were outside the

62 al-Baath Party, "*Kayfa Yatum al-Intisāb li-Ḥizb al-Baath al-ʿArabī al-Ishtirākī*" [How to Join the Arab Socialist Baath Party], al-Baath Official Website, accessed June 1, 2020, https://bit.ly/2x6KfC8 (no longer available).

63 Hanna Batatu, *Syria's Peasantry, the Descendants of Its Lesser Rural Notables, and Their Politics* (Princeton: Princeton University, 1999), 161.

64 In contrast, parachuting is worth up to twenty points, as I understand from meeting people who got them.

65 Wedeen, *Ambiguities of Domination*, 67.

country. This did not prevent the regime from committing massacres when needed, or from arresting others, as seen in incidents such as what happened in Qamishli, or the many massacres that took place after 2011. This highlights that the regime utilized the recalled memories of the massacres and, once they were marginalized, did not hesitate to commit new ones or arrest people. These two aspects were in my view the key component of domination, with the help of the symbolic violence practiced and performed in several cultural products, which I discuss in Chapter 1.

Publishing Houses

Publishing houses played a vital role in shaping Syrians' language and controlling the linguistic public space by distributing very select types of books and increasing the level of censorship on books and magazines. With the help of the popular organizations, propagandists, and publishing houses, the entire publishing field was controlled and censored by the regime, especially due to ʿIrsān, who remained president of the Arab Writers Union for thirty years. Publication of a book or novel could be considered dependent on a writer's loyalty to the regime—though not necessarily to the regime alone, as loyalty could also be offered to ʿIrsān's "gang."[66]

The vast majority of magazines and newspapers were shut down after Assad gained power; only three, which all represented the regime's perspective, were allowed to keep going. As visual and musical spaces were filled with the products of Hafez al-Assad, the same became true for books. While Hafez al-Assad was a historic leader for Syria and deserves many books to be written about him—as the regime assumed—the market was completely inundated with such works.[67] The importance of such books was not the value of their thoughts and philosophy but rather how the writer or reader could benefit from the book. Such a product had

66 A detailed description of the requirements for publishing books is provided in Chapter 2.
67 There have been many books written specifically about Hafez al-Assad, including, for example, *Qāsim al-Ribbdāwī, Ḥāfiz al-Assad wa al-Ḥuryya* [Hafez al-Assad and Freedom]; Karīm al-Shībānī, *Fī Ẓilāl al-Bayʿa* [In the Shadow of the Pledge]; Fūʾād al-ʿIsh, *Ḥāfiz al-Assad: Qāʾid wa Risāla* [Hafez al-Assad: A Leader and a Message]; Ṣafwān Qidsī, *al-Baṭal wa al-Tārīkh* [The Hero and the History]; Byār Salāma, *al-Qunṣul: Jīl Ḥāfiz al-Assad* [The Council: The Generation of Hafez al-Assad]; Jūrj ʾAyn Malik, *al-Taṣḥīḥ al-Thawrī* [The Revolutionary Correction]; Ḥamdān Makārim, *Masīrat al-Wafāʾ wa-al-ʿAṭāʾ* [The March of Bestowal and Faithfulness]; Asʿad Ali, *Famm al-Nabiʿ wa al-Taʾalluq fi Khiṭāb al-Rʾīs* [The Water Source and the Glamour of the Speeches of Hafez al-Assad]; Riyāḍ ʿAwwād, *Qirāʾa fī Fikir al-Qāʾid Ḥāfiz al-Assad* [A Reading in the Thoughts of the Leader Hafez al-Assad]; Farīz Sammūnī, *Suṭūr Khalida maʿ al-Qāʾid al-Amīn* [Eternal Lines with the Trusted Leader]; Aḥmad Qarna, *Ḥāfiz al-Assad Ṣāniʿ Tārīkh al-Umma* [Hafez al-Assad the Arab Nation-Maker]; ʿAbdil Ghanī ʿAshī, *Aḥbabnāk fa-Bāyʿnāk* [We Loved You, Then We Acknowledge You as a Leader].

value in its ability to take the shape of other products, such as cash or social or political positions. For example, it was hugely important for a writer to prove their loyalty to the regime by showing that a publication was about the "eternal leader." Even symbolic violence could be measured as capital by watching others' loyalty to the regime. In addition, such books could be used to obtain higher points and grades in university exams or written interviews for government jobs, as well as ostensibly demonstrating skill in public speaking or debate when mixed with politics. Moreover, using such a book when speaking in social settings helped encourage others to show respect to the speaker. I discuss one such book in detail in the section analyzing books as symbolic products in Chapter 1

Chapter 1
The Silenced

هل تستطيع الكلمات أن تفعل شيئاً؟ هل تخيفهم؟
عبدالرحمن منيف من "شرق المتوسط"

Can words achieve anything? Do words frighten them?

ʿAbd al-Raḥmān Munīf, *Sharq al-Muṭawassit* [East of the Mediterranean]

Domination and its Products

As discussed in the introduction, under the rule of Hafez al-Assad, three main actors—the propagandists, popular organizations, and publishing houses—controlled the symbolic market in Syria and inundated it with products. While other elements also contributed to narrowing the public space, it is impossible to cover here all of the potential centers or focal points that gave instructions for immersing the Syrian public space in these symbolic products. Voluntary participation by individuals was always welcomed if it served the Baath ideology and the objectives of praising the regime and Assad. The consumption of symbolic products, active and passive participation in this, accumulation of symbolic products, and abstaining from performing created structured zones and boundaries in the Syrian public space. Within these organized, but also arbitrary, boundaries of who provided what products to the public space, it is only possible to focus on the products, and to argue that the regime had a structured scheme to channel, regulate, and supervise individual behaviors and actions. In other words, as long as there were agreed values for the benefit and utility of having a specific product in the symbolic market, then individuals would compete to produce them in order to obtain prestigious—not necessarily economic—benefits. This chapter will discuss the emergence and use of these products. By highlighting specific examples of the various types of products, it will demonstrate how they centered on Hafez al-Assad, and how their gradual distribution was used to achieve his ultimate objective of being viewed as a god of Syria and also control the public space by those products as a form of symbolic violence accompanied by the prison violence. It aims not only to prove that symbolic products were used to present Assad as godlike, but also to show the importance of tracing these symbolic products over time in order to study and analyze their development through to 2011—when such products came up against revolutionary ones—and thus better understand the emergence of the revolutionary language. As such, the description of this language and its domination contributes to a thick

https://doi.org/10.1515/9783110767698-002

translation of the layers of meaning of the revolutionary language and its history, as reflected in the slogan of 2011, "Hafez, curse your soul."

It is of course impossible to examine all of the many different symbolic products that might potentially be studied. I have not, for example, included jokes, university and school curricula, literature (pre- and post-2011), blogs and social media, films, and many other products due to the significant amount of research time this would require, which would necessarily exceed the timeframe of interest of my study. I have therefore only included products of similar types to the revolutionary language (i.e. slogans, graffiti, and musical and visual products such as songs and photos). I believe this metaphoric mapping of the products presents a more comprehensible scene about the relationship between Syrian symbolic products before and after 2011.

In this chapter, I focus on the development of symbolic products in the Syrian symbolic market and their content. Tracing several of the symbolic products in the Syrian market shows how the Assad regime, by inundating the market with the help of the propagandists, popular organizations, and publishing houses, was able to gradually achieve its goal of controlling all possible spaces. This in turn influenced everyday language in Syria, and provided control over the products in the Syrian market by giving significant importance to otherwise non-valuable products, such as books praising Assad. These (non-)valuable products thus became significant and could be exchanged for other forms of capital, as manifested in five different types of symbolic products: visual and musical ones, books, slogans, and everyday terms.

Visual Products

Following Assad's *coup d'état* in 1970, the regime sought to intensively occupy all of Syria's public visual spaces, which, in turn, affected the practices of individuals. By "occupy" here I mean using all visual spaces for the purposes of the regime and its symbols. This act of occupation was very similar to immersion in that all visual units in public places (including streets, squares, and government buildings such as schools, universities, and hospitals), and sometimes private ones (such as private offices, medical clinics, shops, and restaurants), were filled with the regime's visual symbols as a sign of individuals' loyalty and the regime's domination, making "Asad's cult [. . .] a powerful, albeit ambiguous, mechanism of social control."[68] Following a plan that may have been developed by Ahmad Iskandar Aḥmad or Assad himself—

68 Lisa Wedeen, *Ambiguities of Domination: Politics, Rhetoric, and Symbols in Contemporary Syria* (Chicago: The University of Chicago Press, 1999), 24.

and against George Ṣaḍḍiqnī's advice[69]—Assad's image was present everywhere in Syria, and "regularly depicted as omnipresent and omniscient."[70] To achieve this, Assad immersed all visual spaces with his image, accompanied by short captions such as "first doctor" (in private or public hospitals), "first peasant" (in the peasant syndicate), "first pharmacist" (in the pharmacist syndicate), "first engineer" (in the engineer syndicate), "first teacher" (in the schools and Ministry of Education), and "first soldier" (in military establishments).[71] To magnify his presence in everyday Syrian life, his image was (and still is) printed on school certificates, notebooks, scholar scarf pins, military uniforms, and in magazines and newspapers, and displayed in the offices of the Baath Party and in car and bus windows. In 1997, his portrait was added to the new 1,000-pound note, and remains there still.

This visual invasion ensured Assad's presence in the collective unconsciousness of Syrians. Wherever they went, they were monitored by his face, and seemed to be surrounded by him. Strikingly, in most of these images, Assad appeared to be gloomy or frowning, as if he were directly, scornfully looking at the viewer.[72] Due to the forced consumption of Assad's images on the market, visual signs in Syria became limited to one practice and objective: seeing Assad. All of them worked hand-in-hand toward the objective of glorifying the country's leader, and the other signs accompanying Assad's portraits lost their layers of meaning. For example, with the intensive portrayal of Assad throughout the entire Syrian public space, Assad and Syria became essentially synonymous. Looking at his image brought Syria to mind and vice versa, operating what was akin to a linguistic distortion of the signifiers and signified in "Syria"—another violence breaking the normal relationship between a signifier and the signified. Under Assad's rule, even linguistic meaning was abused to signify differently. In this sense, Syria had only one signification—Hafez al-Assad—while Assad's image had several layers of meaning to Syrians, including "revolution," the "Baath Party," "resistance," "Pan-Arabism," "Palestine," and "Syria."

Figures 1 and 2 provide two examples to demonstrate this point: first, the logo of the Baath Party that was displayed beside Assad's image in various locations such as notebooks, at events, in universities, and schools; and second, a cloth patch with images of Hafez and his sons. The Palestinian flag was shown more than the

69 See the discussion in the introduction about the role(s) played by specific individuals.

70 Wedeen, *Ambiguities of Domination*, 1.

71 Wedeen, *Ambiguities of Domination*, 1. Wedeen also observes that Assad was described in newspaper photos as the "father," the "combatant," the "first teacher," the "Savior of Lebanon," the "Leader forever," and the "gallant knight."

72 Some people believe that the frowning photos of Assad became especially common after the death of his son Bassel in a car accident in 1994. Those born in the 1980s and later cannot remember ever seeing a smiling Hafez al-Assad in the images presented on the streets or in public visual spaces.

Syrian one at official events, and both flags were sometimes presented simultaneously side by side. If we were to collect up all of the visual symbols and attempt a combined reading, there would be many interpretations connected to Assad, but the other symbols accompanying his image were only associated with one concept.

Figure 1: Baath Party logo.[73]

One possible interpretation of the Baath Party logo displaying the Palestinian flag and map of a borderless Arab world that comes to mind based on the visual symbols in the Syrian public space is that of Assad as Syria and as struggling to defend Palestine, if we take him to be the head of the Baath Party. It suggests that he is the only Arab leader calling for unity, while the Palestinian flag placed in the middle gives the impression that Assad is the only Arab leader who is calling for Palestine to be liberated from Israel: an indirect reference to Assad as the one who would liberate Palestine.[74] This presents Assad not only as an Arab leader but also as an Islamic one, given the implicit reference to Ṣalāḥ al-Dīn, "who wrested Jerusalem from enemy control in 1187."[75]

---

73 Al-Baath Party, official logo, al-Baath Party official website, accessed June 1, 2020, https://goo.gl/MT929K (no longer available; the logo has changed slightly since, but a copy of this version has been kept in the research archive).

74 Wedeen, *Ambiguities of Domination*, 10. Wedeen refers to and makes a comparison between Assad and Salāh al-Din al-Ayyūbī: commenting on a poster produced for the 1991 referendum on Assad's leadership, she observes that the poster invokes "the battle of Hittin. . . [but] might as easily call to mind the *contrasts* between Asad's regime and the legends of Salah al-Din."

75 Wedeen, *Ambiguities of Domination*, 1.

In addition to using his image to claim the Syrian public space, Assad also gradually began to use it to communicate messages to the public. In the early 1990s, when it was time for his son Bassel to gain power, Assad started to combine Bassel's image with his. Bassel was also cleverly presented to the Syrian public space not only as Assad's heir, but also as someone to be admired, in a similar way to Bashar at a later date. Basel was presented to Syrians as a sportsman and equestrian, with the Championship of Equestrianism created for him. This was initially called the Championship of Peace and Love, and then changed to be called Bassel's championship. As the Arabic word for equestrian is *fāris*, which can also mean a knight on horseback, Bassel was also associated with the idea of a chivalrous knight. However, his death in a car accident in 1994 brought an end to the plan to make Bassel his father's official heir, and marked the beginning of preparing Bashar to be the successor instead.[76] Suddenly Syria became Assad's Syria, the property of Hafez al-Assad (and later, of the Assad family). This was seen primarily in welcome banners in border areas. Anyone entering Syria from the Lebanese border crossing, for example, would be greeted by a banner reading "Welcome to Assad's Syria," indicating that they were not visiting Syria but rather Assad's property.

Figure 2: Cloth patch.[77]

76 William E. Schmidt, "Assad's Son Killed in an Auto Crash," *New York Times*, January 22, 1994, accessed July 18, 2023, http://www.nytimes.com/1994/01/22/world/assad-s-son-killed-in-an-auto-crash.html.
77 A cloth patch of the kind that can be found in several shops in Syria. Author photo, personal research archive.

It is impossible to pinpoint the exact moment at which Syria began to be presented as Hafez al-Assad's property in terms of visual products as Assad's Syria, but the literature suggests an approximate period for this. Yassin Haj Saleh notes that "Syria was first described as the Lion's den in the late 1970s [with reference to Assad as the lion], but after a while it became directly known as Assad's Syria."[78] This is consistent with the Assad regime's gradual approach to change, as we will see in relation to songs and slogans. Contrary to Haj Saleh's estimated date of the 1970s, Assad is unlikely to have made this particular change at that time due to the political tensions in Syria, notably with the Muslim Brotherhood and with Rif'at al-Assad, which resulted in the latter leaving the country in 1984.[79] Azmi Bishara claims that Syria started to be called "Assad's Syria" after the defeat of the Muslim Brotherhood in 1982.[80] It is difficult to precisely date the term, but the earliest evidence I have been able to document dates back to 1987, when Syria hosted the Mediterranean Games. At the opening ceremony, which was attended by Hafez al-Assad, Bassel delivered a speech. After the parade was over, the camera panned to the stadium screen, which read, as shown in Figure 3 below, "Assad's Syria cordially greets the participants and guests of the Tenth Mediterranean Games."[81]

Musical Products

Filling the Syrian public space with images was only one aspect of forcing domination on Syrians. Music was also used as a tool, not only to elevate Assad's rank but also to control Syrians and assert the idea that there was no alternative to the Assad family in Syria. In this section I will focus in particular on nationalist music, by which I mean songs dedicated to Syria, the homeland of Syrians, and widely used anthems. The nationalist songs and various anthems are also considered here to be

78 Yasīn al-Ḥāj Ṣāliḥ, "*Dawlat al-Baʿth wa-Sūriyā al-Assad wa-l-Ḥurūb al-Sūriyya*" [The Baath State, Assad's Syria and the Syrian Wars], *Souria Houria*, October 16, 2011, accessed July 18, 2023, https://goo.gl/Rwm4tr.

79 Vivienne Walt, "Enter the Uncle: An Aging Assad Relative Joins the Syrian Fray," *Time Magazine*, November 24, 2011, accessed July 18, 2023, http://content.time.com/time/world/article/0,8599,2100180,00.html. Tensions arose between Hafez and Rif'at after the latter tried to take power from his brother; they eventually reached an agreement for Rif'at to go to exile.

80 Azmī Bishāra, *Sūriyya Darb al-Ālām naḥwa al-Ḥurryah: Muḥāwala fī-l-Tārikh al-Rāhin* [Syria: The Path of Suffering towards Freedom (2011–3)] (Doha: Arab Center for Research & Policy Studies, 2013), 264.

81 General Sports Federation, Syria, "*Iftitāḥ Dawrat al-Alʿāb al-Mutawasiṭiyya 1987*" [The Opening of the Mediterranean Games, 1987], *GSF Sport* (YouTube channel), uploaded February 15, 2017, at 32:58, accessed July 18, 2023, https://www.youtube.com/watch?v=uLy8wsItWrM&t=648s.

Figure 3: Welcoming message in the stadium reading "Assad's Syria cordially greets the participants and guests of the Tenth Mediterranean Games".

symbolic products distributed in the Syrian market. A gradual change took place in the content of such songs between 1961, when Syria was still part of the United Arab Republic with Egypt, and the late 1990s. A longer timeframe is used here to show the differences in musical products and the related political events. This gradual change occurred due to the concurrent gradual change in the dispositions of the Syrian political habitus, creating different symbolic products at different times.

Nationalist Songs

As there are no archives of Syrian music over the past five decades, I conducted Google searches for existing musical products with keywords such as "national song" and "Syria." Using this information, I chose specific periods, songs, and singers to create a timeline for certain musical products. I drew on the testimonies of Khairy Alzahaby and Fādil al-Sibāʿī regarding the songs that they heard when they lived in this period in Syria,[82] and also relied heavily on several older books about the history of music in Syria, which do not provide a critical analysis of Syrian music or songs, but rather describe the history of music in Syria and the Arab world.[83] Based on my research and interviews, I identified three main phases

---

82 Author interviews with Khairy Alzahaby, May 19, 2017; and Fādil al-Sibāʿī, June 6, 2017.

83 Adnān Ibn Dharīl, *al-Mūsīqā fī Sūriyya: al-Baḥth al-Mūsīqī wa al-Funūm al-Musīqyya Munthu Miʾa ʿĀm* [Music in Syria: The Musical Research and Arts of the Last Hundred Years] (Damascus: Maṭābiʿ Alif Bāʾ – al-Adīb, 1969); Majdī al-ʿAqīlī, *al-Samāʿ ʿInda al-ʿArab* [Listening for Music for

of Syrian national songs, during each of which the content and concepts expressed in song lyrics changed in line with the country's political situation. The first period ran from 1958 to 1961, encompassing the United Arab Republic union between Syria and Egypt; the second began with Syria's secession from the union in 1961 and ended with the October War of 1973; and the third and final phase began after the October War and continued through to the late 1990s.

From the late 1950s until separation in 1961, Syria and Egypt were one republic. It was a dream of Arab countries to be united, and Syria and Egypt were the first example of a great union of Arab countries in one state. Most of the songs in Syria during this period were thus about the Arab world as one entity, even if reference to the Arab world was intended to mean Syria. Syria was referred to as one republic with Egypt in numerous songs by artists such as the Fulayfil brothers, Muḥammad ʿAbd al-Wahhāb, Suhayl ʿArafa, Um Kalthum, and ʿImād Hamdī. Hamdī, for example, sang: "Oh my heart, sing again. . . Egypt and Syria are reborn again."[84] These examples presented Syria and Egypt as one country, and continuously praised Jamal Abdul Naser and his Pan-Arab beliefs.

The second period began by describing the phase of Israeli-Arab conflict, such as in the song "Allāhu akbar fawqa kayydi al-Muʿtadī" [God is the Greatest and Greater than the Deception of the Aggressor]. During this period, nationalist songs started to drift away from Pan-Arabism,[85] and the word "Syria" began to become a main element in song vocabulary, as seen in the example of "Sūriyya yā Ḥabibatī [Syria, my Beloved]." This shift from the union with Egypt to referring solely to Syria might be the only concession that George Ṣaḍḍiqnī—who, as discussed above, had been against the mass use of Assad's image—gave to Hafez al-Assad, and happened after he left his ministry. The mention of Syria in the song represents a turning point in Syrian musical history. From then onward, Syria was presented as a *qutr*, meaning not a country but a region of the larger Arab world: *Al-Quṭr al-Sūrī* means a country that is not totally separate but is part of a bigger region or union. In this case, Syria remained under the umbrella of Baathism and Pan-Arabism but

---

Arabs] (Damascus: n.p, 1973); Laylā Maliḥa Fayyāḍ. *Mawsūʿat Aʿlām al-Mūsīqā al-ʿArab wa al-Ajānib* [The Encyclopedia of Famous Arab and Foreigner Musicians] (Beirut: Dār al-Kutub al-ʿIlmiyya, 1992); ʿAbd al-Karīm al-ʿAllāf. *al-Ṭarab ʿInda al-ʿArab* [Ṭarab for Arabs] (Baghdad: al-Maktaba al-Ahliyya, 1963); Fūʾād Rajāʾī and Nadīm Darwīsh, *Min Kunūzinā* [From Our Treasures] (n.p, 1970).

84 "*'Ghannī yā qalbī,' 'iḥnā al-Shaʿbb' – ḥafl Sīnamā Dimashq 24 Fibrāyīr 1958*" ["Sing, oh my Heart" and "We are the People," concert at Damascus Cinema, February 24, 1958], YouTube (video), uploaded April 19, 2017, accessed July 18, 2023, https://www.youtube.com/watch?v=Jmv3o4AkMQ0.

85 "*Allahu Akbar Fawqa Kayydi al-Muʿtadī,*" *MrFoxmalaga* (YouTube channel), uploaded October 6, 2010, accessed July 18, 2023, https://www.youtube.com/watch?v=AWWc51YfLws.

in a different way. This shift, from a united republic into the single country of Syria, represented a change in both name and identity.

During the third and final period, which followed the October War, Assad entered into another war with the Syrian opposition, and nationalist songs slowly changed from referring to Syria as a *quṭr* to Syria as a separate country from the Arab world. Although this shift was obscured by the Baath ideology, songs began to refer to Syria as a country. Then, slowly, song themes and topics became united with Assad and Syria as one symbol. The same act seen in the visual products—images of Assad signifying Syria, but not vice versa—occurred in Syrian nationalist songs, which began to center on the name of Assad. This became increasingly obvious after Assad's victory over the Muslim Brotherhood in the 1980s, following which an enormous number of songs were dedicated to him and performed by the most famous singers in the Arab world. These included, among many others, "Ḥamāka Allāhu yā Asadu" [God Protects You, Assad] by Assala Nasri, "Tislam li-al-Shaʿbb yā Ḥāfeẓ" [Be Safe for Your Own People, Hafez] by George Wassouf, "Abū al-rijāl" [The Father of Men] by Samira Said, "Waʿd al-qāʾid" [The Promise of the Leader] by Musṭafā Naṣrī, and "Yā Ḥāfeẓ al-Assad" [Oh, Hafez al-Assad] by an unknown singer.[86]

As with visual products, the purpose of musical products was to communicate messages from and about the regime, and to dominate the musical scene. While the regime was unable to control all of the musical spaces in Syria, it did hold sway over radio and local television channels, regime festivities and events, schools, and universities, and also connected well-known pop singers with songs for Assad, as we will see later. The messages of the songs suggested that Assad was the only possible leader of Syria. Listening to all of these songs with this information in mind, we can see how Assad was sending gradual hints to the people that he would become the Assad of Syria, its owner and leader. Before Bassel's death in 1994, Hafez al-Assad communicated a new message to Syrians via a new song entitled "Abū Bāsil qāʾidnā" [The Father of Bassel is our Leader].[87] This was a direct indication, in line with the visual products, that Bassel was intended to succeed Hafez.

---

86 Assala Nasri, "*ḤAMĀKA ALLĀHU YĀ ASADU*," YouTube, uploaded May 26, 2011, accessed July 18, 2023, https://www.youtube.com/watch?v=058eGyHgQqA; George Wassouf, "*Tislam li-al-Shshaʿb yā Ḥāfiẓ*," YouTube, uploaded October 10, 2013, accessed July 18, 2023, https://www.youtube.com/watch?v=BlHCaFx3N00; Samira Said, "*Abū al-Rijāl*," YouTube, uploaded March 4, 2018, accessed June 1, 2020, https://www.youtube.com/watch?v=h9bE8QHXKTY (no longer available); Musṭafā Naṣrī, "*Waʿd al-Qāʾid*," YouTube, uploaded June 28, 2015, accessed July 18, 2023, https://www.youtube.com/watch?v=i4ggaK9_jyY; "*Yā Ḥāfeẓ Assad*," unknown artist, YouTube, uploaded September 23, 2012, accessed July 18, 2023, https://www.youtube.com/watch?v=6_2ezoO7SRs.

87 "*Abū Bāsil qāʾidnā*," *AsadSafita* (YouTube channel), uploaded August 18, 2011, accessed July 18, 2023, https://www.youtube.com/watch?v=JgtagVYjLzk.

Analysis of the language of these songs also shows that nationalist songs used Modern Standard Arabic (MSA) more frequently than colloquial Syrian. MSA was used due to the circumstances in Syria as well as the spread of Pan-Arabism as an essential part of Baath ideology. Songs in MSA were thus seen as valuable because they reflected the Baath ideology. This changed in the late 1990s, when new songs about Assad started to be sung in colloquial Arabic, including "Abū Bāsil qā'idnā."

National Anthems

The other component of musical immersion concerns the various anthems heard in Syria. Like most other countries, Syria has a single official national anthem: this was adopted after gaining independence from France in 1946, and performed for the first time on the country's Independence Day, before the Baathists took power in Syria.[88] It does not espouse an ideology but is solely about Syria, and makes no reference to who is ruling the country. Entitled "Ḥumāt al-diyār" [The Guardians of the Homeland], it has four verses, each describing a part of Syria: the first salutes the army as the "protectors of the country," the second is about the country's natural beauty, the third is about what people [citizens] have sacrificed for Syria, and the final one is about its history.[89] Before this was chosen as the official anthem, Syria had three de facto anthems: (1) "Naḥnu al-shabāb" [We are the Youth], lyrics by Bshāra Khurī; (2) "Bilādi al-'Urbi Awṭāni" [Arab Countries are my Countries], lyrics by Fakhrī al-Bārūdī; and (3) "Fī Sabīl al-Majd wa al-Awṭān" [For the Glory of the Countries], lyrics by Omar Abu Rishah.[90] All of these were loved by Syrians, who wanted one or more to be chosen as their national anthem due to their meaning. With the advent of the Baath Party, the Syrian market gradually filled with new anthems that gained widespread popularity. When the Baath Party came to Syria, an anthem was created for it.[91] The popular organizations then fol-

88 Arab Orient Center for Strategic and Civilization Studies, London, "*Qiṣṣat al-Nashīd al-Waṭanī*" [The Story of the National Anthem], accessed July 18, 2023, http://www.asharqalarabi.org.uk/ruiah/qutuf-98.htm.

89 Arab Orient Center, "Story of the National Anthem." I will return to Syria's national anthem in the following chapter.

90 Bshāra Khurī, lyrics, "*Naḥnu al-Shshabāb + Kalimāt*" [We are the Youth + Lyrics], *Kareem El Khatib* (YouTube channel), uploaded November 5, 2014, accessed July 18, 2023, https://bit.ly/3eE-joNW; Fakhrī al-Bārūdī, lyrics, "*Bilādi al-'Urbi Awṭāni*," *Tornadocccp* (YouTube channel), uploaded October 16, 2008, accessed July 18, 2023, https://www.youtube.com/watch?v=VJd4uDBV6P0; 'Umar Abū Rīsha, lyrics, "*Fī Sabīl al-Majd wa al-Awṭān*," YouTube, uploaded December 29, 2012, accessed July 18, 2023, https://www.youtube.com/watch?v=MldpK-9TSgs.

91 "*Nashīd Ḥizb al-Ba'th al-'Arabī al-Ishtirākī*" [The Anthem of the Socialist Arab Baath Party], *AlBaath Palestine* (YouTube channel), uploaded March 9, 2010, accessed July 18, 2023, https://bit.ly/2XRtO5Z.

lowed the same path, creating their own anthems in addition to the official Syrian national anthem.

In contrast to the official anthem, the Baath anthem describes Syria from a Baathist perspective, considering Syria as part of Baath, not vice versa. In terms of vocabulary and themes, the Baath anthem focuses on specific topics such as factories, fields, Baath, Pan-Arabism, revolution, weapons, and youth. Its emphasis on these words highlights the Baathist agenda and ideology and the lyrics do not mention Syria, since it focuses on Arab unity rather than individual Arab countries. The chosen terms suggest that by taking over these terms in the symbolic field, the regime would be able to control the country.

Yet one anthem for the Baath organization was not enough: two additional ones were composed for the Baath Vanguard Organization and the Youth Revolution Federation and competed in the Syrian market as symbolic products. The anthems of the popular organizations use almost the same terminology as the Baath anthem, but are aimed at different consumers: children, teens, and youth. The organizations function similarly to educational institutions in affecting the linguistic practices and the shape of the Syrian political market, in relation to the competition to garner agents for those products. It remains the case that any individuals who memorize the anthem are rewarded with credit, and this is a requirement for members. Even school students who are not members of one of these organizations are involved in frequent repetition of the anthems since many of their activities take place during official school hours, and all students are obliged to attend various ceremonies, chanting the anthems and even passively participating.

Having four anthems at the same time made Syrians unsure of which one represented them. One interpretation of having four anthems is that the Baath Party wanted its anthems to be chanted more than the official national anthem of Syria. I will return to these anthems in Chapter 2, which looks at how they developed after 2011.

Books

Along with the visual and musical immersion of symbolic products in the Syrian market, books were used as a tool to occupy public spaces, and the regime sought to dominate the production of publications. The books I discuss here are those produced by the loyalists and the regime to praise Assad, which could be found in most libraries, cultural centers, universities, and bookshops, and were usually offered free of charge.

As explained in the section on publishing houses, many books were written in praise of Assad, including *Thus Said Assad*, the title of which was clearly meant to invoke Nietzsche's *Thus Spoke Zarathustra* and present Assad as the preacher to his people. As shown in Figure 4, below, the cover page is followed by a Qur'anic verse saying, "God is the best protector and he is the most merciful among merciful ones."[92]

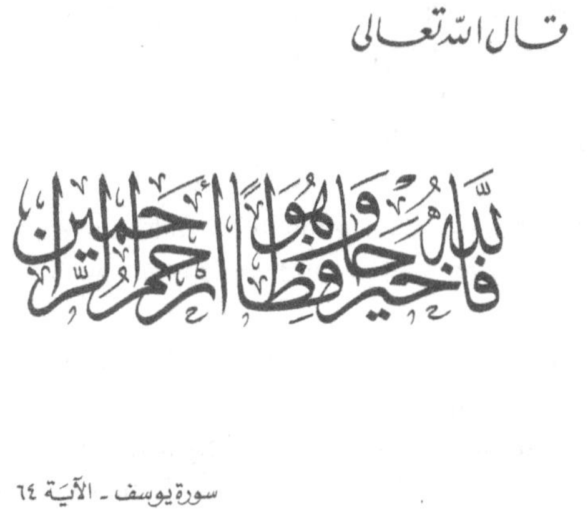
قال الله تعالى
فالله خير حافظا وهو أرحم الراحمين
سورة يوسف - الآية ٦٤

Figure 4: The introductory material of Thus Said Assad includes a Qur'anic verse which translates as "God is the best protector and He is the most merciful among merciful ones".[93]

As is common with individual first and last names in Arabic, the name *Ḥāfez* has a specific meaning as "protector," and is one of the ninety-nine attributes of God in Islamic culture. As we will see with other symbolic products, Assad was thus promoting himself from president to leader and, in this book, to the level of God according to the Qur'an. This is evidenced by the fact that it is written in the Qur'an that he [Hafez] is the best protector, which is the last level of the Assad cult wherein he was seen as a god in Syria (I will discuss this further in terms of language later in the chapter), and means that anyone who believes in God and the Qur'an must worship Assad. I was unable to see any purpose for reproducing the verse in the first pages of this book other than promoting this association between "Hafez" in the Qur'an as God and Hafez in Syria as its godlike figure. The calligraphy shown in this figure has a double meaning: first, it emphasizes the value of Qur'anic verse; and second, it glorifies Hafez as the best protector, like God, as in the original meaning. In context, the Qur'anic verse is said by Jacob to show his confidence that the almighty God will protect his son Joseph: transferred to Syria and its leader, it suggests that the "best protector" will be able to keep Syria safe.

92 Mustafa Ṭlās (ed.), *Hakadhā Qāl al-Assad* [*Thus Said Assad*], n.d., n.p., 3.
93 Ṭlās (ed.), *Hakadhā Qāl al-Assad*, 3.

The second example from the introductory material of *Thus Said Assad*, shown in Figure 5, is about justifying the use of an iron fist, and is taken from the second most important Islamic source, the Hadith. It quotes the Prophet as saying: "Arfaja (may Allah be pleased with him) reported that the Prophet (may Allah's peace and blessings be upon him) said: 'If someone comes to you seeking to undermine your solidarity or disrupt your unity when you have agreed on a man (as your leader), kill him.'"[94] This hadith thus justifies killing someone who tries to disrupt the unity of a people or community. Here the act of killing is a divine duty, presented by Assad to show believers and non-believers alike that they must either accept him as the leader uniting Syrians or face death.

قَالَ رَسُولُ اللهِ صَلَّى اللهُ عَلَيْهِ وَسَلَّم

مَنْ أَتَاكُمْ وَأَمْرُكُمْ جَمِيعٌ عَلَى رَجُلٍ وَاحِدٍ

يُرِيدُ أَنْ يَشُقَّ عَصَاكُمْ أَوْ يُفَرِّقَ جَمَاعَتَكُمْ فَاقْتُلُوهُ

أخرجه مسلم عن عرفجة بن شريح

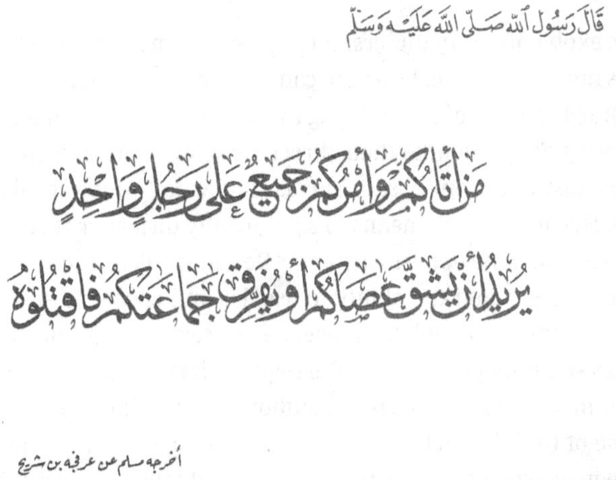

Figure 5: The introductory material of Thus Said Assad includes a hadith, translated into English as: "Arfaja (may Allah be pleased with him) reported that the Prophet (may Allah's peace and blessings be upon him) said: 'If someone comes to you seeking to undermine your solidarity or disrupt your unity when you have agreed on a man (as your leader), kill him.'"[95]

This hadith and Qur'anic verse suggest that if anyone revolts against Assad's power they will be killed, as was the case in the 1980s massacre in Hama, for example. For those who follow Islam, these are not meant to be Assad's words but those of God and of His prophet.

The book itself contains numerous chapters describing what Assad thinks about specific abstract concepts such as "revolution," "freedom," "human being,"

---

94 Sahih Muslim, Book 20, 4565, trans. Hadeeth Encyclopedia.com, accessed July 18, 2023, https://bit.ly/3dng7SS.

95 Sahih Muslim, Book 20, 4565; Ṭlās (ed.), *Hakadhā Qāl al-Assad.*

"democracy," "people," "faith," and "martyrdom," all of which are explained with reference to his speeches and interviews. "Freedom," for Assad, is:

> An innovative and creative freedom in the framework of limits decided by the entire people, otherwise [freedom] becomes chaos and an attack and aggression on every individual and son of the nation.[96]

Although this definition of freedom is very superficial and provides little to no understanding of how Assad really sees the concept, it becomes a valuable saying simply because Assad once said it. Similarly, for Assad, a "revolution" is:

> An innovative explosion [of] people's energy. . . it is an accurate awareness of the main interests [and profits] of the people and a sharp vision for what helps those interests come true.[97]

The importance of Assad's explanations in understanding these terms is not in what he says. The meaning does not matter at all. Freedom can be attained through obedience to Assad, and democracy consists of Assad being the sole candidate in an election. In this sense, such terms do not denote their direct meanings but have other dimensions and take their meaning from Assad. This is similar to "doublespeak," as described by Edward S. Herman, which means to significantly distort or reverse the meaning of a word, performing the opposite action.[98] Doublespeak is frequently combined with "the ability to lie. . . and the ability to use lies and choose to shape facts selectively, blocking out those that don't fit an agenda or program."[99] While the "doublespeak" concept was skillfully practiced by the regime, the very existence of such books was important in strengthening Assad's authority, not because of their content but rather because of their impact on the Syrian public space. While their content was commonly seen as worthless, it could become a valuable product if it helped an individual to be outspoken in praising the regime, pass university oral exams, be seen as performing well in interviews, or score well in written exams by referring to such material in answers to exam questions. An ordinary individual taking part in Syrian daily life in the symbolic market was equipped with the disposition to understand that accumulating such products in the market might be exchanged for, or overlap with, other types of products. For example, *Thus Said Assad* was a symbolic product, but it became an economic one when its author was paid for his work. The same product can take a different form when it is responsible for someone's attainment of a job, a position, or an academic grade. The potential

96 Ṭlās (ed.), *Hakadhā Qāl al-Assad*, 57.
97 Ṭlās (ed.), *Hakadhā Qāl al-Assad*, 27.
98 Edward S. Herman, *Beyond Hypocrisy: Decoding the News in the Age of Propaganda* (Boston: South End Press, 1992).
99 Herman, *Beyond Hypocrisy*, 3.

benefit to an individual of possessing or memorizing such a book made it important as a symbolic product that Syrians competed to have and use in their daily lives.

Slogans

In parallel with the regime seizing power over sensitive market spaces—controlling the visual and musical spaces and publishing houses, and founding popular organizations to inculcate the younger generations—there was a shift in language use in Syrian society. This resulted from new social and political phenomena in the public space, which obliged individuals to repeat practices they were not necessarily convinced of or believed in, and to consume only specific products. The distribution of such products and the change in agent practices led to the emergence of a new language that was represented in slogans and other elements. Through the popular organizations, anthems, music, visual products, and printed works, the political discourse created by the regime was designed to change the nature and mechanisms of Syrians' language construction from childhood through to adulthood. With the help of these factors, slogans were shaped by the dominant discourse that focused on Assad of Syria as president and ultimately established his godlike status. Here, I distinguish between two categories of slogans: those relating to educational and military life, and those relating to Assad personally.

Educational and Military Spaces

As in any other country, schools are one of Syria's most important public spaces. The regime fully appreciated this, and occupied the educational sphere with Baathist slogans and language, using similar practices to those of comparable regimes such as the Soviet and North Korea regimes, through rituals and performances. School students start in the Baath Vanguard Organization during their early years of schooling and move into the Youth Revolution Federation when they are older, and then in university join the Baath Party and/or Student Union. At each stage, they are monitored and observed.[100] These organizations thus play a critical role in students' internalization of linguistic patterns throughout their education. By intruding into the educational sphere, the popular organizations not only controlled the language patterns used in schools but built new educational systems. Like many authoritarian regimes, the Assad regime utilized the mechanism of promoting its propaganda through control and change in language, concentrating on "creating

100 See the previous chapter regarding the approximate number of Baathists in Syria and the required phases of membership in the Baath Party.

new modes of thought."[101] These new modes could be easily created in schools: youth were generally the best audience, as they could internalize Baathist concepts and were easiest to influence. Wedeen analyzes the use of students in rituals in relation to the 1987 Mediterranean Games in Lattakia, which took months of preparation. Here, the voluntary work of pupils was used to show that the regime was "able to compel citizens to enact the choreography movements that iconographically configure worship of the leader, representing his power."[102] This was also part of everyday Syrian life in schools and the military, and accompanied with slogans.

Slogans used in schools or during the rule of Hafez al-Assad have rarely been studied or mentioned. When they are discussed, it is only as a genre or in terms of their general content in praise of Assad, and few or no examples are given. In *Ambiguities of Domination*, for example, Wedeen demonstrates the importance of slogans to Assad, but does not go into detail in framing them. Examples are randomly taken from some interviews. Over the course of the entire book, she only mentions a few slogans, such as: "with spirit, with blood, we sacrifice for you, O Hafez."[103] This slogan and the other examples she notes are not provided in order to offer a study of slogans, but rather to examine "the cult's invocation of a particular rhetorical device," which is well-analyzed but does not set out a broader framework of slogans. Similarly, Salwa Ismail does not provide many examples of slogans or offer a study of them, attempting instead to connect the memories of her interviewees with the content of slogans, especially in school and family life. Her work is thus more focused on everyday life, including certain terms and sentences used by Syrians.[104]

In Syrian schools, the repetition of slogans starts with the morning assembly, where the most important repeated slogan is performed in the shape of a conversation between pupils and the military teacher. It is the slogan of the Baath Party and the pupil uniform is a military one.

أمة عربية واحدة

Military teacher: "One Arabic nation"

ذات رسالة خالدة

Pupils: "With one eternal message."

101 Nikolai Bukharin and Evgenii Preobrazhensky, *The ABC of Communism* (Bookyards, 1922), PDF available from: https://www.bookyards.com/en/book/details/8712/A-B-C-Of-Communism#.Worfa1WnFaQ.
102 Wedeen, *Ambiguities of Domination*, 21–2.
103 Wedeen, *Ambiguities of Domination*, 64–5.
104 Salwa Ismail, *The Rule of Violence: Subjectivity, Memory and Government in Syria* (Cambridge: Cambridge University Press, 2018), 97–129.

أهدافنا:

Military teacher: "Our objectives?"

وحدة، حرية، أشتراكية

Pupils: "Unity, freedom, and socialism."

عهدنا

Military teacher: "Our pledge?"

أن نتصدى للامبريالية الصهيونية وأن نسحق عصابة الاخوان المجرمين العميلة

Pupils: "To confront the Zionist imperialism and crush the treacherous Criminal Brotherhood gang."[105]

As is evident here, from the beginning of their formal education, students are thus introduced to political terms and concepts—and unquestionable obligations—that they may never have heard of before, such as "freedom," "socialism," "unity," "Zionism," "imperialism," and "Muslim Brotherhood." At the same time, it is considered taboo for parents to explain such concepts to their children, for several reasons. One obstacle is the conflict between their collective memory of what they have witnessed under the Baath regime and the regime's own use of these terms and concepts: it is, for example, very challenging to explain the meaning of "freedom" when this is clearly absent, and it is difficult to see how socialism can have any meaning for a six-year-old student. Ismail describes this experience among her interviewees: "Ruba and Mustafa recalled the sense of unease in the compulsion to proclaim statements that they did not understand or with which they could not identify."[106] Still more hazardous would be an attempt to describe the Muslim Brotherhood, and potentially convey an explanation other than the one approved by the regime. This is why military teachers have the responsibility of explaining these terms to students and spreading the military ethos in schools, including through military uniform checks, military discipline, and military-style hair and makeup. The daily repetition of such terms without awareness of their meaning results in individuals superficially performing obedience to these duties. The Baath Party slogan, like others, is performed as an act of imitation without attention to meaning: as Wedeen describes, individuals behave "as if" they believe in them. This is why such slogans face problems with performativity; they make no substantive changes and

105 Here "Criminal Brotherhood" is meant to refer to the Muslim Brotherhood: the word "Muslim" was removed from the slogan and replaced with "Criminal."
106 Ismail, *Rule of Violence*, 103–4.

betray that their performance is an act. They are merely rituals. Over time, students perform the slogans clumsily while fumbling and stuttering. The performance then lacks the dynamism and affect of a speech act. It fails in its direct objective, which is to give meaning to these terms, but achieves a further goal that is more important than comprehension: the obedience that Assad's regime forcibly (and successfully) requires from Syrians.

Outside of academic life and educational spaces, the military has its own space in which slogans were also used to control soldiers' military life in Syria.[107] Since most Syrian men do military service, the military space is one in which the regime could easily disseminate its discourse, particularly as the period of military conscription typically follows on from the end of university or other schooling. Similar to other spaces like schools, streets, and governmental offices, military spaces were also filled with many symbolic products like slogans. Slogans used in the military were composed according to Baath ideology. For example, soldiers were forced to repeat the following slogan while exercising:

زمجر. . . زمجر بغضب. . . وأنتقم من أعداء بلادي.

[Roar. . . Roar angrily. . . take revenge on my country's enemies.]

If we compare this to the Baath Party's main slogan of "one Arabic nation," it is clear that the slogan is absurd and performed without meaning, and ignores the questions that it raises: who are the enemies of the country, and what revenge should be taken? According to the Baath Party's main slogan, the enemies of the country are imperialism, Zionism, and the Muslim Brotherhood. How can revenge be taken against those enemies if no trace of them can actually be found in Syria? All of these slogans are thus merely empty performances provided by soldiers who do not understand them, which makes their performance closer to a ritual. In addition to slogans like this, during Hafez's rule, murals praising him were found everywhere on the walls of military units, such as one with the famous poem by Muḥammad Mahdī al-Jawāhrī:

---

107 As of 2008, the Syrian military had 215,000 active men. Military service in Syria "is obligatory for 24 months for all men from 18 to 42," as per legislative decree number 30 (2007), though some exemptions are offered, for example for those who are the only male in their family. For more information, see: Anthony H. Cordesmann, *Israel and Syria: The Military Balance and Prospects of War* (Westport: Praeger Security International 2008), 166; Syrian Parliament, "*al-Marsūm al-Tashrīʿī Raqam 30 li-ʿĀm 2007 Qānūn Khidmat al-ʿAlam*" [Legislative decree number 30 in 2007 for military service], Syrian Parliament official website, accessed July 18, 2023, http://parliament.gov.sy/arabic/index.php?node=201&nid=4921&.

سلاماً أيها الأسد.
سلمت وتسلم البلد.
وتسلم أمة فخرت.
بأنك فخر من تلد.
وانك انت/ موعدها.
وفجر غدٍ وما يعدّ.

[A smart salute [peace be upon you] to you, Assad.
Remain protected intact with the country.
A nation which is proud of your noble progeny shall remain protected intact.
That nation is proud to have you [Assad] as its long-awaited promise.
And the dawn of tomorrow.]

In order to analyze this verse, it is important to understand the event that occasioned this praise: Al-Jawāhrī composed this verse about Assad after he survived an assassination attempt in 1980, and it is a linguistic document that marks the beginning of the cult of Assad.[108] This poem resulted in “a long stay in Damascus and a personal meeting with Assad” for al-Jawāhrī.[109] What is striking about these lines is the way in which they present Assad: not only as a leader but also as a godlike figure. Above I have provided two possible translations for *salām* [peace]: it can be interpreted first as a salute to Assad, but also in the sense of greeting for holy figures like prophets and saints—for example, when referring to the prophet Muḥammad, one should add “peace be upon him.” It thus provides a linguistic indicator of how language describing Assad had changed. At the time, Assad was so pleased with the poem that he brought its author to Damascus to inscribe it for eternity on all the military barracks’ walls. The examples shown above, in particular the latter one, paved the way for Assad to gradually be presented as immortal in slogans during the three decades of his rule of Syria, as I will discuss in the following section.

Assad and Forever(ism)

Within its monopoly over the military and educational milieus, the impression given by the regime’s slogans is that over time, they had a specific development on the level of content and the language that was used. Classical Arabic was preferred by the Hafez al-Assad regime, especially in slogans, murals, banners, songs, and photo captions, since Assad and his circle saw themselves as the authentic

---

108 *Zaman al-Wasl*, *“Yawm Muḥāwalat Ightīyāl Hafez al-Asad 26 Ḥizīrān 1980”* [The Day of the Assassination Attempt on Hafez al-Assad: June 26, 1980], June 27, 2014, accessed July 18, 2023, https://www.zamanalwsl.net/news/article/51087.

109 *Arab Encyclopedia* s.v. “al-Jawāhrī, Muhammad Mahdi,” accessed July 18, 2023, http://arab-ency.com.sy/detail/3663.

source of Arabism, or at least as synonymous with it. Patrick Seale notes this in his book, mentioning Assad reciting the "classical poetry of Hassan Ibn Thabit"[110] and as "insisting on a high standard of Arabic prose in all statements or letters issued under his signature, and himself making time to read classical Arabic."[111] Slogans referring to Hafez al-Assad can be divided into several phases. In the first phase of his rule, the slogan strategy was limited to the Baath ideology, without focusing on the president. One example of such a slogan is:

اليد العاملة هي العليا في دولة البعث

[The upper hand is the one that works in the Baath state.]

This focus on work is consistent with the socialist ideology of the Baath Party. What characterizes Baath slogans is that they lack any musicality or rhythm. They were composed only to convey direct and clear concepts from the regime. Later, there was a clear shift in slogan development, from Baath ideology to a focus on Assad, for example in:

نحن فداء الوطن والعروبة والقائد المفدى

[We sacrifice ourselves for the country, Arabism, and our ransomed (beloved) leader.]

The translation of the Arabic word *mufadda* as beloved leader does not convey the full meaning of the Arabic phrase. It means a beloved leader for whom people will sacrifice everything or, more specifically, a leader who deserves the sacrifice of his peoples' blood and souls. Here it thus invites Syrians to exchange their lives and blood for Assad's, prolonging his life at the expense of their own if necessary. The order of the slogan is also notable: it first mentions the homeland, followed by Arabism, and then the leader, reflecting the gradual move in slogans to include Assad as part of the homeland and Arabism.

While focusing on Arabism and the idea of an Arab nation, slogans began to display a mixture of Baath ideology and Assad himself as a leading figure in Arabism. The following slogan, for example, represents Assad as a symbol of the Arab nation. Here and in the slogans that follow, the Baath ideology is in the background and Assad is presented as a leader, preaching its ideas to the entire Arab world:

حافظ أسد رمز الأمة (الثورة) العربية

[Hafez... Assad... the symbol of the Arabic nation [revolution].]

110 Patrick Seale, *Asad: The Struggle for the Middle East* (Berkeley: University of California, 1990), 13.
111 Seale, *Asad*, 180.

This blend did not last for long, as the concepts of Arabism and homeland were eliminated from slogans to focus solely on Assad, for example in:

بالروح والدم نفديك يا حافظ

[Hafez, we redeem you with our soul and blood.]

This shift made Assad the focal point and decentered Baath ideology from the Syrian political scene, paving the way for a new type of slogan worshipping Assad while excluding Baath ideology and Syria. For example:

الأسد إلى الأبد (إلى الأبد يا حافظ الأسد)

[Assad forever.]

This slogan demonstrates a new development in Assad-related themes and represents the third phase, immortality, in which Assad rules forever. The strength of this slogan is that it does not explicitly specify whether it refers to Hafez al-Assad himself or the Assad family as a whole. During this period in the 1990s, following the victory over the Muslim Brotherhood, this slogan was meant to present Assad as an immortal leader. The importance of this linguistic indicator is that it marks the beginning of Assad as a god to Syria: a person who would never die and could act without restraint or consequence. Believing in this truth, the regime worked hard to describe Assad as more godlike, as seen for example in *Thus Said Assad.*[112] The process of presenting Assad as godlike thus took time and was conducted gradually, moving him from president, to leader, eternal leader, and, as in the following example, comparing him to the Prophet Muḥammad:

قائدنا إلى الأبد الأمين حافظ الأسد.

[Our Leader forever; the honest [trusted, faithful] Hafez al-Assad.]

The lines before and after the Sura, in the example below, provide the contextual meaning of *al-amīn* [trusted, faithful] that Assad wanted. This word is used only once in the Qur'an and refers to the Prophet Muhammad.[113] As such, the logic is clear that Assad is the prophet of Syria, whom Syrians should believe in because he is "endowed with power," just as the Sura below states in describing *al-amīn.* This change was well-planned and inserted gradually into the Syrian market. It was

112 Ṭlās (ed.), *Hakadhā Qāl al-Assad.*

113 It should also be noted that *al-amīn* is the highest rank in the Baath Party when it is written as *amīn al-ḥizib* [The Party Secretary or the Party Leader]. However, putting the definite article "al" in Arabic after the adjective opens up its interpretation as the only party leader or as an attribute of the prophet in the Qur'an.

not possible for Assad to use slogans like "forever" from the beginning; instead, the regime gradually elevated Assad to the status of a saint or prophet during this phase.

19. إِنَّهُ لَقَوْلُ رَسُولٍ كَرِيمٍ
20. ذِي قُوَّةٍ عِندَ ذِي الْعَرْشِ مَكِينٍ
21. مُّطَاعٍ ثَمَّ أَمِينٍ

[19 This is the speech of a noble messenger.
20 Endowed with power, eminent with the Lord of the Throne.
21 Obeyed and honest.][114]

Assad's rise to the rank of prophet and the gradual escalation of his importance was strengthened not only by calling himself *al-amīn* but also through the use of certain Islamic terms and imbuing himself with attributes associated with the prophet Muḥammad at the time of his re-election. The attributes of Assad in slogans and language used by the regime were accompanied by a holy term meant to be an additional description he wanted to be associated with himself: *bay'a*, an Islamic term that means "a pact between a ruler and the ruled. The ruler undertakes management based on justice and is responsible for Muslims' interests."[115] The first *bay'a* was with the Prophet Muhammad and is known as "*Bay'at al-'Aqaba.*"[116] Using Islamic terms in the Syrian linguistic sphere reflected the position that Assad wanted to occupy, in the same class as prophets and saints, as confirmed by calling himself *al-amīn* in the previous slogan. The similarity between *bay'a* and *al-amīn* is that both of these terms were used for the Prophet Muhammad.

This can be seen by tracing the terms the regime used to describe the act of "electing" Assad through his speeches after the "elections" in 1978, 1985, 1992, and 1999.[117] In his 1978 speech, Assad said, "I am very glad you re-elected me."[118] This was a period in which he was just establishing the roots of control over Syrians that

114 The Qur'an, At-Takwir, Sura (81:19–21), trans. *OneUma.net*, September 20, 2018, accessed July 18, 2023, https://www.quranful.com/.

115 Ṭāha Ḥusayn, *al-Majmū'a al-Kāmilah al-jiz' 4* [The Complete Collection: Vol 4] (Beirut: The International Company for Books, 1982), 59.

116 Muhanna Al-Hubil, "assīrah alnabawyah Wa al'ālam alJadīd Taṣḥīḥ almafāhīm [The Life and Traditions of the Prophet and the New World: Correcting New Terms], *Al Jazeera*, December 13, 2016, accessed July 18, 2023, https://goo.gl/1pgDXi.

117 Muhammad 'Abdū Al-Ibrahim, "Assad's Speeches," President Bashar al-Assad unofficial website, accessed September 20, 2018, http://www.presidentassad.net/index.php?option=com_content&view=categories&id=88&Itemid=476 (no longer available).

118 Hafez al-Assad, "Khitāb al-Sayid al-Ra'īs Hafez al-Assad ba'da 'dā'ihi al-Yamīn al-Dustūrī Amām Majlis al-Shsha'b 1978" [Address Following the Constitutional Oath Upon Re-election to a Second Term in Office], March 7, 1978, People's Assembly, President Bashar al-Assad unofficial

would allow him to remain the leader forever, and thus a time in which it was still impossible for him to claim he was Syria's only possible leader. In 1985, however, Assad described his *iʿādit intikhāb* [re-election] as an *istiftāʾ* [referendum], not an election.[119] This indicated that he wanted to remain its leader forever and represented a gradual development in the terms used by the regime, since there is a significant difference between an election and a referendum. Here, it is important to be aware that in 1985, Assad had just finished eradicating the Muslim Brotherhood from Syria and was thus finally in a position to declare himself the prophet of Syria without threatening his kingdom. Finally, in Assad's 1992 speech, he used *bayʿa* as a new term clearly and explicitly in place of election or referendum, stating at the beginning that the "people's *bayʿa* was precious in its meaning and great in its content where millions took to the streets, squares, and cities to declare to the entire world that national unity in Syria became an integrated political, economic, and social system."[120] This was Assad's first public use of the term, and indicates the internal and external stability of his political situation. It marks the point at which Assad had completely dominated Syria discursively.

Assad did not stop with the rank of prophet: his attributes instead reached the level of a god in a chant for Syrians loyal to Assad:

حلك يالله حلك يطلع حافظ محلك

[Oh, God! It is time to give your place to Hafez.]

I have personally heard this sentence repeated by Assad supporters, but there is no independent source documenting such a slogan about Hafez. These lines were, however, continually present in the Syrian unconsciousness.[121] It is therefore unsurprising to find another slogan that says:

---

website, accessed September 20, 2018, http://www.presidentassad.net/index.php?option=com_content&view=article&id=655:7-3-1978&catid=211&Itemid=476 (no longer available).

119 Hafez al-Assad, "Kalimat al-Sayid al-R'īs Hafez al-Assad bi-Munāsabat Tajdīd al-Bayʿa 1985" [Speech After the Pledge], 1985, President Bashar al-Assad unofficial website, accessed September 20, 2018 http://www.presidentassad.net/index.php?option=com_content&view=article&id=572:11-2-1985&catid=267&Itemid=493 (no longer available).

120 Hafez al-Assad, "al-Sayid al-R'īs Hafez al-Assad bi-Munāsabat Adā'ihi al-Qasam al-Dustūrī li-al-Dawra al-Riʾāsya al-Rābiʿa 1992" [Speech After Taking the Constitutional Oath Following His Fourth Election and Celebrating the Twenty-Ninth Anniversary of the Glorious Eighth of March], March 1992, President Bashar al-Assad unofficial website, accessed September 20, 2018, http://www.presidentassad.net/index.php?option=com_content&view=article&id=624:12-3-1992&catid=274&Itemid=493 (no longer available).

121 While I have been unable to find an independent source that includes this slogan, I have personally heard it repeated by Assad supporters, and several of my interviewees concurred that these lines had a constant presence in the Syrian unconsciousness.

لم يتبق لنا إلا أنت

[[There is] nobody left for us but you [Hafez]]

This slogan is very similar to the famous Islamic profession of faith: "There is no god but God." As in the other spaces discussed in this research—visual products, musical products, and books—Hafez al-Assad thus gradually became godlike in the language of slogans, via a visual and linguistic path from being Syria's president to its only leader, a leader and symbol of the Arab world and its symbol, to garnering the characteristics of saints and prophets through "foreverism." Finally, he appointed himself the god of Syria, as demonstrated by slogans and other products.

Table 1: The gradual development of Hafez al-Assad's slogans.

Slogan (English translation)	Importance	Slogan (original Arabic)
The working hand is the highest in Baath state	Baath ideology; non-musical slogan	اليد العاملة هي العليا في دولة البعث
We sacrifice for our Beloved Leader	Non-musical slogan; Assad is mentioned	نحن فداء الوطن والعروبة والقائد المفدى
Hafez al-Assad, the symbol of the Arab Nation	No Syria or Homeland; only Hafez al-Assad as the leader	حافظ أسد رمز الأمة العربية
Hafez, we redeem you with our souls and blood	People sacrificing their souls for the leader	بالروح والدم نفديك يا حافظ
Hafez al-Assad forever	Immortality	إلى الأبد يا حافظ الأسد
Our trusted leader, the trusted one, Hafez al-Assad	Prophet status	قائدنا إلى الأبد الأمين حافظ الأسد
Oh, God! It is time to give your place to Hafez	Godlike status	حلك يالله حلك يطلع حافظ محلك
[There is] nobody left for us but you [Hafez]	Godlike status	لم يتبق لنا إلا أنت

Everyday Terms

With the Syrian symbolic market immersed in Assad's symbolic products, everyday Syrian language produced new, previously unused terms that drew their meanings from the Syrian context, with all of its social, cultural, and political dimensions. This creation of new words and terms that had no prior equivalent in Syrian colloquial or Arabic language is a strong indicator of how deeply Assad's symbolic products affected Syrian everyday life.

Bījū *and* Marsīdīs

In most of the world, Peugeot and Mercedes are car brands like any other—but this was not the case in Assad's Syria. In the early Baath era, and particularly during Assad's rule, Peugeot cars represented not just a means of transport, but authority. At the time, a Peugeot car owner was necessarily either an employee of the state or a member of the rich bourgeois/landowner class.[122] Peugeot cars were the first to be imported to Syria by the government and were mostly distributed to government departments, which then gave the vehicles to important workers, primarily *mukhābarāt* [security/secret police] agents or those who had proven to the regime that they were loyal to the bone. Peugeot cars spread throughout *mukhābarāt* detachments in small and large cities, conveying the sense that these drivers were ready to arrest people at any time. When there was an arrest campaign, most of the cars used were Peugeot. These cars thus developed a bad reputation and became a source of fear for Syrians. As one journalist noted, whenever a "Syrian sees one of them, it is better to take a different street."[123]

While Peugeot cars were associated with the wealthy or the *mukhābarāt*, Mercedes vehicles were associated with high-ranking officers in the regime. It was assumed that the person inside them was an important person in the government, notably because the windows were usually tinted black, keeping the passenger hidden, and because of the aggressive and bold way such cars were generally driven. Mercedes cars were called *shabaḥ* [ghosts], because the person inside the car was, like a ghost, typically unseen. These vehicles were used as a tool for buying officers' loyalty or satisfaction and a key instrument of bribery and a sign of corruption. What distinguished these government cars was that they came with a free monthly fuel card. Officers usually used one car and sold the fuel cards to other car owners to earn additional income.

As such, Peugeot and Mercedes cars have a very different contextual meaning in Syria compared to the rest of the world. The social facets of the words were added to them, and the terms *Bījū* and *Marsīdīs* were coined for Peugeot and Mercedes in the Syrian linguistic realm. Even though Mercedes cars are sometimes called *shabaḥ* in other Arab countries, the meaning is different in Syria. These words have a different functional role in the Syrian market, in which the economic product of a car was turned into a product that caused fear and was associated with violence. These vehicles are an instrument of repression and violence, serving either as a

122 Samīr Sʿīfān, "al-Niẓām al-Sūrī wa ʿIqdat al-Sayyārāt" [The Syrian Regime and the Complex of Cars], *Zamān al-Waṣil*, December 6, 2015, accessed July 18, 2023, https://www.zamanalwsl.net/news/article/66728/.

123 *Enab Baladi*, "Thalāth Sayyārāt Yakrahuhā al-Sūryūn wa-Tufaḍiluhā al-Mukhābrāt" [Three Cars Syrians Hate and Secret Police Prefer], August 8, 2016, accessed July 18, 2023, https://www.enabbaladi.net/archives/100679.

symbol of *mukhābarāt*—indicating someone's arrest—or the presence of an important person or *shabaḥ*. The prevalence of an alternative term for a specific car brand not only indicates the peculiarity of the Syrian context but also the presence of additional layers of meaning as an extension to the meaning of *shabaḥ* already associated with the vehicle, as we will see below.

Shabbīḥa

It has generally been accepted that *shabbīḥa* is the word for riders of *shabaḥ* (Mercedes cars), as discussed above. Lexical research of this word in the Syrian context reveals many layered meanings, the reasons for them, the origin of the word, and what it was derived from. It is important to note that while the term *shabbīḥa* was also used in the 2011 protests, there it had a different meaning due to the different context. I will therefore return to this term with regard to the changes that occurred under the rule of Bashar al-Assad and its development in 2011 in later chapters.

The linguistic root of *shabbīḥa* is the verb *shabaḥa,* which means to stretch, straighten, or extend one's body or something else. *Shabaḥa Fulān* means to stretch a person's body for lashing. The imperative form in the Syrian dialect is *ishbaḥ fulan,* which means "hang him by his hands." It is probable that the concept of *shabbīḥ* developed from *shabaḥ* [ghost] because people look like ghosts to detainees due to their exhaustion and blurred vision caused by torture. The agent (subject/doer) of the verb is not *fa'il* but *shabbīḥ*, which stresses excessive confirmation of the action. It is *ṣīghat mubālagha* (the hyperbolic form), which means a mode that shows exaggeration. It is derived from the three letter verbs into different forms such as *fa'īl, fa''āl, fa'ūl. Shabbīḥ* means a person who does a lot of *tashbīh* actions. Despite the richness of the linguistic meaning of the term, the social context adds more depth to it.

In its Syrian social context, the general meaning of *shabbīḥa* is a group that breaks the law under the protection of the creator of the law (the regime). Such groups smuggle goods from neighboring countries. There are two different types of *shabbīḥa*: the first group smuggles goods—such as foreign brands and products the regime does not provide cheaply enough—into Syria from border cities, while the second group uses cover from the regime to smuggle illegal goods like cigarettes and drugs, as well as engaging in other activities.

The first group of smugglers are locals and do not have the negative connotation of those associated with the regime. Smuggling was frequently the only way to earn a living in poor villages, such as Maḍāyā and Sirighāya, which suffered from impoverishment, marginalization, and a lack of development projects, thereby encouraging locals to work as smugglers. Smuggling created new markets for foreign products, including hunting rifles, hunting knives, electric devices, fireworks, and almost anything else not available through the official Syrian market. Despite the fact that they were smugglers acting in a similar way to *shabbīḥa*, Syrians did not refer to this group as *shabbīḥa* because they were seen as a source for extraordinary goods, especially given the lack of high-quality foreign goods in Syria. The regime did not condone the illegal market but, due to their mountainous village locations and the difficulties of reaching those areas, preferred to combat smuggling at the border and leave the villages alone. While this group were technically *shabbīḥa*, they did not have a bad reputation and brought cheaper foreign brands to Syrians. In addition, they maintained a peaceful relationship with locals and did not hurt civilians. They were thus differentiated from the second group of *shabbiha*, and seen in a more positive light. because they did not harm the public but offered them goods that would have cost people more if they bought them on the official market.

The second group of *shabbīḥa* are gangs organized directly or indirectly to achieve the interests of the regime, which may include kidnapping, killing, or terrorizing people, in addition to financial interests. The *shabbīḥa* gangs were created by the regime itself, and did not have the popular support of the first group. They worked for money and through loyalty to the regime. In addition to this, they did not offer anything to the people but were a constant source of danger. Some claim that the first *shabbīḥa* group was "founded by Hafez al-Assad himself before he took power in 1966 and it had the name of the al-Kata'ib al'Umālyya [Labor Brigade]."[124] The Labor Brigade helped ensure the success of Assad's *coup d'état* and was later disbanded to show that he was the protector of the country. The first *shabbīḥa* group founded under Hafez al-Assad's reign was Sarāyā al-Difāʿ [Defense Brigade] led by Rifʿat al-Assad (Hafez's brother).[125] The *shabbīḥa* phenomena was largely invisible outside of the coastal cities: it was not entirely absent elsewhere, but was limited since the center of *shabbīḥa* was Lattakia. What is striking are the titles that *shabbīḥa* call their bosses: *muʿalim* [teacher/master] or *khāl* [uncle].[126] These ambiguous terms make it unclear who is being referred to, which may be an important officer in the regime or even higher depending on the tone and the situation in which the term is used. If a person claims to have orders from *muʿalim*, it begs the question as to which *muʿalim* and could definitely reach the biggest

---

124 Hadīr al-Zahhār, "*Shabbīḥat Firaq al-Mawt li-niẓām al-Assad*" [Shabbīḥa. . . The Death Squads of the Assad Regime], *al-Ahram*, May 5, 2012, accessed July 18, 2023, http://www.ahram.org.eg/archive/Journalist-reporters/News/149487.aspx.

125 Yassin al-Haj Saleh, "*Fī al-Shabbīḥa wa-al-Tashbīḥ wa Dawlatihimā*" [On Shabbīḥa, Tashbiih, and Their Country], *Souria Houria*, January 30, 2012, accessed July 18, 2023, https://goo.gl/vRHraK.

126 Mamdūḥ ʿIdwān, *Ḥaywanat insān* [Human's Animalization] (Damascus: Mamdūḥ ʿIdwān Publishing House, 2007), 134.

mu'alim of all: Hafez al-Assad. In the chapters that follow I will return to this term and how it changed before and during the revolution.

The Discursive Apparatuses of Bashar al-Assad

The accumulation and distribution of symbolic products remained as it had been under Hafez al-Assad after his death in 2000.[127] He was the president who would not die, described in all discursive products as the eternal leader of Syria. There was a great deal of tension on the streets of Damascus, as all shops suddenly closed before the official announcement. What would happen? Who would become president after him? Was it true that Rif'at Assad would return to Syria and become president, or would there be a *coup d'état*? All of these worries and rumors circulated among Syrians. Regardless of the political aftermath of the death of Hafez, the rumors and fears of Syrians, and the details of the parliamentary assembly amending the constitution to allow Bashar to succeed Hafez as president,[128] the symbols of Syria—visual, musical, and in all other possible ways of using language as a medium—anticipated that Bashar al-Assad would be the new president of Syria after Hafez's death. Images of Hafez al-Assad and his sons were everywhere in Syria. At the beginning of what I have termed the visual immersion, only Hafez al-Assad was present, but the president later communicated a message to Syrians by including his sons Bassel and Bashar with his images in all public and private facilities. The presence of Bashar al-Assad alongside his father in all visual spaces prior to Hafez's death made it clear that Bashar was in place to become the new president of Syria after the death of his elder brother. In the following section, the same symbolic products that were illustrated above will be retraced to examine the relationship between the periods of Hafez and Bashar al-Assad.

Theoretically speaking, it is impossible to create a quick shift or a radical change in the dispositions of individuals that affects their practices, perceptions, and attitudes and change the rules of competing in the market. Habitus is "a set of dispositions which incline agents to act and react in a certain way."[129] Practices, perceptions, and attitudes are created by these dispositions. According to this defi-

---

127 *Al-Jazeera*, "*Wafāt al-Ra'īs al-Sūrī Hafez al-Assad bi-Azma Qalbyya*" [The Death of Syrian President Hafez al-Assad from a Heart Attack], June 11, 2000, accessed July 18, 2023, http://www.al-jazirah.com/2000/20000611/fr2.htm.

128 CNN, "Syrian President Hafez al-Assad Dead at 69," June 10, 2000, accessed July 18, 2023, https://edition.cnn.com/2000/WORLD/meast/06/10/assad.03/.

129 Pierre Bourdieu, *Language and Symbolic Power*, trans. Gino Raymond (Cambridge, MA: Harvard University Press, 1993), 12.

nition, the practices, perceptions, and attitudes of the individual are created by the dispositions generated within the habitus. The most important attribute of these practices is that they are performed consciously or unconsciously as a result of the regular relationship between individuals and the surrounding institutions, which results in the distribution of products in the symbolic or economic markets. Individuals are gradually enculturated due to acts of internalization and externalization and engagement with the institutions in the social context in which they live. This drives individuals to compete in the symbolic market and to accumulate products. The most important characteristics of individual dispositions in the field are generative, durable, transposable, and structured, which means that they reproduce themselves, last, and are transposable with its products. A habitus that contains such dispositions does not change quickly. When Bashar al-Assad came to power, the Syrian symbolic market did not establish entirely new practices and strategies that would lead to a change in practices on one hand and a change in the relationship between individuals and the market on the other. Among others, the institutions, tools, and discursive apparatuses discussed in my analysis of Hafez al-Assad's rule were key factors responsible for inclining individuals to acquire new knowledge and strategies. After Bashar inherited the kingdom that his father had created over decades in Syria, the competition for the same symbolic products remained, but was slightly different.

Development, Rather Than Sudden Change to, the Old Discursive Domination

The graph of the number of political prisoners discussed in the introduction shows how the first few years under Bashar al-Assad saw more prisoners released and fewer new arrests.[130] Syria was entering a new phase with a new president whom they hoped—and whom many believed— would not be like his father. Bashar was handed a Syria that had been tamed and dominated by the old discourse that his father had created via visual, musical, and book products, providing a landscape and a setting that was already constructed and ready to be built over and improved. Tracing the same discursive apparatuses that created Hafez al-Assad's discursive domination shows that very little changed. In terms of the media sphere and the Ministry of Information, there were few changes. Even the Ministers of Information were not as notable as previous ministers, because their predecessors had established the ground rules for taking over all of Syria's possible visual, musical, and personal spaces. The struggle thus moved from the regime institutions and

130 See Appendix 1.

individuals to an internal conflict between the ministers themselves. For example, according to Lamā Aḥmad (the daughter of Iskandar), "Salmān [the first Minister of Information during Bashar al-Assad's rule] burnt and destroyed Aḥmad Iskandar Aḥmad's archives in the Radio and Television General Authority because of the patronage he had when Hafez al-Assad was alive."[131]

Nationalist songs glorified the country's new leader, describing him as the hope of Syria. Bashar did not need to gradually work on his cult status to become godlike for Syria, but rather inherited it from his father. He thus inherited even the symbolic domination, albeit with slight changes and developments. First, he was often described based on his physical appearance, and since he was only in his thirties when he ascended to the presidency, the initial discourse described him as Syria's "hope." Gradually, he inherited his father's titles. Second, he was more modern than his father, because he had lived in England and had direct contact with the "developed world," and was thus also dubbed "the leader of the march of modernization and development in Syria." While Hafez al-Assad was the leader of the Corrective Movement of Baath, his son represented the revival of this revolution through the modernization of Syria.

It was not difficult to take over and develop musical products for Bashar. Most of the symbolic products that used the name Assad remained but referred to Bashar instead of his father, and "Hafez" could easily be removed from symbolic products or changed to "Bashar" if necessary Unlike during his father's rule, after Bashar came to power a number of very famous pop singers glorified the president in various songs. Others, including George Wassuf, Wael Kfoury, Najwa Karam, Wadih El Safi, Assi El Hallani, Fares Karam, Rouwaida Attieh, sang for him, all attempting to present him as Syria's only hope of becoming a modern country.

One notable example of the symbolic products or immersion discussed above was performed by Bashar al-Assad during his 2007 election campaign. While Hafez al-Assad did not directly use the word *bay'a* in most of his election victory speeches, Bashar explicitly did so, without hiding behind terms like "referendum" or "election" as his father had done. As discussed above, despite the term's religious background, it is a normal act that authoritarian regimes do regardless of how they label the act of extending their rule over a country.

The significant leap made by Bashar al-Assad was in terms of the genre and content of songs. Most of the songs about him were in the Syrian dialect and in the form of pop or modern music, whereas the nationalist songs of his father had been in classical Arabic and used a military rhythm. In addition to new modern rhythms, colloquial Arabic was also used instead of MSA. Colloquial Arabic had started to be

131 Author interview with Lamā Aḥmad, May 16, 2017.

used during the last years of Hafez's rule but its use was empowered by Bashar's symbolic products. In terms of their content, all of these songs celebrated Bashar as a natural extension of Syria with reference to the Baath Party. Lebanese singer Najwa Karam, for example, directly refers to this in one of her songs about Baath and the Assad family:

بدنا نحافظ بدنا نجاهد بالروح البعثية... ونسلم بشار القائد رايتنا السورية

> [We want to protect and struggle with Baath spirit... and hand over the Syrian leadership to Bashar, the leader] [132]

As is clear here, the verb *nḥāfiz* [protect] is a pun that refers to Hafez as its stem is Hafez, the father of Baath in Syria. The second part refers to full obedience in giving Bashar leadership of Syria. It is important to note here that the singer is Lebanese, and that she was singing for Syria on the occasion of the new *bay'a* of Bashar al-Assad. The lyrics of the song also refer to handing over leadership, which is an act similar to *bay'a.*

The most famous song about Bashar was by Shahad Barmada and promoted by Syriatell, one of the Assad family's companies. The song is called "Minḥibbak" [We Love You].[133] The distribution of such songs in the Syrian market marked a turning point in the way Bashar was presented, in contrast to his father. The way that the song was performed, particularly during Bashar's "election" campaign, was rich with meaning. A video on YouTube shows an incredible performance at the al-Maḥaba Festival (also called the Festival of Bassel, literally meaning Festival of Love) in Lattakia Stadium involving hundreds of dancers, very similar to the opening of the Mediterranean Games in 1987. The stadium setting and hundreds of participants suggest the song is forcing them to comply with the authority of the regime in a very direct and simple way to show obedience. But while the 1987 Games involved a highly disciplined performance focusing on a military parade and a show of strength, the performance of "Minḥibbak" involves processions representing all communities, professions, and youth and children saluting and dancing to the song: policemen, doctors, students, old women (representing mothers), and others dance, representing the Syrian people, as expressed in the lyrics. The song starts by repeating the song refrain "we love you," then defines the pronoun "we" in the song: "We

---

132 Najwa Karam, "*Najwa Karam Bashar al-Assad al-Qā'id – Ugnya li-Bashar al-Assad Bidnā Nḥāfiz*" [Najwa karam Bashar al-Assad the leader – A Song to Bashar al-Assad, we want to protect], *Tarek Karam* (YouTube channel), uploaded April 2, 2007, accessed July 18, 2023, https://www.youtube.com/watch?v=W_v5tkrIoeU.

133 Shahad Barmada, "*Minḥibbak*" [We Love You], *Addounia Shabeha* (YouTube channel), uploaded February 14, 2013, accessed July 18, 2023, https://www.youtube.com/watch?v=PFPz0dlVT6s.

are your family. . . We are your people. . . We love you." It then goes on to explain who the Syrians are that love Bashar:

> The coming generation, the people of civilization, we are the people who appoint your slogan as ours, we are the teachers, the civilization devotees, the country guardians, farmers, workers, and doctors.[134]

Comparing the themes of this pop song about Bashar to the Baath anthems, for example, the major difference is the pop genre, since the objectives of the Baath Party, as outlined in its anthems, are all mentioned in "Minḥibbak." The fact that a pop song could function as a nationalist song and be sung in the Syrian dialect, in contrast to the Baath songs in classical Arabic, represented a major development in Syria. Another indication of the shift is that the song itself does not mention the Baath Party at all, suggesting that Baath was merely an instrument for the Assad family to take power during Hafez al-Assad's time. "Minḥibbak" became not only the best-known song about Bashar but was also used as the slogan of his "election" campaign.

At the visual level, nothing changed. Like his father before him, Bashar al-Assad used all possible ways of promoting himself. He was the leader now, not the son of the leader, and the images of Hafez were thus gradually replaced with images of Bashar or both images were put side by side. In 2007, the slogan "We love you" was everywhere, and the campaign launch provided a useful opportunity to remove the old images of Hafez al-Assad. Radio, television, internet, and cellular networks promoted "We love you," and there were TV competitions, various celebrations and concerts also called "We love you."

Banners in bus stops spread displaying a fingerprint—referring to voting—on a background of the Syrian flag. This sent an indirect message that, like his father before him, Syria was Bashar and Bashar was Syria. The words "We love you" did not disappear with the election but became engraved in the Syrian unconsciousness. This highlights the importance of visual products in controlling Syrians. In 2007 there was an "election/referendum" in Syria in which only one candidate stood: one in which there was no competition. Taking its lead from Hafez al-Assad's regime, the visual and musical products used in the campaign were tools for the new regime to control Syrians. The domination started by killing and torturing, and then became symbolic, reminding the public of the violence but refraining from inflicting more physical violence when it was not necessary. What proved its efficiency was the system of rewards for using those products.

---

134 Barmada, "*Minḥibbak.*"

Another dynamic of individual reactions to this event was the atmosphere of competing to show loyalty and faithfulness to the regime, which can be seen as a struggle in Bourdieusian terms. The symbolic products represented here by banners, and later by slogans, were goods that people used to show their privilege. They demonstrated their loyalty through displaying banners pledging to vote for the leader. Personally, I observed banners being hung in front of shops by their owners. Some described them as a kind of protection spell, and these products were indeed used for protection from the regime, reflecting a kind of symbolic violence—though the existence of some people genuinely loyal to the regime and engaging in such behavior with no other motive, cannot be excluded. This participation in the direct or indirect production of such election-related products encouraged people to participate in order to gain something symbolic such as proving loyalty or, in some cases, even more. In this sense, and indirectly, the regime used symbolic products to force an obligation on Syrians and control them at the same time. For example, one of the banners I saw was presented by the al-Dhahabī business. If it was not present the next year or during the next national event, then the establishment would endanger its own survival, since it was obliged to show obedience to the regime in order to remain in business. When an individual or a group of people starts competing in the market struggle, they cannot simply stop, for if they try to do so, it is understood as disobedience to the regime. However, participation in this voluntary and well rewarded competition is also not obligatory. What all of these banners had in common is the fact that they not only described Bashar positively, but also told people what to do, for example with the slogan "yes is our word." This is what the campaign was all about: obliging or coercing Syrians to vote "yes" for the sole candidate and thereby imply that all Syrians wanted Assad to continue in power. Aside from the system of rewards, this environment made anyone opposed to Assad feel isolated, since the regime created an atmosphere around the campaign that these products were created by the people themselves—and this was in fact the case, given that if people did not participate, they would lose out badly, especially if they owned a shop or business.

In terms of publishing houses and the book market, the publishing process largely remained unchanged under Bashar al-Assad, although some new publishing houses were founded. Publishing a book required approval from the Ministry of Information, and an "Authorization of Deliberation" from the Arab Writers Union president ʿAli ʿUqla ʿIrsān, who remained in his position until 2005.[135] Importantly,

135 "*Ṣināʿit al-Kutub fī Sūriyya wa Tārikhuhā Wāqiʿuhā wa Āfāquhā*" [The Book Industry in Syria: Its History, Reality and Prospects], *al-Baath*, July 3, 2006, accessed July 18, 2023, http://www.aljaml.com/node/17471.

the Authorization of Deliberation could not be obtained if the work dealt with any of three interrelated concepts: religion, politics, and sex. Although there were numerous private publishing houses, self-censorship by such publishers meant that they did not cross the red lines indicated above.

The red tape of the Authorization of Deliberation to censorship and the approval of the Ministry of Information were not the only obstacles to publishing a book. The Arab Writers Union presidents that followed ʿIrsān, such as Ḥussīn Jumʿa (2005–14) and Niḍāl Ṣāliḥ (2014), also used it as a kiosk to publish and sell books by a close group of writers. Under the authority of the union, "they [could] appoint an officer as a writer or print for him a faked book too."[136] Dozens of books were published to praise and glorify Bashar to take the place of those published for or about his father. These included, for example, a book entitled *The Greatest of the Twentieth Century* that included Bashar al-Assad, despite the fact that he had begun his political life in the twenty-first century.[137] A military general also wrote a book about Bashar entitled *Thus Said al-Assad* in 2018.[138]

The role of the Baath popular organizations was ridiculed by Baath members, and official rituals, like the party slogan or party anthems, continued to be performed without meaning. To provide an example, at one of the regular meetings I attended in 2007 for non-full members (those who had joined the party recently) or new starters, after chanting the slogan the meeting consisted solely of chatting about pop songs and the best television series that people were watching. Another example drawn from my personal experience is the comic sketch in season 4 of the TV series Buqʿat Ḍawʾ ["Spotlight"] in 2004, called "Manābir" ["Platforms"]. This consists of five short sketches depicting political speeches and meetings in the Syrian political landscape under Baath Party rule, all of which showed how political meetings and speeches in Syria were trivialized.[139] These sketches showed how absurd

---

136 Ali Nāṣir, *"Manshūrāt Itiḥād Kuttāb Sūriyya bi-Rubʿ Dolār"* [The Publications of the Arab Writers Union for a Quarter USD], *Al-Araby al-Jadeed*, September 11, 2014, accessed July 18, 2023, https://goo.gl/S4JVtC.

137 *Syrian Arab News Agency*, *"ʿUẓamāʾ al-Qarn al-ʿIishrīn: Kitāb Yuṣalliṭ al-Ḍawʾ ʿalā 14 Shakhṣyya ʿĀlamyya min Ayqūnāyt al-Niḍāl ḍidda al-ʾIstiʿmār"* [The Greatest of the Twentieth Century: A Book Sheds Light on 14 International Figures from the Icons of Anti-Colonization], April 4, 2017, accessed July 18, 2023, https://www.sana.sy/?p=535189.

138 *Tishreen*, *"Thus said al-Assad,"* May 19, 2018, accessed June 1, 2020, http://tishreen.news.sy/?p=155603 (no longer available).

139 *Buqʿat Ḍawʾ*[*Spotlight*] Season 4, 2004: The five sketches: *Buqʿat Ḍawʾ Khiṭāb* 1 [Spotlight: Speech 1], *Sham Drama* (Youtube Channel), March 27, 2018, accessed July 18, 2023, https://www.youtube.com/watch?v=JKTFpK4Zc7E; *Buqʿat Ḍawʾ Manābir 2* [Spotlight: Platform 2], *Sham Drama* (Youtube Channel), April 18, 2017, accessed July 18, 2023, https://www.youtube.com/watch?v=ku9BwYpfEGY; *Buqʿat Ḍawʾ Manābir 3* [Spotlight: Platform 3], *Drama al-Fan* (Youtube Channel), January 3, 2017, it was to have any political activity in Syria represent the Baath political engagement in society. It was difficult for the generations that had been raised with these organizations since childhood not to belong to these organizations or recognize the symbolic products because most people had been coerced into joining them. This did not mean people were banned from joining them, but under Bashar's rule it was clear that the work of those organizations was absurd based on the symbolic products they released. Anthems were repeated and people went to the meetings of these organizations to obtain personal benefits, such as a job or a promotion. Meetings started and ended with the slogan, but in between people simply gossiped about everyday life: further proof that people were performing the rituals of Assad without believing in or finding meaning from them.

The Products

The domination of the Syrian linguistic and discursive public space by the regime's discourse paved the way for the extension of Hafez al-Assad's linguistic realm. Terms like *Bījū*, *Marsīdīs*, and even slogans became associated with new experiences and knowledge. This was for numerous reasons, the most important factor being the opening up of the Syrian economy, which changed the horizon of thinking due to the presence of the internet and new, modern Western goods. Everyday language, such as terms for cars, for example, developed along with economic openness. Under Hafez's rule *Marsīdīs* and *Bījū* were the best known, but after 2000 new vehicles were imported, such as the Hummer. Hummers were particularly imported for the Assad family or their close associates. The very famous Hummer displayed on the Mazzeh high road in Damascus with images of Bashar and the Syrian flag provides an example of how terms developed, from *Bījū* and *Marsīdīs* to Hummers and other vehicles that were used only by people close to the regime. Audi cars were imported only for retired officers, who could eventually sell them if desired. Despite the presence of new vehicles, the Marsīdīs shabaḥ car remained the best known, along with the *mukhabarat Bījū.*

The term *shabbīḥa* became more prevalent across all Syrian cities. Before Bashar al-Assad came to power, *shabbīḥa* were not seen in the big cities other than Lattakia, but were present only in the Lattakia region or in the border region when

accessed July 18, 2023, https://www.youtube.com/watch?v=S3MEaXVc1Uc; Buqʿat Ḍawʾ Manābir 4 [Spotlight: Platform 4], Drama al-Fan (Youtube Channel), January 21, 2017, accessed July 18, 2023, https://www.youtube.com/watch?v=VG1MMu14SpQ; *Buqʿat Ḍawʾ Manābir 5* [Spotlight: Platform 5], *Scoop Drama* (Youtube Channel), November 17, 2018, accessed July 18, 2023, https://www.youtube.com/watch?v=VG1MMu14SpQ.

engaged in smuggling operations. But during the early years of the Bashar regime, *shabbīḥa* gangs became more empowered and began to intervene in the daily lives of Syrians. The new economic openness increased the work of *shabbīḥa*, and new people were explicitly announced to Syrians as *shabbīḥa*, such as Fawwāz al-Assad or Muḥammad Tawfīq Assad (known as the Sheikh of the Mountain). Syrian drama and television programs began to use *shabbīḥa* as an everyday term. The earliest occurrence of *shabbīḥa* in Syrian drama I have been able to find was in 2006, in a series called ʿArabyāt (meaning "from around the Arab world" or derived from *ʿaraba* [vehicle]), while the earliest occurrence of the verb *shabaḥa* was in 1996, in a series called Abū al-hanā (the name of the series' main protagonist). In these series, the terms were used to mean someone who was outside the law and able to break it easily.

The End of "Forever"

The regime's method of using slogans changed significantly after Bashar came to power. The regime's cumulative linguistic experience had increased, and Syrians in the political habitus improvised new strategies to take part in market competition. The old slogans remained, but with slight changes or insertions. A slogan like "Assad forever" could still be used but with reference to Bashar. The name Hafez was directly replaced with Bashar in the slogan "*bī-l rūḥ wa-l damm nafdīka yā Bāshshār*" [With soul and blood. . . [we] redeem you Hafez/Bashar]. Colloquial Syrian was used extensively in political discourse, which had never been the case before Bashar's rule. Hafez al-Assad loved poetic language and the most eloquent language possible in almost every speech or slogan used by Syrians. In contrast, Bashar al-Assad tried to make changes to the language, while retaining classical Arabic for use at official events or among intellectuals. His attempts to use colloquial Syrian in interviews were successful, especially his expressions like *Sūriyya ʾAllāh hāmīhā* [God protects Syria].[140] This sentence was only said once before spreading and being seen on banners everywhere on the streets. The use of colloquial Arabic instead of MSA gave people the opportunity to express themselves outside of the classical Arabic Baath protocol for the first time. People composed simple colloquial slogans.

---

140 Bashar al-Assad, "*Min Kalimat al-Sayyid al-Raʾīs Bashar al-Assad fī-Jamiʿat Dimashq*" [Mr. Dr. Bashar al-Assad's speech at Damascus University], 2005, *cflitchannel* (YouTube channel), uploaded March 6, 2011, accessed July 18, 2023, https://www.youtube.com/watch?v=wiP0LIL0U1M.

To trace the slogans used under Bashar al-Assad, I drew on YouTube videos of Bashar's public speeches to support my personal observations of participating in several such events during my life in Syria in university or at school, before attempting to record any new slogans that were chanted for him. As mentioned above, most of the slogans were in colloquial Arabic, and new slogans were composed that had not been used in Hafez al-Assad's time. For example, when Bashar visited Hasaka in Syria in 2002, people chanted *Allāh yeḥmīk yā Bashar* [God bless you, Bashar] when he talked about the promised reforms.[141] People in Hasaka chanted *Sha'bb blāadī kulū Ynādī Bashar ḥāmī blādī* [People of Hasaka are crying "Bashar is the protector of my country"] when he talked about the people being in solidarity with him. When he talked about external threats, the people cried *Yā Bashar lā tehtamm 'andak sha'b bi-yishrab damm* [Don't worry, Bashar, you have a blood-drinking people], which is associated with manhood and bravery.[142] On his visit to Homs, the people cried *Allāh yeḥmīk yā Bashar* [God protect you, Bashar]. In one of his meetings with the Arab Lawyers organization, an individual stood up and said *Lamm yatabaqqā lanā illā ant* [[there is] no one left for us but you, Bashar], a declaration that directly maps onto some of the latter developments in the slogans used for Hafez al-Assad.[143] The "forever" slogans were slightly changed. Instead of "Forever Hafez al-Assad," it became *Allāh Sūriyya Bashar wa-bas* [God, Syria, and only Bashar]. This latter slogan denotes the new inseparable triad of Bashar, Syria, and God.

As described above, all these slogans were repeated in all of Bashar's visits to Syrian cities after he took power in Syria. All of the slogans consisted of normal praise or prayers for Bashar. While I will not explore the question of the spontaneity of these slogans in any great depth, there are two reasons to believe that they were indeed spontaneous. The first is that they were very simple and uncomplicated in their construction, and represented responses to speeches that contained numerous promises, particularly at the beginning of the Bashar era, when people honestly thought that he might represent real change for Syria. The second is that Bashar al-Assad used dialect in some sentences and expressions in an attempt to be more of a man of the people than his father. There may well also have been people inside the crowd accompanying the presidential parade who were prepped to shout

141 Bashar al-Assad, "*Kalimat al-sayyid al-Ra'īs Bashar al-Assad fī-Muḥāfaẓat al-Ḥsaka*" [Speech of Mr. President Bashar al-Assad in Hasaka], *cflitchannel* (YouTube channel), uploaded February 27, 2011, accessed July 18, 2023, https://www.youtube.com/watch?v=Nb9tvI34Usk.

142 The literal translation here is "Don't worry, Bashar, your people can drink blood for you."

143 Bashar al-Assad, "*Kalimat al-Ra'īs al-Assad fī-Mu'tamar al-Muḥāmīn al-'Arab*" [President Assad Speech at the Arab Lawyers Conference, January 21, 2006], *Syria RTV* (YouTube channel), uploaded July 15, 2013, accessed July 18, 2023, https://www.youtube.com/watch?v=2qbeOI-Ew4Y.

out slogans that others would then repeat. Meeting Assad in person represented a good opportunity for individuals to be creative in improvising new slogans in order to either prove their loyalty to the regime or benefit from their performance on other occasions. It is important also here to note that many people in the crowd were required to turn up because they worked for government bodies: if they did not show up and participate, they might be published or risk interrogation. During the first decade of his rule, Bashar Assad went out in public, in the streets, ate in normal restaurants, took pictures with people, shook hands, and kissed and hugged people who asked him to. His shift from classical Arabic to colloquial Arabic coincided with him becoming closer to the language and consciousness of the people, both physically and linguistically. When he was slow to make progress toward reforms, people made excuses for him and claimed that he was a good person, but that those around him were preventing him from achieving his goals—a claim also made during his father's rule.

Conclusion

This chapter provides an introduction to the language of the Syrian revolution and to understanding how a new language might subvert the Assad family's discursive domination. It shows that the language of the regime was rooted in the complex details of the social and cultural life of Syrians, which were reflected in the language of discursive domination. This symbolic domination can be seen in the general increase in the number of released prisoners, and the general decrease in the number of arrests, from the end of Hafez al-Assad's regime through to the early years of rule under Bashar al-Assad. This raises questions about the methods or mechanisms used to control Syrians. One apparent way of dominating Syrians discursively was by the symbolic market, constructed and regularized by the regime, institutions, and individuals. This was seen in the intensive immersion of symbolic products at the visual and musical levels, bolstered by support from the ministers of Information and the takeover of Syrian publishing spaces. These factors secured discursive control over the public space. This discursive domination of the symbolic market produced specific products and affected the language of Syrians, resulting in the emergence of terms such as *Bījū* and *Marsīdīs* and the new meaning in the Syrian space of words like *shabbīḥa*. Tracing these symbolic products shows that Hafez al-Assad's ultimate objective was to symbolically make himself the god of Syria, and that he gradually succeeded in doing so. The examples provided—slogans, music, portraits, and books—show that Hafez reached a godlike status in the final stages of this process: He was God, and would not die. After him, Bashar al-Assad inherited the dominated kingdom and continued to work on symbolic domination, without resorting to killing during his first decade of rule unless it was a necessity. Bashar maintained the methods of domination developed by Hafez but did not create new ones: a point that will be discussed in detail in a later chapter and may have been a reason for the beginning of the Syrian protests in 2011. The fear and violence that the regime constructed in Syria was divided into two parts: the actual violence committed, and the resulting imagined violence. The committed violence was inflicted by Hafez, and Bashar tried to ease the imagined violence by representing himself as a new person different from his father the dictator. I will return to the symbolic products of the Assads, the father and the son, in Chapter 3 to explore the relationship between these background products and the symbolic products of the revolution, and in the concluding part of the book, will look at how this thick language was transferred to English. Was the revolutionary language an extension of the language of the regime, or did it constitute a break with everything that came before it?

Chapter 2
They Can Speak Up Now

-عرقوب: أن نتخيل؟
-السياف: مسموح.
-عرقوب: أن نتوهم؟
-السياف: مسموح.
-عرقوب: أن نحلم؟
-السياف: مسموح. . . ولكن حذار!
-عرقوب : أن يتحول الخيال الى واقع؟
-السياف: ممنوع.
-عرقوب: أو يتحول الوهم الى شغب؟
اللسيّاف: ممنوع.
-عرقوب: أو تتحد الاحلام وتتحول الى أفعال؟
-السياف: ممنوع.
سعد الله ونّوس من "*الملك هو الملك*"

ʿArqūb: To imagine that?
Headsman: That is permitted.
ʿArqūb: To fancy that?
Headsman: That is permitted too.
ʿArqūb: To dream that?
Headsman: That is permitted but be careful with it!
ʿArqūb: That imagination does not turn into reality?
Headsman: That is strictly forbidden.
ʿArqūb: Or that fancy becomes riots?
Headsman: That is strictly forbidden.
ʿArqūb: Even if dreams get united and assembled into accomplished fact?
Headsman: Also forbidden.

Saʿdallāh Wannūs, *al-Malik Hūwa al-Malik* [The King is the King]

The Emergence of a New Oppositional Language

In order to trace how a slogan like "Hafez, curse your soul" emerged in 2011, and to understand the context in which it then appeared, it is essential to understand why what Lisa Wedeen refers to as acting "as if," and the prison violence and its echoes on governing Syrians, as described by Salwa Ismail, stopped having the effect of enforcing Syrians to show their compliance with the Assad regime after 2011. In conjunction with acting "as if" and the prison violence, the symbolic violence of the regime through discourse, symbols, slogans, music, and books became less effective, represented in Syrians' protests in 2011. In this chapter, I will show how pro-revolu-

https://doi.org/10.1515/9783110767698-003

tionary Syrians liberated three key areas of the country's public space—the visual and musical spaces, and book publishing—in order to develop a new language that defied the regime. By looking at various examples of this language, we can see the interaction between the dominant and oppositional discourses along with their complex layers of meanings, thick histories, and thick contexts. As such, explaining the language of the regime is a similar act to the thick translation that is included throughout this book for several elements of the revolutionary language that intersected with that of the regime.

As the Syrian public space and discursive apparatuses began to take a different shape from those of the past, they became the center of struggles for control, as regime opponents and supporters sought to fill the available spatial units.[144] These struggles can be measured by the dissemination of symbolic products, represented by different musical and visual products to those established over decades by the regime. Producing slogans, music, graffiti, and banners became a metaphorical struggle or fight between the regime's supporters and those of the revolution in order to have more of the public space. All of these struggles over the Syrian public space were contestations between regime supporters and pro-revolutionaries that represented a metaphorical war in the symbolic market from the perspective of pro-revolutionary actors.

This metaphorical conflict erupted into a military war and eventually, after 2013, into a proxy war that continues to be fought today. The metaphorical war over symbolic products affected Syrians' practices, ultimately changing their use of language. The shift in practices led to changes in the schemas, attitudes, and perceptions of individuals in relation to the ways they were treated, behaved, and managed their daily lives, as clearly observed through changes in language use after March 2011. This language can be described not only as revolutionary in political terms—given the political goal—but also in terms of its creativity and novelty, with new terms and expressions introduced to reflect the revolutionary situation and loosen the grip of the regime's symbolic and discursive violence.

Dominant and Oppositional Discourses

Similarly to what was discussed in the previous chapter, in March 2011 the new ascending power, represented by revolutionary activists, began to pursue liberation from regime institutions and syndicates as a step toward liberating the Syrian

144 By "spatial units" here I mean all of the possible spaces that the regime or pro-revolutionaries could fill, whether musically, visually, or with other types of symbolic products.

public space. One of their actions was to replace all of the regime-sponsored unions and syndicates with others described as "free"—such as the Union of Free Syrian Students, Free Syrian Writers Union, Free Syrian Artists Union, Free Syrian Journalists Union, and many others in different fields.[145] New parallel organizations were also created in the political sphere, including numerous movements, political gatherings and groups. All of these new concepts, ideas, and institutions started with demonstrators and pushed Syria toward the development of a new discourse, as protestors sought to achieve their emancipation by replacing the Baath organizations and restructuring their dynamics. The existence of two discourses, at least in 2011,—the dominant one of the regime, and an oppositional one—invites us to compare the symbolic products of the revolution with those of the regime. Approaching this issue from the perspective outlined by Michel Foucault in *The History of Sexuality* offers a useful way to understand the relationship between dominant and opposing discourses and the circumstances that facilitate their emergence or disappearance.[146] Comparing the revolutionary and regime languages reveals the mechanisms by which these discourses worked to co-opt, subvert, and re-subvert each other. While these languages appeared to be in opposition, they in fact complemented each other, and depended on each other to produce this new discourse.[147]

Drawing on the work of Foucault, and related to the idea of dominant and opposing discourse, miriam cooke calls dissident art a "commissioned" form that Syrians were permitted. This means that the opposing discourse, the one managed by the regime, was the internal one. It was possible to criticize the regime, but only within specific limits and targeting specific people. There were many examples of

---

145 *Syria Untold*, "Union of Free Syrian Students," May 6, 2013, accessed July 18, 2023, https://syriauntold.com/2013/05/06/union-of-free-syrian-students; Nizār Muhammad, *"Ittiḥād al-Kuttāb al-Aḥrār: Aqlām Sūriyya bi-Wajh al-Niẓām"* [The Free Writers Union: Syrian Pens Against the Regime], *Al-Jazeera*, May 1, 2014, accessed July 18, 2023, http://bit.ly/3857DNi.; Free Syrian Journalists Union, "About Us," Official Facebook page, accessed July 18, 2023, https://www.facebook.com/Etehad.alahrar.syria/; Free Syrian Journalists Union, official website, www.ufsyrians.com (no longer available). The Free Syrian Journalists Union was founded on June 21, 2011.

146 Michel Foucault, *History of Sexuality*. vol. 1, trans. Robert Hurley (New York: Pantheon Books, 1978).

147 In *The History of Sexuality* Foucault elaborates more on the emergence of opposing and dominant discourses and helps explain how an opposing discourse emerges. While he is explicitly interested in sexuality, this understanding can also be applied to political discourse. As we will see, an opposing political discourse is born from the dominant one via various practices. Narratives about sex and sexuality work in a similar way to narratives about the 2011 Syrian revolution. In both examples, there is a dominant authority that censors some knowledge and prevents the spread of other types of knowledge.

what can be called the internally commissioned, opposing discourse against the regime itself. Under the rule of Hafez al-Assad, the TV programs al-sālib wa-l-mūjab [The Plus and the Minus] and Idhā ghannā al-qamar [When the Moon Sings] criticized the regime directly and cleverly.[148] Opposition writers such as Sa'd Allāh Wannūs and Mamdūḥ 'Idwān, among others, were also allowed to publish. Under Bashar al-Assad, the Syrian opposition dared to be more clearly and directly critical of the regime than the latter had ever thought possible, resulting in the 2000 "Statement of 99," the 2001 "Statement of 1000," and the 2005 "Declaration of Damascus."[149] The 2011 Syrian revolution was clearly an extension of this internal oppositional discourse, affected by the limited spaces of freedom permitted by the regime, and later separating from it to become independent in the early months of the protests. But what was the relationship between this internal discourse and the dominant one?

This new discourse, represented in this chapter by symbolic products, consisted of many aspects and elements that shaped what I call the revolutionary language. Using the same components as the language of the regime, pro-revolutionaries were able to create their own language and discourse that was ultimately able to enter the symbolic market with many symbolic products. In this chapter, I show the similarities and contrasts between both languages, focusing first on pro-revolutionary products, and then presenting equivalent products or those produced via interaction with pro-regime supporters. Drawing a map of the language of the regime—represented by slogans, banners, and visual products—helped me to trace the revolutionary language, since this latter was not created or composed by a specific individual, but by the same social, political, cultural, and economic context that created the language of the regime, and was disseminated by the newly formed political committees, the media, and publishing houses. Together, these elements created a subversive language that was designed to replace the regime language at the visual and musical levels, as well as to create and recycle new books.

Pro-Revolutionary Political Committees

From the experiences of other Arab protests in 2010–11, such as those in Egypt, Libya, and Yemen, Syrian dissidents knew that the best way to regain the public space was not only to demonstrate, but also to create political apparatuses to organize political protests on the ground. New political committees thus emerged in the Syrian public

---

148 Tawfīq Hallāq (Syrian author), interview with the author, May 21, 2017.
149 'Abd al-Razzāq *'Īd, Wa-Yas'alūnaka 'an al-Mujtama' al-Madanī: Rabī' Dimashq al-Maw'ūd* [And They Ask You About the Civil Society: The Buried Damascus Spring] (Beirut: Dār al-Fārābī, 2004), 134–5.

space to organize activists as well as to decrease the legitimacy of regime institutions, in particular after mass defections from the military, educational institutions, and syndicates. These new revolutionary organizations were offered to people as alternatives to those of the regime. In an act designed to subvert the Baath popular organizations, activists created "popular committees" that reflected the revolution's ideals as a way to replace regime organizations, including coordinations, groups, movements, political groupings, revolutionary councils, the Syrian Revolution Coordinators Union, Local Coordination Committees of Syria, and the Free Student Union,[150] which were all designed to support revolutionary activism.[151]

These newly formed political committees were given names similar to those used by the regime, but were designed to avoid the widespread uselessness of the latter. The goal was not to replicate regime institutions but to create new, revolutionary ones. The importance for activists of founding new political and activist groups stemmed from the absence of knowledge about such political terms under the Assad regimes, which fought or blocked knowledge of political activism. In other words, for Syrians, political terms or labels were either not used at all or were explained in line with the regime's ideology, as in the book *Thus Said Assad*—discussed in Chapter 1—in which everyday Syrian terms were explained in a bid to ensure that the people would understand politics in the way Assad wanted them to. In addition to this, there was no clear justification for calling one group a *ḥirāk* [mobility] or another a *ḥaraka* [political group]. The absence of the use of such terms in previous decades resulted in Syrians using a variety of labels for their groups in an *ad hoc* manner without reference to how terms might differ denotationally from one another, as they lacked the experience of being organized politically in any way other than the highly chaotic and arbitrary manner of the Baath Party.

Revolutionary political terms were coined in a way that reflected the context of the protests. These political terms were devised not only to create political committees, but also to express their roles and functions in the field. Every political committee started by declaring itself as part of the revolutionary field by issuing a founding statement explaining its expectations and attitudes. This was made necessary by the fact that protestors were initially unable to hold organized demonstra-

150 Coordination(s) is a political term that was translated from *tansīqiyya* by activists as "coordination." A full explanation can be found in Chapter 3.

151 After March 2011, the Syrian public space was full of new terms for the political work of activists on the ground, including *ḥirāk, ḥaraka, tansīqiyya, jam'iyya, majmū'a, majlis thawrī,* and *lijān*. On the differences between these terms, one activist told me that "They are the same and they serve the revolution" (Amjad [a founder of the Local Coordination Committees of Syria], interview with the author, November 27, 2015).

tions, although they kept an eye on political events and the reaction of the international community to the protests, especially after the arrest of children in Daraa, which ignited demonstrations in many places.[152] The presence of these founding statements in the public space played an important role in unifying activists in specific political and revolutionary groupings as well as organizing revolutionary acts. Even more important still was a new term used for the first time in the Syrian public space, *tansīqiyya* [coordination], which will be discussed at greater length in the section on new revolutionary terms in Chapter 3.

Liberating the Syrian Public Space Through Audio, Visual, and Written Media

After March 2011, the public space was affected by the introduction of new media, including social media, alternative media outlets, online websites and blogs, and newspapers (both printed and digital). This played an important role in shaping the Syrian public space, since it meant that for the first time since the 1963 Baath Party *coup d'état*—after which all newspapers were shut down and syndicates were dissolved and later replaced with Baath syndicates—the media outlets and apparatuses of the regime had to compete with others.

The news agencies and providers that formed post-2011 did not call themselves news organizations or correspondents from the outset, but rather evolved in phases. They typically began as pro-revolutionary social media pages unofficially reporting on events and news for specific locations where the page administrators were living. Due to the lack of any credible pro-regime media and the challenges for international journalists in covering Syria, these social media pages were gradually adopted as alternatives to the official media and as reputable sources of information. Revolutionary media outlets like Sham News, Ugarit News and others thus started on social media and then gradually became separate news agencies.[153] Following the early months of the protests, many Facebook pages for revolutionary political organizations eventually became media outlets covering events due to the decline in revolutionary activities caused by the increased militarization. Coordinations, or *tansīqiyya*, were developed to meet the requirements of demonstrators and the revolutionary field. Each coordination had several offices that were each responsible for only one type of revolutionary act, such as the media office, political

152 Jamāl al-Bārūt. "*Al-'Aqd al-Akhīr fī-Tārīkh Sūryya: Jadaliyyat al-Jumūd w- al-Iṣlāḥ* [Syria in the Last Decade: The Dialectic of Stagnation and Reform] (Doha: Arab Center for Research and Policy Studies, 2012), 184.

153 *Sham News Network*, "*Man Naḥnu*" [Who We Are], *Sham News Network* official website, accessed January 5, 2020, accessed July 18, 2023, http://bit.ly/2UuWy4j; *Ugarit News*, Facebook, accessed December 20, 2014, http://ow.ly/stvbB (no longer available).

office, and organizational office. Over time, the media offices of these coordinations became quasi-professional media centers and were seen as authentic sources of news due to the ban on foreign journalists from entering Syria—a step taken by the regime to try and prevent Syrian protestors from reaching the international audience.

In addition to the news media, new websites, blogs, and forums emerged for public discussion of the situation in Syria. These played a vital role in shaping the revolutionary language by presenting and discussing articles and explaining everything related to the protests, including the revolutionary language used in slogans and on banners. Such websites included Kibrīt [Match], Sūriyyatna [Our Syria], al-Ṣafḥāt al-Sūriyya [The Syrian Pages], Nūr Sūriyya [The Light of Syria] and al-Mundassa [The Female Infiltrator].[154]

This last website, al-Mundassa, was the best known and was widely used as a place for debate and discussion of events in the first three years post-March 2011. It is generally considered to be one of the first websites on which the daily events of the revolution and its protests were discussed, and according to an interview with co-founder "Eddi" and information on the website, its first post—considered to mark the beginning of the website—was published on April 11, 2011.[155] These websites were both affected by and affected the revolutionary language in turn. In my interview with her, Eddi explained that "the idea of the website came due to its necessity in 2011, but it was not new," and added that "the Syrian public sphere witnessed, before the 2011 protests, many blogs—even though posts were published anonymously. The period before 2011 witnessed a huge shift in activism in the world of blogs in Syria."[156] During the period before 2011, "Syrian bloggers faced repression in many cases. A dozen bloggers are reported to have been arrested in the period from 2006 onward, and some are still in detention."[157] The most famous bloggers were Tal al-Mallohi, who was arrested in 2009 after writing a letter to Bashar al-Assad, Razan Ghazzawi, whose blog was called Razaniyāt, and Āyāt ʿIṣām.[158] In his study of Syrian bloggers, Yenal Göksun observes the shift that

---

154 *Kibrīt* (blog), accessed July 18, 2023, https//kebreet.wordpress.com; *Sūriyyatna* (blog), accessed July 20, 2017, www.souriatnapress.net (no longer available); *al-Ṣafḥāt al-Sūriyya* (official website), accessed July 18, 2023, www.Syria.alsafahat.net; *Nūr Sūriyya*, official website, accessed July 18, 2023, www.Syrianoor.net. Unfortunately, the official website of *al-Mundassa* no longer exists, but I have a copy of it in my personal research archive.

155 Eddi (co-founder and editor of *al-Mundassa*), interview with the author, May 20, 2017.

156 Eddi, interview.

157 Sarah Jurkiewicz, *Blogging in Beirut: An Ethnography of a Digital Media Practice* (Bielefel: Transcript Verlag, 2018), 63.

158 Tal al-Mallohi, *Mudawwinatī* [My Blog], http://talmallohi.blogspot.com/; Tal al-Mallohi, *Rasāʾil* [Letters], http://latterstal.blogspot.com/; Tal al-Mallohi, *Filasṭīn* [Palestine], http://palestinianvillagcame with the emergence of a new blogging phenomena in Syria: “Mass demonstrations and public campaigns led some Syrians to start blogging. . . to document the revolution and to create awareness about the current situation.”[159]

Unlike blogs, however, al-Mundassa was open to all and specifically aimed at all Syrians, regardless of the author or topic. It was also characterized by both the variety of topics discussed and the type of language used. Users came from different classes of Syrian society, and even supporters of the regime were allowed to express their opinions in articles. A close examination of the website archive shows that thirty to fifty new articles were published every week. After two years of protests, the number of articles gradually decreased until the website stopped publishing due to militarization of the revolution. According to Eddi, “participants were from different backgrounds. It was possible to find Islamists, liberals, secularists, and non-religious people.”[160] The website’s description of its editorial line stated that “comments on posts are open to loyalists and those who are against the regime. Freedom of speech is secure as long as it is presented in a respectful way.”[161] Its section titles were mostly in colloquial Syrian Arabic or Modern Standard Arabic (MSA) using the same words as colloquial Syrian, for example: “Fann al-Thawra” [The Art of the Revolution], “Ṣūra wa-kam Kalima” [A Photo and Some Words], “mushārakāt maftūḥa” [Open Participations], “Bi-l-Ṣawt wa-l-Ṣūra” [Voice and Photo], “Dassa ʿāl-Māshī [A Quick Hint], and “Kartūniyyāt” [Cartoons].[162] The website’s design and language reflected both the overall Syrian context and its changes. Using colloquial Syrian Arabic was an attractive aspect and active topic of discussion on the website and in my view one reason that the website was popular, since users did not need to be fluent in MSA. Some of the website’s articles were about the use of colloquial Syrian Arabic instead of MSA, which was seen as a development of the revolutionary language. For this reason, some readers and authors were surprised by it, while others supported it.

When Bashar al-Assad came to power, a few new regional print media outlets were introduced. In 2011, however, within the first few months of the protests, the number rapidly grew to fifteen. Soazig Dollet quotes a study by Enab Baladi on

---

es.blogspot.com/; Razan Ghazzawi, *Razaniyāt* (blog), accessed July 18, 2023, https://razanghazzawi.org/articles/. Tal al Mallohi is known for her three blogs noted above, accessed July 18, 2023. Unfortunately, I was unable to find any existing blog for Āyāt ʿIṣām.

159 Yenal Göksun, “Cyberactivism in Syria’s War: How Syrian Bloggers Use the Internet for Political Activism,” in *New Media Politics: Rethinking Activism and National Security in Cyberspace*, ed. Banu Baybars-Hawks (Newcastle: Cambridge Scholars Publishing, 2015), 56.

160 Eddi, interview.

161 *Al-Mundassa*, archived documented version.

162 Please see Figure 34 for a screenshot of the design of the *al-Mundassa* website.

the new Syrian press, reporting that as of March 2015, "268 publications (newspapers and magazines) have been launched since March 2011."[163] This huge number shows the struggle to disseminate more symbolic products to gain more units in the Syrian public space. As noted above, contributing to such work was risky, since the regime reacted by arresting the journalists, distributors, writers, and anyone else who helped create them.

A few months after the beginning of the protests, the Syrian public space witnessed the release of new magazines and newspapers, in contrast to the previous four decades during which only three main newspapers were circulated in the country (supplemented by a few local ones produced by the regime itself). These new print media were published under special conditions (secrecy) and distributed as an act of resistance, since distributing nongovernment publications was highly risky. Print publications—newspapers and magazines—were intensively distributed to resist and subvert the knowledge and news that had previously been released and controlled by the regime. According to the Syrian Print Archive website, over fifteen new pro-revolutionary print media outlets were launched in Syria in the period between March 1, 2011 and December 31, 2011.[164] The dramatic

---

163 Enab Baladi, "*al-I'lām al-Sūrī al-Badīl*" [The Syrian Alternative Media] (March 2015), quoted in Soazig Dollet, "The New Syrian Press: Appraisal, Challenges, and Outlook," CFI The French Media Development Agency, September 2015, accessed July 18, 2023, https://www.cfi.fr/sites/default/files/etude_presse_syrienne_EN.pdf. Unfortunately, the original study by *Enab Baladi* is no longer accessible: https://www.cfi.fr/sites/default/files/etude_presse_syrienne_EN.pdf.

164 These fifteen print media outlets were: *Āzādī Ḥuriyya* [Freedom], first released April 24, 2011 and written in Kurdish and Arabic, accessed July 18, 2023, https://syrianprints.org/ar/issues?agency=292; *Bukrā Sūryā* [Tomorrow Syria], first released May 23, 2011, accessed July 18, 2023, https://syrianprints.org/ar/issues?agency=6; *al-Thawra al-Sūriyya* [The Syrian Revolution], first released July 26, 2011, accessed July 18, 2023, https://syrianprints.org/ar/issues?agency=136; *Akhbār Mundass* [Infiltrator News], first released August 12, 2011, accessed July 18, 2023, https://syrianprints.org/ar/issues?agency=9; *Īqā'āt* [Rhythms], first released August 14 2011, July 18, 2023, https://syrianprints.org/ar/issues?agency=177; *Ḥuriyyāt* [Freedoms], first released August 23, 2011, accessed July 18, 2023, https://syrianprints.org/ar/issues?agency=24; *Ṣawt al-Thawrah al-Sūriyya* [The Voice of the Syrian Revolution], first released August 31, 2011, accessed July 18, 2023, https://syrianprints.org/ar/issues?agency=219; *Aḥrār Sūriyya* [The Free (People) of Syria], first released September 2, 2011, accessed July 18, 2023, https://syrianprints.org/ar/issues?agency=7; *al-Badīl* [The Alternative], first released September 2, 2011, accessed July 18, 2023, https://syrianprints.org/ar/issues?agency=11; *Sūriyyatnā* [Our Syria]. first released September 26, 2011, accessed July 18, 2023, https://syrianprints.org/ar/issues?agency=13; *al-Ḥaqq* [The Right], first released October 30, 2011, accessed July 18, 2023, https://syrianprints.org/ar/issues?agency=15; *Sūriyya Biddhā ḥuriyya* [Syria Wants Freedom], first released October 30, 2011, accessed July 18, 2023, https://syrianprints.org/ar/issues?agency=25; *Aḥfād Khalid* [Descendent of Khalid], first released December 9, 2011, accessed July 18, 2023, https://syrianprints.org/ar/issues?agency=3; *Jusūriyyā* [Bridges Syria], first released December 10, 2011, accessed July 18, 2023, https://syrianprints.org/ar/issues?agency=197 (the title is a portman-

growth over this short period shows how Syrians directly subverted the regime control's over the print media and started to reject those produced by the regime in the public space.

Comparing the periods before, during, and after Baath control over Syria illustrates the huge difference in the number of print media outlets. Before the Baath Party came to power, there were seventy-four magazines and newspapers in circulation, but only three during its period of control: the newspapers Al-Baath, Tishreen, and Thawra.[165] The situation changed under Bashar al-Assad's reign, but all new print outlets were subject to regime censorship. It is important to note that there were some print media outlets during the Baath period that were not well-known, including *al-Jamāhīr* [The People] (1965) in Aleppo, *'Uruba* [Arabism] (1973) in Homs, *al-Fidā'* [The Sacrifice] (1973), *al-Wiḥda* [The Unity] (1984), and al-Furāt [Euphrates] (2004)."[166] All of these were regional, and printed by the government publisher, the al-Wiḥda Organization.

Musical Products

Along with newspapers and blogs, music was a vital element that shaped the new Syrian public space and helped to create the new revolutionary language, notably through the various characteristics displayed in these songs that had not previously been present in the lyrics of patriotic Syrian songs.[167] As discussed in the previous chapter, under Hafez and Bashar al-Assad, patriotic songs were initially mainly composed in MSA, and later shifted toward mostly using colloquial Syrian. Here I present examples of patriotic songs released within a one-year period, from March 23, 2011 to March 23, 2012, following the main timeframe for studying the other symbolic products of the revolution. Although there is no complete list of songs, the collected songs of the revolution provided in the appendix—based on internet searches and interviews with most of the song producers where possible—aspires to be the largest available.

teau combining *jusūr* and *Sūriyyā*); and *al-Ghad* [The Syria Tomorrow Movement], first released December 26, 2011, accessed July 18, 2023, https://syrianprints.org/ar/issues?agency=19.

165 Syria Ministry of Culture, "Book Industry in Syria;" *Al-Baath* [Baath newspaper], official website, accessed July 18, 2023, http://newspaper.albaathmedia.sy/; *Tishreen*, official website, accessed July 18, 2023, http://tishreen.news.sy/; *Thawra*, official website, accessed July 18, 2023, http://thawra.sy/.

166 *al-Furāt Newspaper*, official website of al-Wiḥda for Journalism and Print Media, accessed June 1, 2020, http://furat.alwehda.gov.sy/node/243076 (no longer available).

167 In order to maintain the pronunciation and musicality of certain songs and slogans, they have been transliterated as they are pronounced, rather than converted to MSA.

A brief look at the list of revolutionary songs shows that they started being released a week after the protests began,[168] with the first revolutionary song uploaded to YouTube a few days after the Daraa protests.[169] This was "Yā Sūrī Kassir al-Quyūd wa-Intafiḍ" [Hey Syrian, Break the Chains and Rise Up].[170] The day after, on March 25, 2011, an older song connected to Palestine, called "Allāhu Akbar Yā Sūriyya Kabbirī" [Oh, Syria; God is the Greatest], was uploaded to serve a revolutionary purpose.[171] The day after this, Waṣfī Maʿṣarānī produced two songs, "Darʿānā Tunādī" [Our Daraa is Calling] and "Sūrī Anā" [I Am Syrian].[172] On March 27, 2011, Samih Choukair released a song called "Yā Ḥīf" [Alas/Shame On You].[173] This song was the most famous of the pro-revolution songs, to the extent that many people still

---

168 Please see the Appendices for the list of the revolutionary songs and the link to listen to the songs. The songs are hosted by Creative Memory of the Syrian Revolution. https://creativememory.org/ar/collections/eylaf-bader-eddin-songs.

169 Since this section about songs relies heavily on YouTube, I considered gauging song visibility by the number of views and/or comments, but ultimately decided not to do so since the number of views on YouTube is inaccurate for many reasons: first, some songs have since been deleted by YouTube or by the channel owner; second, not every visitor leaves a comment, which means the number of comments does not correspond to the number of views; and third, whether or not the original YouTube video has been deleted, people frequently upload videos to different YouTube channels under different titles, taking away views from the original video and making it difficult to work out when a song was first released. I therefore attempted to contact song producers to confirm the actual release date of songs, or otherwise took the oldest one as the original release date if this was not possible. Finally, these songs were uploaded to other websites in addition to YouTube, such as Facebook and SoundCloud, all of which presented the same challenges with songs and accounts being deleted, and the number of views and comments thus not providing accurate information.

170 "Yā Sūrī Kassir al-Quyūd wa-Intafiḍ," unknown artist, *lovesyria11* (YouTube channel), uploaded March 22, 2011, accessed July 18, 2023, https://www.youtube.com/watch?v=74I0tFnSwvg.

171 "Allāhu Akbar Yā Sūriyya Kabbirī," unknown artist, *EnsanGroup* (YouTube channel), uploaded March 26, 2011, accessed July 18, 2023, https://www.youtube.com/watch?v=t1sfnQJy8UQ&has_verified=1.

172 Razān Siryya, "Waṣfī Maʿṣarānī Mirʾāt Ghināʾiyya li-Waqiʿ Sūriyya" [Waṣfī Maʿṣarānī is a musical mirror for the current situation of Syria], *Radio Sawa*, March 27, 2015, accessed July 18, 2023, https://arbne.ws/3f2c3bJ; Waṣfī Maʿṣarānī, "Darʿānā Tunādī," *freedar3a* (YouTube channel), uploaded March 24, 2011, accessed July 18, 2023, https://www.youtube.com/watch?v=OS5DQpe6FtE; Waṣfī Maʿṣarānī, "Sūrī Anā," *freedar3a* (YouTube channel), uploaded March 25, 2011, accessed July 18, 2023, https://www.youtube.com/watch?v=Z_pdYeXtwrI. Waṣfī Maʿṣarānī is a Syrian singer and composer who began his singing career at the beginning of the Syrian revolution. One of his parents is from the Czech Republic, and he was there when he produced his first song.

173 Tāriq Ḥamdān, "Samīḥ Shuqayr.. Ghurfa Ṣaghīra fi-al-Manfā" [Samih Choukair. . . a Small Room in Exile], *Al-Araby*, September 17, 2014, accessed July 18, 2023, https://bit.ly/3cYIINQ; Samih Choukair, "Yā Ḥīf," *terrrromtak* (YouTube channel), uploaded March 27, 2011, accessed June 1, 2020, https://www.youtube.com/watch?v=Us4_fvsugOw (no longer available). Samih Choukair is a believe it was the first. A day after Choukair's song was released, a song called "Āhin 'alā Waṭanī al-Ḥabīb" [Oh, My Beloved Country], which had previously been used by the regime in the context of the Israeli occupation of Palestine, was also adapted for revolutionary purposes.[174] After this, Yahya Hawwa produced a song called "Shū Minḥibbik yā Blādī" [Oh, My Country, We Love You So Much] in colloquial Syrian.[175] The revolutionary musical path then returned to MSA, as new songs were produced or older ones reused. From mid-April until the end of May 2011, most songs were in colloquial Syrian. In late May, the pro-revolution group al-Mundassūn al-Sūriyyīn al-Aḥrār [The Free Syrian Infiltrators] and Fajr Sūriyya [The Syrian Dawn] were founded.[176] A month later, Al-Dibb al-Sūrī [The Syrian Bear] was founded.[177] At the beginning of June 2011, Abṭāl Mūskū al-Aqwiyā' [Moscow Strong Heroes] produced a new song called "Biddnā Na'abbī al-Zinzānāt" [We Want to Make the Detention Centers Full].[178] After this, most songs were produced by individual singers under their real names. The willingness of singers to identify themselves represented a new step for revolutionary music, breaking the wall of fear.

Close examination of the information above shows that singers were afraid to perform under their real names in the early phase of the revolution, with the majority of songs at that time performed anonymously or produced under the name of an unknown band with only a YouTube channel or a Facebook page. At most, a Facebook account under a fake name might claim responsibility for a song. I tried to investigate some anonymous bands to examine the production of songs in

---

famous Syrian singer and songwriter. He has performed dozens of songs about Syria and is considered to have left an indelible mark on the development of patriotic songs in the Arab world.

174 Salīm 'Abd al-Qādir Ma'sarānī, songwriter, *"Āhin 'alā Waṭanī al-Ḥabīb,"* produced by Sham Media UK, *freesyriasongs* (YouTube channel), uploaded March 28, 2011, accessed July 18, 2023, https://www.youtube.com/watch?v=GSjS_SbPhsM&t=15s.

175 Yahya Hawwa, "Yahya Hawwa, Voice of the Syrian Revolution," interview by Omar Shahid, *The Guardian*, February 17, 2013, accessed July 18, 2023, https://bit.ly/2SecmGT; Yahya Hawwa, *"Shu Minḥibbik yā Blādī," DearSyria* (YouTube channel), uploaded March 31, 2011, accessed July 18, 2023, https://www.youtube.com/watch?v=ciVE4YusXZo. Yahya Hawwa is a Syrian singer who lived in Amman, Jordan. His family was killed in the Hama massacre in 1982. He became a famous pro-revolutionary singer after 2011.

176 Al-Mundassūn al-Sūriyyīn al-Aḥrār, official YouTube channel, accessed July 18, 2023, https://www.youtube.com/user/AlMundaseenBand/about. While songs released by Fajr Sūriyya carry the name of the band, there is no information about the band itself, and the group has no social media accounts.

177 Al-Dibb al-Sūrī, official YouTube channel, accessed July 18, 2023, https://www.youtube.com/user/syrianbear04.

178 While the song *"Bidnā Na'abbī al-Zinzānāt"* was purportedly released by Abṭāl Mūskū al-Aqwīyā', nothing is known about the band itself. Songs were uploaded under the group's name, but it has no social media accounts and no material about the band itself can be found online.

2011, and was able to reach al-Mundassūn al-Sūriyyīn al-Aḥrār and Nuṣṣ Tufāḥa. Al-Mundassūn al-Sūriyyīn al-Aḥrār was working on composing from Saudi Arabia and Jordan. According to Muʿādh Naḥḥās, a lyricist for the band, "We were working in different geographical places. I am a university lecturer in KSA [the Kingdom of Saudi Arabia], and we had some members of the band here. The rest were in Jordan, recording."[179] It was previously unknown who was behind the band's songs. The other group, Nuṣṣ Tufāḥa, was famous for encouraging pacifist activities in Syria at the end of 2011 and the beginning of 2012. The leader of the band, Muḍā' Maghribī, was working from the occupied Golan Heights. He told me how the band started: "As a Syrian who lived [through] both suppression of the Israeli occupation and of the regime, I had an artistic project that the revolution in Syria helped me to start with." He continued, explaining that the name of the group, "Nuṣṣ Tufāḥa means half an apple. It is a metaphor for the precious fruit that helped people in Golan Heights to earn their living and a metaphor for the regime taking Syria and the Israeli Occupation taking our lands."[180] This band composed its songs in collaboration with many people from different Arab countries, before the songs were edited in Maghribī's studio and released on YouTube.

Rap music was a new phenomenon in the Syrian public space compared to other Arab countries, emerging there in the 1990s.[181] Before 2011, rappers in Syria were not well known, and rap was not a widespread musical genre because of "political repression and restrictions on freedom of expression."[182] After March 2011, however, rap was one of the first musical genres to begin producing pro-revolutionary songs. Beau Bothwell traces music played on radio stations about the Syrian nation, providing numerous examples of how songs broadcast on radio stations reflected the image of the nation from different perspectives, depending on the radio station itself. For example, by comparing the lyrics of the song "Against the Country" with Assad's speech in 2012, he finds the relationship between the lyrics and Assad's speech is "extensive, and touch[es] on the standard points of Assad's rhetoric."[183] Bothwell's study sheds light on how the political rhetoric of

---

179 Muʿādh Naḥḥās (lyricist for al-Mundassūn al-Sūriyyīn al-Aḥrār), interview with the author, July 4, 2017.

180 Muḍā' Maghribī (leader of Nuṣṣ Tufāḥa), interview with the author, July 11, 2017.

181 Jennifer L. Roth-Burnette, "Syria," in *Hip Hop Around the World: An Encyclopedia*, 2 vols., eds., Melissa Ursula Dawn Goldsmith and Anthony J. Fonseca (Santa Barbara: Greenwood, 2019), 683.

182 Torie Rose DeGhett, "'Record! I am Arab': Paranoid Arab Boys, Global Ciphers, and Hip Hop Nationalism," in *The Hip Hop & Obama Reader*, eds. Travis L. Gosa and Erik Nielson (Oxford: Oxford University Press, 2015), 100.

183 Beau Bothwell, "Minnḥbbuk (ya Baṭa): Musical Rhetoric and Bashar al-Assad on Syrian Radio During the Civil War," in Tyranny and Music, eds. Joseph E. Morgan and Gregory N. Reish (Lanham: Lexington Books, 2018), 170.

the regime was used in music on the radio. In addition to this, my personal archive and fieldwork interviews show that rap was used in Syria to express the revolutionary discourse from the beginning of the protests. The first rap song related to the protests, "al-Bayān Raqam Wāḥid" [Proclamation Number 1] by MC RoCo,[184] was released on March 28, and was followed by a second song on May 9, 2011, "al-Bayān Raqam Itnīn" [Proclamation Number 2].[185] Music was then released by a new rapper, MC Revi. Despite the fact that rappers were making music related to the revolution, it took time for rap to develop and improve as a musical genre. This is why most famous rap bands, such as the Lāji' al-Rāb [Refugees of Rap], did not begin to produce their most important revolution-related songs until mid-2012.

Looking more closely at the list of songs I have compiled that were released between March 2011 and March 2012, they can be categorized into three main groups, based on their relationship with those of the regime. The first group consists of songs that were created by supporters of the revolution to respond to or compete with an older song created by the regime; the second group conversely of songs produced by the regime in reaction to a song or songs produced by revolutionary supporters; and the final group of the national anthems and alternative versions that were developed post-March 2011 as a means to subvert, protest, or contest regime control. I will provide an example from each group and discuss it.

Revolutionary (Musical) Responses to the Regime

These groups of songs show that both sides affected each other, and confirm the importance of regime songs to the emergence of certain revolutionary ones.[186] This does not mean, however, that both sides promoted the same content. It was a metaphorical war conducted through symbolic products, with each side trying to overturn and subvert the symbolic products of the other. For example, one of the most important ideological songs for the regime was "Syria, My Beloved," which was released at the time of the October War against Israel.[187] The importance of this song lies in its utilization of regime-claimed victories in 1973, and its historic impact on Syrians was

184 MC RoCo, *"al-Bayān raqam wāḥid," syrianagent2011* (YouTube channel), uploaded May 8, 2011, accessed June 1, 2020, https://www.youtube.com/watch?v=AipWcjJ3Z9Q (no longer available).

185 MC RoCo, *"al-Bayān Raqam Itnīn," syrianagent2011* (YouTube channel), uploaded May 9, 2011, accessed June 1, 2020, https://www.youtube.com/watch?v=A_tffmJakMA (no longer available).

186 Please see Appendix 4, Group 2 for song lyrics.

187 Najāḥ Salāma and Mufhammad Salmān, *"Ūghniyat Sūriyā yā Ḥabībatī al-Aṣliyya wa-l-Kāmila"* [The Song of Syria, my Beloved], complete original version, *SyrianHistoryX* (YouTube channel), uploaded November 26, 2009, accessed July 18, 2023, https://www.youtube.com/watch?v=yKli72vT-PmE. This particular war, part of the Arab-Israeli conflict, is known by various names in different languages and regions. The Assad regimes refer to it as the Liberation War while Israel calls it the

reflected in its reuse by the regime after March 2011. Covers of the original version, with no changes to the lyrics or musical composition, were performed in the first few months of the revolution by artists including the famous Syrian singers Nassif Zeytoun, Ḥusām Madaniyya, and Shahd Barmada.[188] I was unable to establish the exact release date of the original song, but the earliest cover version I found online was uploaded on July 17, 2011.[189] The pride in the (imagined) victorious past encapsulated by this song led to its reproduction in order to create solidarity between the regime and Syrians.

The regime's revival of the song enhanced the desire of pro-revolutionary activists and Syrians to counter the claims expressed by the regime in the song. A response to the ideological and historical messages that the regime wanted to send through the song came via a revised version of the song released by pro-revolution activists. Two bands, Ghurabā' Grūb [Strangers Group] and al-Mundassūn al-Sūriyyīn al-Aḥrār, created different versions of the song.[190]

The Ghurabā' Grūb version, called "Nashīd Sūriyā yā Ḥabībatī" [The Anthem of Syria, my Beloved] was produced by Al-Majd, a Saudi channel that supported the revolution. The origin of the channel provides an indication as to its ideological leanings and perspective. In particular, it espouses Islamist ideology, and this attempted involvement in the creation of revolutionary products shows how external Islamist actors intervened in Syria early on in the protests with the support of foreign governments. What distinguishes this song from other revolutionary songs is that it was performed as an Islamic anthem, despite this being unclear from the

Yom Kippur War. It is known elsewhere in Arabic as the October War, which is the term I have chosen to use.

188 *al-Fann*, "Nassif Zeytoun," accessed July 18, 2023, https://bit.ly/2SqFSJw; ʿAmmār Al-Daqir, "Ḥusām Madaniyya Ḥubb al-Najāḥ" [Ḥusām Madaniyya, the Love of Success], *al-Thawra*, February 19, 2016, accessed April 20, 2020, https://bit.ly/2WeSxjV (no longer available); *Layalina*, "Shahad Barmada," accessed July 18, 2023, https://bit.ly/3f8iEkW. Nassif Zeytoun is a Syrian pop singer who became famous after winning the *Star Academy* competition in 2007, Ḥusām Madaniyya is a Syrian pop singer who became famous after reaching the final stage of *Super Star Academy* in 2006, and Shahad Barmada is a Syrian pop singer who became famous after reaching the final stage of *Super Star* in 2006. She also sang the song used for Bashar al-Assad's election campaign, "Minḥibbak" [We Love You].

189 Nassif Zeytoun, Ḥusām Madaniyya, and Shahad Barmada, performers, "Sūriyā yā Ḥabībatī," cover of song originally by Najāḥ Salāma and Muhammad Salmān, *magnonetsawtnassif* (YouTube channel), uploaded July 17, 2011, accessed July 18, 2023, https://www.youtube.com/watch?v=ZXQw-Gd50Buk.

190 Ghurabā' Grūb, "*Nashīd Sūriyā yā Ḥabibatī*", *syriagloryn* (YouTube channel), uploaded January 12, 2011, accessed July 18, 2023, https://www.youtube.com/watch?v=q2x4g7p1Afs; Al-Mundassūn al-Sūriyyīn al-Aḥrār, "Sūriyā Yā Ḥabībatī", *syriagloryn* (YouTube channel), uploaded March 10, 2012, accessed July 18, 2023, https://www.youtube.com/watch?v=0cGo09crarc&has_verified=1.

lyrics.[191] This can be seen through calling it an "anthem," not a "song," referring to the Islamic content of the musical product itself. The absence of musical instruments and women's voices in the song indicate that the song was performed in line with the policies of the Islamic Al-Majd channel.[192] However, despite the ideology of its producer, the song lyrics have no overt Islamic elements.[193] The rhythm and melody of the Ghurabāʾ Grūb version, which calls on Syrian cities to protest against Assad, are similar but not identical to the original song. This song depends primarily on subverting the original by replacing all of the original lyrics of the song. This shows a wholescale change of content. Subversion in this case is thus built on the appropriation of the song without strictly following the number of verses or the original composition of the lyrics, contrary to al-Mundassūn al-Sūriyyīn al-Aḥrār.

The song "Sūriyā yā Ḥabībatī" by al-Mundassūn al-Sūriyyīn al-Aḥrār is another example of a song that was re-released by the revolutionary side. In contrast to the version by Ghurabāʾ Grūb, the aim of this cover was to imitate the song used by the regime while playing with the words: not only to distribute an alternative to the original, but also to create a new song in terms of both form and content. Changing the song title to "Sūriyā yā Ḥabībatī" [A New Song of Syria, my Beloved] was one step toward subverting the song itself as a regime product. This is important because the claims of the regime to have won victory over Israel in the October War are present in the lyrics of the original version—even though the Assad regime's "forces were forced out of the parts of the Golan they had initially recovered, and lost more territory on the road to Damascus."[194] In addition to this, "Asad by consequence made headway politically among his own people,"[195] by showing Syrians that "the war of October was a liberation war against the Israeli Occupation."[196] Despite the fact that the Syrian army did not win the war, Hafez al-Assad embarked on an enormous cam-

191 *Nashīd* in Arabic has two meanings in relation to music. The first relates to national anthems that describe a country and the second is an Islamic one. *Nashīd Islāmī* [an Islamic anthem] is a type of song that uses almost no musical instruments, only a *daff*, and does not include any female voices.
192 For more information about the Al-Majd channel, see: Qanāt al-Majd al-Faḍāʾiyya Tuthīr Jadalan fī al-Wasaṭ al-Iʿlāmī al-Suʿūdī [al-Majd Channel Sparks Controversy in Saudi Medi]. Al-Sharq al-Awsat, April 6, 2004, accessed July 18, 2023, https://archive.aawsat.com/details.asp?article=227163&issueno=9261#.Y4i8snbMI2w.
193 A *daff* is a type of drum that the Prophet Muhammad used to listen to. For this reason it has a religious association only if used in this context.
194 Hanna Batatu, *Syria's Peasantry, the Descendants of Its Lesser Rural Notables, and Their Politics* (Princeton, NJ: Princeton University Press, 1999), 202.
195 Batatu, *Syria's Peasantry.*
196 Syria Ministry of Defense, "The General Results of the Liberation War of October," official website of the Ministry of Defense, accessed July 18, 2023, http://www.mod.gov.sy/index.php?node=554&cat=886.

paign to celebrate this "victory." After decades of teaching Syrians about this victory in schools and universities, the majority of the population educated in Baath establishments (schools, universities, and popular organizations) believed in the Syrian army's ability to achieve victory over Israel. This is why the song had a major influence on the Syrian unconsciousness and explains why it was the first to be remade. It is also a clear example of how Assad's regime was able to ignore facts and turn defeat into victory, at least metaphorically. This demonstrates the importance of Assad's symbolic products in telling the public exactly what it should know. Symbolic products such as this song controlled public perception by directing how events were received and perceived, particularly given the few media outlets at that time.

A comparison between the original song and the cover version by al-Mundassūn al-Sūriyyīn al-Aḥrār finds that both have the same length and rhythm and almost the same number of words, but with totally different content. It is very creative how slight changes are inserted in order to reverse all of the song's meaning, target, and to contradict the propaganda of the regime. The revolutionary song contradicts and opposes all of the elements of the original; a contradiction highlighted by the use of similar grammar and tenses between the original and cover versions of the song. While the first line of the original song says "Oh, Syria, you gave me back my dignity and freedom," the new song says, "Oh, Syria, give me back my dignity and freedom." The different tense not only reflects the subversive nature of the new song but also represents a response to the original version, showing that nothing under the Assad regimes has returned freedom or dignity to Syrians. The original song claims that the Tishrīn War [October War] gave Syrians their dignity and freedom, but the new song poses a question: where is dignity and freedom in Syria in 2011? To put it differently, the recomposition of the original song after 2011 has an intertextual relationship with the older text. In song terms, when "any cultural practice which provides a relatively polemical allusive imitation of another cultural production or practice" occurs, it is called a parody in order to contextualize intertextuality.[197] The cover song does not merely repeat the original but offers new information from after the October War, the context of the original version. Another example of subversion at the lexical level is the reversal of the vocabulary of war and violence to that of peace and love, with each side displaying its ideology and mentality. Where the original version has "war" in the second line, the cover refers to "love," and the line "the bullet of the rifle is the factory of freedom" becomes "the bullet of the rifle will not kill the freedom," which depicts the demonstrators as victims and shows pacifist resistance to the regime's violence. The lyrics also directly condemn the regime's brutal suppression of peaceful demonstrations. Even where the two versions use

197 Simon Dentith, *Parody: The New Critical Idiom* (London: Routledge, 2000), 9.

the same words, the context and thus what is signified is different. For example, both versions mention the term "revolution," but in the original song this means the two *coup d'états* of the Baath revolution: the first when the party took power (1963), and the second when Hafez al-Assad came to power by what he called the "Corrective Movement" (1970). The original version of the song, as typical of the regime's speeches, calls for solidarity with the March Revolution of Baath, or the definition of revolution given by Assad in *Thus Said Assad*, while the new song calls for the (new) revolution of 2011.[198] One of the functions of the new version is thus to correct and offer a new interpretation of the old terms used by the Baath Party in Syria.

If we take a closer look at the changed words, we can see that the new song casts aside all of the ideological terms of the Baath Party. For example, instead of "Now I am an Arab," as in the original version, the cover song has, "Now I long for my tomorrow with freedom in my hand"—a shift from the Pan-Arabism sought by Baathism to the future as the objective of the revolution. While these phrases are not direct antonyms in lexical terms, they symbolically represent the opposite of each other: while the first speaks of gaining freedom in the name of Pan-Arabism, the second speaks of freedom in the sense of "I own my tomorrow," which refers to emancipation and agency in decision-making. Replacing "Baath" with "people" also suggests that only Syrian people should rule the country, not Baath, reflecting the end of one-party rule as one of the reforms demanded by demonstrators. In both the original and cover versions, "Sūriyā yā Ḥabībatī" reflects competition over symbolic products between the pro-regime and pro-revolution sides.

This song is not the only example of the revolutionary side remaking or creating similar new symbolic products for public consumption but there are several songs that followed the same strategy to subvert Assad's music. In the previous chapter, I discussed the use of the famous song "Minḥibbak" in Bashar's 2007 election campaign,[199] and this was also remade, with an emphasis on the "m," adding an *alif* (ā) after it to make it a negative particle and turn the title into "Mā Minḥibbak" [We Do Not Love You]."[200] There was also another cover of the same song as it was performed by Shahad Barmada but subtitling it and putting an emphasis on the ī in Minḥibbak to become Mīn Ḥabbak, meaning "Who loved you [Bashar]?"[201].

---

198 Ṭlās (ed.), *Hakadhā Qāl al-Assad*, 27.

199 See the previous chapter for more information on Bashar al-Assad's 2007 "Minḥibbak" campaign, in Chapter 2.

200 "*Yā Bashar mā Minḥibbak*" [Hey, Bashar, We Do Not Love You], unknown performer, YouTube, uploaded June 8, 2011, accessed July 18, 2023, https://www.youtube.com/watch?v=FCjL3zwO4h4.

201 "'Ughnīyat mīn Ḥabbak yā-Bashar [The Song of Who Loved you Bashar]". Abu Fidā' al-Ḥamwī (Youtube channel), uploaded August 29, 2011, accessed July 18, 2023, https://www.youtube.com/watch?v=h-BvYKIzcD8.

Subtitling the song was one way of changing its goal and this was also supported visually by displaying photos. For example, the song says "We are the future generation who love you [or in the subtitles who do not love you], and it shows a photo of Assad as a butcher saying "Welcome to the future generation, Bashar loves you so much."[202]

Regime (Musical) Responses to Revolution Products

The second group of songs were those remade by the regime as alternatives to revolutionary symbolic products: just as those who supported the revolution remade older pro-regime songs, pro-regime factions did the same with certain revolutionary songs that were popular hits in the Syrian public space. One example of this is Samih Choukair's famous revolutionary song "Yā Ḥīf." The regime did not change the title but added to it to create a "patriotic version" of "Yā Ḥīf."

"Yā Ḥīf" is a Syrian expression used to express sorrow or injustice. In Arabic, *ḥayf* used by itself means "injustice and unfairness."[203] The whole expression, with the addition of *yā* can be expressed as "alas," meaning being deceived, astonished and surprised, but can also be translated as "shame on you [person or object]." In the view of Samih Choukair, "the title and this expression was chosen because it summarizes everything that happened in Syria. It compresses the whole Syrian tragedy from 2011 till now."[204] The song is about the beginning of the protests and the arrest of children in Daraa, and can be seen as a story containing the literary elements of setting, plot, narrative point(s) of view, and theme. The setting is Syria and the time is 2011. As is evident from the lyrics and translation displayed in Table 2, the plot is presented through a description of the events, as Choukair details the shooting of protestors, the arrests of children, people chanting "freedom" in the streets, and the reaction of the regime. The song uses multiple points of view: a main narrator is used for the prelude and epilogue, and the remainder of the song is performed from the first person "we" and "us." The whole song is a confession to a mother [*yummā*], a word that is used seven times in this short song. The phrase "*yā ummī*," or "oh, my mother," is an expression used when one has been deeply deceived, is surprised by something, or is complaining about something. In addition, addressing these lyrics to "mother" highlights the confidentiality of a dialogue that usually happens in private, rather than publicly in a song. This technique of addressing his mother gives Choukair's narration of events a greater degree of authenticity.

202 Abu Fidā' al-Ḥamwī, "'Ughnīyat mīn Ḥabbak yā-Bashar."

203 *Al-Maany Online Dictionary*, s.v. "*ḥayf*," accessed July 18, 2023, https://www.almaany.com/ar/dict/ar-ar/%D8%AD%D9%8A%D9%81/. This word is pronounced differently in colloquial Syrian.

204 Samih Choukair, interview with the author, June 4, 2018.

Table 2: "Yā Ḥīf" by Samih Choukair, in Arabic with English translation.

"Yā Ḥīf" (Original), by Samih Choukair	ياحيف (الثورة) سميح شقير
Alas/What a shame/Shame on you.	يا حيف اخ ويا حيف
Bullets rained on the unarmed people, alas.	زخ رصاص على الناس العزّل يا حيف
Oh, how do you [the regime] arrest those in the flower of youth?	وأطفال بعمر الورد تعتقلن كيف
While you are the son of my country, you are killing my children. Your back is to your enemy and you attack me with your sword.	وانت ابن بلادي تقتل بولادي. وظهرك للعادي وعليي هاجم بالسيف
This is what happened. . . Alas. . . Oh, my mother. . . Alas. . . This is Daraa. . . Alas.	وهذا اللي صاير يا حيف. بدرعا ويا يما ويا حيف
Oh, mother, when the youth heard that freedom was getting close, they went out to chant for it.	سمعت هالشباب يما الحرية عالباب يما. طلعو يهتفولا
Oh, mother, they saw rifles. . . They said they were our brothers. They would not fire on us.	شافو البواريد يما. قالو اخوتنا هن ومش رح يضربونا
Oh, mother, they shot us with live bullets. . . We died at the hands of our brothers in the name of the security of the country.	ضربونا يما بالرصاص الحي. متنا. بايد إخوتنا باسم امن الوطن
Who are we? Ask history. Let history read our page.	واحنا مين إحنا واسألوا التاريخ. يقرا صفحتنا
One word of freedom, oh mother, trembled the jailors' authority.	مش تاري السجان يما كلمة حرية وحدا هزتلو اركانو
When the crowd chanted, he became like a stung person, throwing fire on us.	.ومن هتفت لجموع يما اصبح كالملسوع يما. يصلينا بنيرانو
It is we who said, "the one who kills his people is a traitor. . . no matter who he is."	واحنا اللي قلنا اللي بيقتل شعبو. خاين. يكون من كاين
The people are like fate, when they ask, they should be answered. The people are like fate. . . the hope is obvious.	والشعب مثل القدر . من ينتخي ماين. والشعب مثل القدر .والامل باين

"Yā Ḥīf" had a huge impact on the Syrian public space, as reflected in the regime's numerous attempts to remake the song. Several versions were produced by the regime at different times, each claiming to be the original version of "Yā Ḥīf," in spite of the fact that the song was composed and first released by Samih Choukair. The first regime version was performed by Sām Ḥisīkū, but with a different rhythm and composition, on April 14, 2011,[205] while the other three were released on July

205 Sām Ḥisīkū, performer and song writer, *"Ughniyat Yā Ḥīf al-Jadīda (al-Fannān Sām Ḥisīkū)"* [The New Song of Yā Ḥīf], uploaded April 14, 2011, accessed July 18, 2023, https://www.youtube.com/watch?v=7-IYdS6dABc.

9 by an unknown singer, on August 11 by Sumar al-Hamwi, and on August 26 by Nawar Haidar.[206] This added description of the new cover of the song as "original" is an indicator that it is a reclaiming process and an act of reshaping the songs by trying to recreate a more powerful product to compete in the public space.

The first response from the regime side was also titled "Yā Ḥīf."[207] The song is two minutes and seventeen seconds long and was produced eighteen days after the one by Choukair. This rapid response and the poor production quality suggest that the song was not initially funded by the regime itself. For example, the sound engineer did not delete the few seconds when the singer cleared his voice before starting to sing, or during the song. The sound quality is also very poor, which makes it clear that the song was not produced in a studio. In terms of themes and lyrics, it does not have the same form as the original version of "Yā Ḥīf" because the length is not the same, the words do not have deep meanings, and it does not use figures of speech. While Choukair's version has thirteen lines that chronologically explain what happened in Daraa, Ḥisīkū's song has only four lines. It was written to criticize Choukair and his art and to cast aspersions on the intentions of demonstrators, accusing them of burning the Syrian flag and attacking the army. The description of the song on YouTube reads "This song is dedicated to every Syrian who loves his country and hates traitors"[208]—a direct message intended solely for supporters of Bashar al-Assad. To put it differently, as seen in the previous chapter regarding the visual products relating to Hafez al-Assad and the signification between Syria and the leader, Syria becomes an attribute of the leader and not vice versa. When the word Assad (Hafez or Bashar) is used, one of its meanings is Syria, but Syria has only one meaning, which is Assad. The depth of meaning and collective layers of "Syria"—as a place, home, culture, language, etc.—are thus limited to mean only "Assad," while "Assad" takes on the function of "Syria" and becomes a richly layered

206 "Tughdur ahlak wa-aṣḥābak khā'in yā Ḥīf" [If You Betray Your People, You Are a Traitor, yā ḥīf], unknown artist, YouTube, uploaded August 19, 2013, accessed July 18, 2023, https://www.youtube.com/watch?v=PhYCGT2Zris&has_verified=1; Walīm al-Ḥasan (Syrian singer/songwriter), interview with the author, June 6, 2018; Sumar al-Hamwi, performer, "Ughniyat Yā Ḥīf bi-Iṣdār Jadīd 2011 Sūriyya" [Yā ḥīf Song in a New Edition, 2011 Syria], *ZeinasoftNetwork* (YouTube channel), uploaded August 11, 2011, accessed July 18, 2023, https://www.youtube.com/watch?v=7bdSFWRsBY0; Nawar Haidar, "Yā Ḥīf," personal research archive. According to my personal research archive, the cover song allegedly by al-Ḥasan was first uploaded on July 9, 2011, but the only remaining copy on YouTube was uploaded on August 19, 2013. While the song originally uploaded on July 9, 2011 listed Walīm al-Ḥasan as the singer/songwriter, he denied responsibility for it in my interview with him on June 6, 2018.
207 Ḥisīkū, "Yā hīf."
208 Ḥisīkū, "Yā hīf."

word with many levels of interpretation.[209] For example, when this song mentions the national army, Syrian flag, homeland and Syria, the singer means Assad, rather than the obvious meanings. As in the previous chapter, the YouTube video displays many photos that express the line that is being sung: for example, the line "Alas for a broken art"[210] is accompanied by a photo of Choukair, and when the Syrian flag and homeland are mentioned in the song, the video shows a photo of Bashar al-Assad (Figure 6). The background of the photo is Damascus, as seen from Qāsiyūn Mountain, along with the Syrian flag. This photo illustrates how Assad means Syria, the flag, and the country, highlighting the plurality of "Assad" as a concept.

Figure 6: A photo of Bashar al-Assad from the video to Sām Ḥisīkū's "Yā ḥīf." The text underneath reads: "We will not forgive those who conspired against you, our leader, Bashar."

The second regime version of "Yā Ḥīf" was released on July 9, 2011, and also has fewer lines compared to the original. Just as Choukair's song explains the protests chronologically, this version provides a chronological view from the regime perspective. It starts by criticizing demonstrators for protesting and blames them for killing civilians and the military, claiming that soldiers were killed while protecting protestors. The singer then goes on to explain what happened, singing: "When people heard 'freedom,' they took advantage of it. The plan for freedom was to sabotage the homeland on behalf of the enemies." The song uses the same technique of addressing the singer's mother, and ends by telling her that pro-Assad marches are taking place all over the country. He indirectly includes a slogan in the song, saying "I swear to God that we will sacrifice ourselves for you [Bashar al-Assad]," and describes Assad as "the Damascene rose who is the hope." Like the first pro-regime version, in

209 Please see the visual section on Hafez al-Assad in the previous chapter.
210 Ḥisīkū,"Yā hīf."

this song "Assad" as a concept has multiple layers of meanings, while Syria is limited to meaning only Assad. As such, the song refers not to marches in support of the country, but to pro-Assad marches: according to the logic of the regime, Syrians who want to support the army or their country should support Bashar al-Assad.

The other two versions of "Yā Ḥīf" were released in August 2011 by Sumar al-Hamwi and Nawar Haidar. These two versions are typical covers of the original in terms of their rhythm, music, style, and composition. Only the content differs, with both songs doing away with the story elements of the original and creating a form of speech that is, like the previous song, generally written from the regime's perspective.

The third version of "Yā Ḥīf" is not a personal attack on Choukair—it does not mention or attack him—but directly contradicts the story of the original song. The title of the video claims that it is a new version of the song, which also demonstrates the struggle to replace revolutionary songs with those of the regime or to express the regime's vision. In it, Sumar al-Hamwi criticizes Arab leaders, claiming that they act as though "Syria is the enemy and Israel is the guest."[211] The song uses examples of Muslim personalities who supported the revolution to assert that it was funded by them, mentioning Yūsuf al-Qaraḍawī and ʿAdnān Muḥammad al-ʿArʿūr.[212] These names are meant to provide evidence that the revolution is calling for Salafism and killing the army. The end of the song does not mention Bashar al-Assad, but does praise the army and Syrian people: here again, ultimately meaning the leader. The video displays a single photo in the background, showing a map of Syria and a quote from Bashar al-Assad that says, "God protects Syria," as a reference to the presence of Assad in the song. This sentence and the maps of Syria confirm again the unity of symbolism in which "Syria" refers to Bashar al-Assad, who is the army, the unity of Syria, and Syria itself.

Terms like conspiracy, Salafism, sedition, and others are used in the same way as the regime-sponsored news media in the fourth, and final, version of the song by Nawar Haidar, which is also entitled "Yā Ḥīf." Like the other versions, it refers to the protestors as "traitors," and draws on the same repetitive themes as the other versions about the non-peaceful demonstrations. At the end of the video, similarly to the other pro-regime versions, a photo of Bashar al-Assad is shown.

Regarding his decision to record a cover of "Yā Ḥīf," Sumar al-Hamwi told me that: "I felt that it was my duty to defend my country. Nobody made me do it, and, unfortunately, I was threatened on Facebook, by SMS, and by phone for producing

211 Al-Hamwi, "Yā Ḥīf."

212 Yūsuf al-Qaraḍawī and ʿAdnān Muhammad al-ʿArʿūr are two sheikhs known for their support of the Syrian revolution. The former is based in Doha and known for his affiliation with the Muslim Brotherhood, while the latter is a Salafist based in Saudi Arabia.

a true version and a national one of 'Yā Ḥīf.' I did not want to compose a new song. I wanted this song to be a clear message to Samih Choukair."[213] Nawar Haidar also argued that "the events [in the original version of Yā Ḥīf] did not take place as the song described. It is not possible for our army to kill people. Let's be honest. There were honest demonstrators, but many people took advantage of what happened in 2011."[214] The songs themselves seem to suggest that these new versions of "Yā Ḥīf" were produced individually and voluntarily, without any force or coercion from the regime. I asked Nawar Haidar and Sumar al-Hamwi if anyone in authority had pushed them to create these cover versions, but they answered "no," and that "it was a national duty" for them to do so.[215] This can be seen as an example of the symbolic violence under the Assad regimes, with these performers likely seeking a reward or wanting to prove their loyalty to the regime. When I looked into how the regime supports the production of patriotic songs, I was told by Syrian composer Walīm al-Ḥasan that: "the regime does not finance most of the songs, and it pays only a little for producing national songs." While there is definitely a certain amount of enjoyment for a singer/songwriter practicing their craft, the real reward "was to obtain protection and privileges from the regime itself."[216] Releasing a song in praise of Assad could be understood as not directly resulting in a material reward, but as potentially providing access or a gateway to other benefits that can bring financial support. For example, such performers might be exempted from military service or given a secret police card providing them with other types of privileges. Similarly to slogans and banners, individuals' dispositions in the symbolic market teach them to take advantage of it to exchange any products with symbolic benefits. This behavior of immersing the field with symbolic products not only benefits individuals but also provides domination over the social setting in which those products are marketed.

Both sides were thus competing to create patriotic songs. Some singer/songwriters might have done so for potential rewards or to prove their loyalty, while the release of other songs can only be seen as a revolutionary act. Patriotic and pro-regime songs were not randomly released in the Syrian public space; there was a struggle between the pro-regime and pro-revolution factions. This struggle can be seen in the tensions between each side offering or subverting alternative songs. It is important to note that, while both sides imitated each other's songs, the revolutionary songs were more innovative in terms of their form and even themes, and were able to make widespread use of genres like rap due to the greater amount of lyrical freedom for singers. Singers and composers also cooperated in a different

213 Sumar al-Hamwi, interview with the author, June 5, 2018.

214 Nawar Haidar, interview with the author, June 6, 2018.

215 Haidar, interview; al-Hamwi, interview.

216 al-Ḥasan, interview.

way, since their lack of financial resources freed them from the control of producers and sponsors, creating new dynamics in the Syrian musical space. This was one reason why bands such as Nuṣṣ Tufāḥa and al-Mundassūn al-Sūriyyīn al-Aḥrār were working online or outside of Syria to create revolutionary songs.

National Anthems

This competition to subvert all of the regime's symbolic products ultimately extended to the official national anthem, which was not perceived by demonstrators as a national symbol because of the regime's adaptations to it. As explained in the previous chapter, despite the existence of an official Syrian national anthem, the Baath Party created three parallel quasi-national anthems for Syria.[217] After March 2011, the supporters of the revolution began to oppose and reject all of the symbolic products of the regime, an act of opposition that extended to the symbols of Syria as a state. Other would-be national anthems were thus created with the intention of these replacing—or at least competing with as musical symbolic products—the official national anthem, as a symbol of the regime.

Table 3 below presents the official national anthem of Syria, "Humāt al-Diyār" [Guardians of the Homeland], in Arabic with an English translation. The anthem highlights the role of the Syrian army in protecting the country, focusing on the idea that the existence of a nation is achieved via a strong army.[218] "Ḥumāt al-Diyār" has four verses. The first begins by calling the army the "guardians of the homeland" and describing the glory of Syria as an untouched den of Arabism and the sun of suns; the second verse extols the landscape and beauty of nature; the third verse refers to sacrifices made by Syrians throughout history; and the final one praises the history of Syria, referencing Al-Walīd Ibn ʿAbd al-Malik[219] and Harūn al-Rashīd.[220] The anthem was composed, performed, and recognized as the official anthem of Syria before the Baath Party came to power. This raises an important question: if the anthem was not a symbolic product from the Baath regime, why it was opposed after 2011?

---

217 For more on the regime's national anthems, see Chapter 1.

218 For more historical information on the Syrian national anthem, see Chapter 1.

219 Yūsuf al-ʿAysh, *Al-Dawla al-Umawiyya* [The Ummayyad State], 2nd ed. (Damascus: Dār al-Fikr, 1985), 243. The rule of Al-Walīd Ibn ʿAbd al-Malik has been described as the Golden Era of the Umyyad Caliphate.

220 Muhammad al-Khuḍarī Bek, *Muḥaḍārāt fī-Tārīkh al-Umam al-Islāmiyya: al-Dawla al-ʿAbbāsiyya* [Lectures on the Muslim Nations: The Abbasid State], ed. Najwā ʿAbbās (Cairo: Al-Mukhtār, 2003), 101. The period of Harūn al-Rashīd's rule has been described as the Golden Era of the Abbasid Caliphate.

Table 3: "Humāt al-Diyār," the official Syrian national anthem, in Arabic with English translation.

Guardians of the homeland	النشيد الوطني السوري الرسمي[221]
The guardians of the homeland! To you is our salutation!	حماة الديار عليكم سلام
Nobles spurn humiliation	أبت أن تذل النفوس الكرام
The den of Arabism is sacredly unapproachable	عرين العروبة بيتٌ حرام
The throne of suns is too high to assault.	وعرش الشموس حمى لا يضام
Damascus gardens are constellations	ربوع الشآم بروجُ العلا
so much like high skies	تحاكي السماء بعالي السنا
It is a sun-lit land	فأرضٌ زهت بالشموس الوضا
Matchless land and sky it is!	سماءٌ لعمرك أو كالسما
Hopes are pinned; hearts beat fast	رفيف الأماني وخفق الفؤاد
over a country-unifying flag	على علم ضم شمل البلاد
It certainly has the black color of its people's eyes	أما فيه من كلّ عينٍ سواد
And so it has martyrs-blood ink	ومن دم كلّ شهيد مداد؟
Fierce pride and glorious past	نفوسٌ أباةٌ وماضٍ مجيد
Souls are offered lavishly	وروح الأضاحي رقيبٌ عتيد
Al-Waleed and Al-Rasheed belong to us	فمنا الوليد ومنا الرشيد
We have every reason to rule and build the country	فلم لا نسود ولم لا نشيد

The process of creating a parallel national anthem designed to replace the official one did not begin until protestors were convinced that this anthem could not represent Syrians. The main reason for rethinking the national anthem and subverting it or replacing it with another stemmed from its opening words: "Humāt al-Diyār" starts by praising the army as the "guardians of the homeland," but the Syrian national army—converted by the Assad regimes to one dedicated to protecting the Assad throne—committed atrocities in the early months of protests in 2011. Protestors did not initially attack the army by denouncing it in alternative anthems, but instead tried to bring it over to their side, as had happened in Egypt and Tunisia, including by designating Friday, May 27 as a day to speak to the "guardians of the homeland"[222]—an event that was promoted via social media and visual products. On this day, Syrians took to the streets and chanted the Syrian national anthem to remind the army that it was meant to protect Syrians, not kill them. Many people

221 Khalīl Mardam Bey, lyrics, and Muhammad Fulayfil, composer, *"al-Nashīd al-Waṭanī al-Sūrī Ḥumāt al-Diyār,"*, n.p. 1936, official national anthem of the Syrian Arab Republic, unknown performer, uploaded January 30, 2015, accessed July 18, 2023, https://www.youtube.com/watch?v=hxq9-q64O08.

222 *Al Jazeera*, *"Sūriyyūn Yunaẓẓimūn Jumʿat Ḥumāt al-Diyār"* [Syrians Organize the Friday of the Guardians of the Homeland], May 27, 2011, accessed July 18, 2023, http://bit.ly/2vukACg.

were injured or killed in these demonstrations.[223] The national anthem was chanted in several protests in different cities to show that demonstrators shared the same essential patriotic feelings and objectives as all Syrians. Later on, certain activists known as the "flower activists," including Ghiyāth Maṭar among others, were tortured to death because they gave flowers to soldiers.[224] The army did not respond to the demonstrators' call: there were several reasons for this, but the most important one was that high-ranking officers were ultimately more loyal to the Assad regime or were themselves from the Assad family.

The first alternative version of the national anthem was chanted in September 2011, at a demonstration in Ṭaybat al-Imām, a town close to Hama. This kept the same rhythm but used different lyrics (Table 4).[225]

Table 4: The subverted Syrian national anthem as performed in a protest, in Arabic with English translation.

National anthem used in the demonstration (first modified version)[226]	النشيد الوطني السوري المعدّل في المظاهرة
The regime's guardians are tyrants and ignoble. You killed proud souls.	حماة النظام طغاة لئام قتلتم، أهنتم نفوساً كرام
You bombarded houses, spilled blood, masked suns and spread darkness.	قصفتم بيوتًا أرقتم دماء حجبتم شموساً، نشرتم ظلام
The Levant land suffers and prays God lifts the affliction.	ديار الشآم تعاني الأذى تناجي العظيم لرفع البلا
The tyrant army is ignoble and hateful. It antagonizes the people and the sky.	فجيش الطغاة لئيم حقود يعادي العباد يعادي السما
The pulse of the land and its prayer are against the tyrant who wreaks havoc.	ونبض البلاد وكل الدعاء على ظالم عاث فيها الفساد
The tyrant sold the people's land, befriended the invaders, displaced the people and spilled blood.	أما باع أرضاً ووالى الغزاة وهجر شعباً، أراق الدماء

223 *France 24*, "*Qatlā wa-Jarḥā fī Jumʿat Ḥumāt al-Diyār*" [Deaths and Injuries on the Friday of the Guardians of the Homeland], May 27, 2011, accessed July 18, 2023, http://bit.ly/39uGEeI.
224 *Procare Press*, "*Ghiyāth Maṭar. . . Ahdāhum al-Wūrūd fa-Raddū ʿalayhi bi-l-Iʿtiqāl wa-l-Taʿdhīb ḥattā al-Mawt*" [Ghiyāth Maṭar . . . He gifted them flowers, they answered by arresting him, torturing him to death], September 11, 2019, accessed July 18, 2023, http://bit.ly/37kS9Ur.
225 This is the earliest date I was able to find on which a modified version of the national anthem was used as a chant in a demonstration to attack the Syrian army.
226 "*Ughniyat Ḥumāt al-Diyār ʿAlaykum Damār. . . Qataltum Ahantum Nufūsan Kirām*" [The Song of the Guardians of the Homeland, Destruction Be Upon You; You Killed, Humiliated Proud People], performed at a demonstration, YouTube, uploaded September 7, 2011, accessed July 18, 2023, https://www.youtube.com/watch?v=iUBlxXbN9d8.

Table 4 (continued)

Proud men and long-standing people confronting the enemies firmly.	رجال أباة وشعب عريق بوجه الأعادي شديد عتيد
Some are wounded or martyred. Certainly the expulsion of tyrants is near at hand.	فمنا الجريح ومنا الشهيد فدحر العتاة قريب أكيد

This version has the same form as the original anthem in terms of its length, number of lines, and rhythm. Its content, however, is totally different and is meant as an attack on the army. Instead of the "guardians of the country" in the first verse, demonstrators referred to "the regime's guardians," insinuating that the national army is merely a killing machine for the regime and loyal to the regime above their country. The rest of the anthem describes Assad as a traitor, referring to his selling of Golan to Israel,[227] and contradicting the regime's story of being in a continuous war against Israel. The new version of the anthem used in the protest ends by praising the demonstrators for representing all Syrians and promising that they will defeat the tyrant soon.

It is important to note here that replacing the anthem in Ṭaybat al-Imām was a key indicator of disruption to public belief in the anthem as a national icon and symbol for Syrians because it represented the regime. The new version was a huge success, with other cities following Ṭaybat al-Imām's lead and singing the same version. In my view, by chanting the modified anthem protestors did not aim to replace the original but rather to criticize the regime and bring attention to what the army was doing to demonstrators.

A few months after this demonstration, on January 17, 2012, al-Mundassūn al-Sūriyyīn al-Aḥrār released a new version of the Syrian national anthem (Table 5).[228] This anthem is similar to the official one, but its four verses focus on the grievances of Syrians. It seeks to drum up enthusiasm and to produce an emotional response in specific parts, with phrases like "the wound of the freewomen is unforgettable" and "the purity of Levant has been profaned." The composer knew that the word "free-

227 Hashem Othman, *Tārīkh Suriyya al-Ḥadīth: ʿAhd Hafezal-Assad 1971–2000* [The Modern History of Syria: Hafez al-Assad Regime 1971–2000] (Beirut: Riad El-Rayyes, 2014), 45. The accusation that the Golan Heights was sold is frequently made because in 1970 Hafez al-Assad was the Syrian defense minister. Following Syria's defeat in the 1967 Six-Day War, Israel captured the Golan Heights and continues to occupy a significant portion of the area.

228 Al-Mundassūn al-Sūriyyūn al-Aḥrār, *"Nashīd Ḥumāt al-Diyār – Tanfīdh wa-Ghināʾ Firqat al-Mundassūn al-Sūriyyūn al-Aḥrār"* [Free Syrian Infiltrators, the Anthem of Guardians of the Homeland], produced and sung by Al-Mundassūn al-Sūriyyūn al-Aḥrār, YouTube, uploaded January 17, 2012, accessed July 18, 2023, https://www.youtube.com/watch?v=u4ixV2Itm1U&t=9s.

women" would have more of an impact than saying "free people" because it has three levels of interpretation here: first, harm to women calls directly for gallantry and chivalry; second, it shows that the revolution is for both men and women; and third, it can have an Islamic connotation depending on the context in which the term is used, as we will see in the section on translation of this word in Chapter 4. Another change comes with the concluding lines: while the official anthem praises al-Walīd, the Umayyad Caliphate, the new one refers to Khālid Ibn al-Walīd in an implicit reference to the city of Homs where he is buried, then considered to be "the capital of the revolution."[229] Muʿādh Naḥḥās, who composed the lyrics, told me that: "It was a spontaneous anthem. I wrote the lyrics in a Facebook post, and then a friend asked me about producing it as an anthem." He added: "The idea came to my mind after I saw a demonstration [where they were] chanting the anthem [and] inverting its meaning." When I asked whether this was the same demonstration discussed in the above paragraph, Naḥḥās answered, "I do not remember. There were many demonstrations that chanted the inverted national anthem of Syria."[230] He confirmed to me that the new version of the anthem lyrics was not intended to replace the official anthem, but rather to rally demonstrators and increase their enthusiasm on the ground.

Table 5: The second new version of the national anthem, in Arabic with English translation.

National anthem (second modified version)	النشيد الوطني الثاني (المندسين)
Guardians of the homeland, peace be upon you. People want to topple the regime. Shedding the people's blood is forbidden. Come and defend a subjugated people.	حماة الديار عليكم سلام- الشعب يريد إسقاط النظام دم الشرفاء عليكم حرام- فهبّوا لنصرة شعب يُضام
The Levant land suffers from sorrow and grieves to God for the hardened injustice. The wound of the freewomen is unforgettable and the purity of Levant has been profaned.	ربوع الشآم تعاني الأسى- وتشكو إلى الله ظلماً قسا فجرح الحرائر لا يُنتسى- وطهر الشآم قد دُنسا
With determination, certainty, an honest heart, and a challenging peacefulness against tyrants, we will eradicate the unjust from everywhere and make patience our provision.	بعزم اليقين وصدق الفؤاد- وسلمية تتحدى العتاد سنقتلع الظلم من كل واد- ونجعل من صبرنا خير زاد
A people loyal to a glorious past made strong through [their] ordeals. The people will keep the pledge, Ibn al-Walid, and will restore glory to the Umma.	فشعب وفيّ لماضٍ مجيد- وفي النائبات قويٌّ عتيد سيبقى على العهد يا بن الوليد – وللأمة المجد سوف يعيد

229 Usāma Abū Zīd, "*Homs ʿĀṣimat al-Thawra al-Sūriyyīa Khāliya min Ahlihā*" [Homs, the Capital of the Revolution for Syrians is Abandoned], *Al Arabiya*, March 20, 2016, accessed July 18, 2023, http://bit.ly/2vqlkIz.
230 Muʿādh Naḥḥās, interview with the author.

Looking beyond the 2011 timeframe of this study, two further anthems were released. The first of these was “Sūriyya Nashīd al-Aḥrār” [The Anthem of the Free], with lyrics and music composed by Malek Jandali.[231] Although this was not intended to become an official national anthem, in an interview with the composer he told me that: “I would be happy to have this anthem as the Syrian national anthem, but it is down to Syrians to choose an anthem for themselves.”[232] The last anthem starts with “The Martyr of the Homeland,” released in 2013.[233] This anthem was structured similarly to the one by Jandali, but was intended to replace the official one. The producer of the anthem told me that: “it is a new anthem that unifies the Syrian opposition and the Free Syrian Army. Any armed group that does not adopt it as its anthem should consider itself a militia.”[234] Since the timeframe of this study ends in 2012, I will not analyze or discuss other would-be national anthems that were released outside of this period. It is important to note, however, that production of such anthems stopped after 2013 (see Figure 7, below).

March 27, 2011
•The official national anthem is used in demonstrations
September 7, 2011
•New version of the national anthem is used in a demonstration
January 17, 2012
•The first alternative national anthem is released by al-Mundaseen
April 11, 2013
•The Anthem of the Free People
July 16, 2013
•A third national anthem is released

Figure 7: Attempts to replace the national anthem.

As seen in Figure 7, the symbolic product that took the longest to be changed or replaced during this period was the national anthem. In contrast to the other types of symbolic products, the production of a (potential) new national anthem was controversial among people in Syria, and was exploited by the regime to argue that protestors lacked the feeling of belonging to Syria and were traitors. The regime

231 Malek Jandali, *“Sūriyya Nashīd al-Aḥrār,”* YouTube, uploaded April 11, 2013, accessed July 18, 2023, https://www.youtube.com/watch?v=x2uSPsJr0vE.

232 Malek Jandali, *“Malek Jandali maʿ Nawras Yakan fī Salām”* [Malek Jandali, interview with Nawras Yakan fī Salām], *Nasaem Radio,* uploaded March 16, 2015, accessed July 18, 2023, https://www.youtube.com/watch?v=WdGO7-ypLm4.

233 al-Ḥabīb al-Muṣṭafā Band, *“Nashīd al-Thawra al-Sūriya: al-Nashīd al-Waṭanī al-Sūrī al-Jadīd 14–7–2013”* [The Syrian Revolution Anthem: The New Syrian National Anthem 14–7–2013], YouTube, uploaded July 16, 2013, accessed July 18, 2023, https://www.youtube.com/watch?v=kz261jIKUqc&t=12s.

234 Ghassān Abū Luz, *“Shahīd al-Diyār .. Nashīd Waṭanī Jadīd li-Tawḥīd al-Muʿāraḍa”* [The Martyr of the Country: A New Anthem for Unifying the Opposition], *Al Arabiya,* July 16, 2013, accessed July 18, 2023, https://goo.gl/ayFm6n.

argued that protestors wanted to change all of the country's national symbols, ignoring the existing ones of the Baath regime. Despite the creation of different versions of the national anthem (and new potential national anthems) intended to subvert or replace the official anthem, such anthems did not become as well known as the revolutionary or regime songs. They merely sought to emphasize to the regime that all of its icons would be destroyed once it was defeated.

Visual Occupation

Activists and demonstrators not only sought to express themselves musically, but also understood that the winner of the struggle over these products would be able to control more of the symbolic public space. Controlling, here, is very similar to the act of occupying, covering, or filling the public Syrian space with all kinds of favorable symbolic products—especially visual ones that are visible to the larger public and indicate who (or which perspective) is controlling the area. The act of "winning" or controlling the visual space was, however, temporary in regime-controlled or pro-regime areas. One objective of controlling the visual space was to broadcast it to the media, which meant that such control was not meant to last for more than few seconds. Pro-revolution visual items were generally only present in pro-regime areas for no more than a day, at the most, depending on how difficult it was to erase the visual materials that were planted. Regime television channels broadcast live almost continually, displaying areas of calm in big cities or quiet neighborhoods to imply that life was normal and dissent non-existent. At the same time, the regime's cameras had to wait for the secret police to finish clearing away revolutionary visual materials before they could resume broadcasting and show that everything was stable. Activists, meanwhile, used social media and channels like *Al Jazeera* to show that the opposite was true. As with other symbolic products, the pro-revolution and pro-regime factions struggled to remain more visible in the Syrian public space in order to signal who had control over more visual spaces.

One way of acquiring more physical and symbolic space was through the use of people's bodies. Both sides thus competed to acquire more space by using any visual materials people could carry—such as flags, banners, and photos—in demonstrations (see Figures 8 and 9).[235] As seen here, both photos essentially depend on the concept of using bodies to fill space, creating a visual sign of ownership through physical occupation. The revolutionary performance, its performativity and graffiti, will be explained in separate parts in the following chapter.

---

235 *"Raf' Aṭwal ʿAlam fī Sūriyya"* [Raising the Longest Flag in Syria], *Discover Syria*, June 15, 2011, accessed July 10, 2018, http://www.discover-syria.com/image/_px/ds-15611syrFlag20.JPG (no longer available).

Figure 8: Bodies visually controlling spaces at a pro-regime march in Damascus.

Figure 9: Demonstration in Hama.[236]

In this case, a demonstration, together with all of its components, was considered a visual product in the public space that Syrians competed to acquire, whether they were pro- or anti-regime. One step toward emptying the visual space of traces of the regime began in the first few months of the protests with a systematic campaign to

236 Friday of Khalid Bin al-Walid, July 22, 2011, personal research archive.

destroy symbols of the Assad regimes and family such as statues and images.[237] The revolutionary act of filling the visual space was an act of the destruction and creation of new spaces. For example, all of Assad's titles adorning places such as schools, mosques, squares, and streets were removed and replaced with titles like "Freedom School," "Martyr Square," and "Revolution Street." The only reaction by the regime was to tighten security and to put statues under surveillance, or to remove them if protests increased, as in cities like Hama, where in early June 2011 the regime anticipated that demonstrators would destroy Assad's statue, and thus removed it to keep it out of their hands. Protestors later went to the former site of the statue and put a donkey in its place to humiliate Assad and destroy his image in Syria.[238]

Another way of acquiring more visual space in the Syrian public space was by distributing pamphlets and leaflets. Although this was a temporary act that only lasted a few minutes, it became very annoying for the security police, who were under instructions to remove or clean up any pro-revolution materials while also being tasked with securing and guarding public spaces. The technique of disseminating leaflets evolved over time by making the flyers smaller, and thus irritating the secret police by making it harder to clean up the pieces of paper from the floor or ground. Because of the difficulty of cleaning up a whole street, it was considered a victory for demonstrators—a victory over the visual space—if they saw leaflets on the ground that had not been removed.

Another visual product worthy of mention in this section was displaying the revolutionary flag. As another act of temporarily acquiring new spaces, raising the revolutionary flag over the regime's public institutions constituted a silent act of resistance designed to claim more public visual space from the regime. Usually, the flag was removed and replaced by another regime symbol to signal that it had won the competition. The timing and method of deploying a visual item played an important role in giving it greater longevity. One YouTube video, for example, shows the flag being displayed during the very early morning hours when no one was on the street.[239] Other items remained in place longer because they were difficult to completely remove from the streets, such as the small leaflets.

---

237 I mention here only the revolutionary side because all of the Syrian public spaces were controlled and filled by the regime's visual products. The equivalent act of destroying revolutionary spaces by the regime will be discussed separately later.

238 *"Izālat Timthāl Ḥāfez al-Assad fī-Ḥamā"* [Removing the Statues of Hafez al-Assad in Hama], YouTube, uploaded August 24, 2011, accessed July 18, 2023, https://www.youtube.com/watch?v=P-pDgow5Iu4Y. Image *"Ḥafl Tanṣīb al-Ḥimār Badal Ṣanam al-Maqbūr Hafezfī-Ḥamā"* [The Inauguration Ceremony of a Donkey Instead of the Idol of Hafez], YouTube, uploaded October 29, 2011, accessed July 18, 2023, https://www.youtube.com/watch?v=gPEJ9rJst8A.

239 *"Dimashq – al-Mazza Rafʿ ʿAlam al-Istiqlāl ʿalā Utustrād al-Mazza"* [Damascus – al-Mazza Hoisting the Independence Flag at Mazza Highway], President Bridge, Mazza Highway, Damas-

What is highly noticeable visually and in terms of performance are the similar aspects between both sides. There were, for example, no differences in the performances or visuals in the two photos showing the longest flag in Syria and the pro-regime march, despite the fact that one was in protest. These photos also highlight competition between pro-regime and pro-revolution factions to display the longest Syrian flag in Syria. Another similar practice was seen in the act of releasing balloons, which was performed by both sides.[240] In YouTube videos, the balloons hold a banner that says "the conspiracy has failed," a reassuring sentence for pro-regime supporters, while a pro-revolution video shows the release of "freedom balloons," each one attached to a small banner with the name of an arrested activist.[241] Both sides thus imitated each other with similar visuals but different content.

Book Publishing

Academic and nonacademic books were rapidly published after a few months of protests. Due to the publishing situation in Syria, most books were published outside the country and funded by universities, cultural organizations, and study centers; they were written in Arabic, but also increasingly in English. The Arab Center for Research and Policy Studies was one of the first study centers to offer researchers the opportunity to focus on Syria. Its first study of Syria was a five-part series written by Jamāl al-Barūt and published from October 23, 2011.[242]

The partial absence of publishing houses and published materials paved the way for academics, journalists, and experts to write about Syrian events from outside Syria itself. While the local revolutionary side tried to produce as much output as possible, it achieved little due to a lack of experience, financial resources, and the foreign language skills needed to convey events. This resulted in the online circulation and distribution of books that were banned in Syria before March 2011 skyrocketing during the first few months of the protests—particularly prison lit-

---

cus, January 30, 2012, uploaded January 30, 2012, accessed July 18, 2023, https://www.youtube.com/watch?v=mUCSXdcelI4.

240 *"al-Masīra al-Milyūniyya al-Sūriyya al-Kubrā"* [Great Syrian Million People March], uploaded June 21, 2011, accessed July 18, 2023, https://www.youtube.com/watch?v=vfJkGnNBZNk&t=51s; *Sham News Network*, *"Shām al-Lādhiqiyya Bālūnāt al-Ḥurriyya bi-Jumʿat Aṭfāl al-Ḥuriyya"* [Sham News Network, Lattakia, Freedom Balloons on the Friday of the Freedom Children], uploaded June 11, 2011, accessed July 18, 2023, https://www.youtube.com/watch?v=9ZIstBaRN8c.

241 *"al-Masīra al-Milyūniyya al-Sūriyya al-Kubrā"*; *"Shām al-Lādhiqiyya Bālūnāt al-Ḥurriyya bi-Jumʿat Aṭfāl al-Ḥuriyya."*

242 Jamāl al-Bārūt, *Jadaliyyat al-jumūd wa-l-iṣlāḥ* [The Dialectic of Stagnation and Reform] (Doha: Arab Center for Research and Policy Studies, 2012).

erature, initially available only in Arabic.[243] Before the protests, accessing banned books in hardcopy or electronic form was impossible, due to the risks involved with possessing them. People were largely uninterested in the censored books available in the public space under the Assad regime due to their lack of hope of political change. Post-March 2011, this attitude changed, and there was a wave of people interested in raising awareness. The use of VPN proxy programs to enable people to use the internet safely to browse or download files without being monitored thus spread rapidly.[244] In fact, social media websites had been blocked before 2011, but in February of that year, Syrian authorities suddenly allowed the public to access social media without the need to use a VPN, since this made it easier for the regime to track activists online.[245]

The small number of books published in 2012 and their reception by the public highlight the shortage of uncensored publications out of Syria and in Syria with the new dynamics that the cultural field witnessed by the sudden shifts of the Syrian space. For example, the first print novel about the revolution was published in 2012 by ʿAbdallāh Maksūr,[246] and Samar Yazbek published her work one month later.[247] Both of these novels were diaries of the revolution. The first publishing house that was considered to be working against the regime was Bayt al-Muwaṭin, which published a collection called Shahādāt Sūriyya [Syrian Testimonies] with contributions

243 A co-founder of Abu Abdo al-Bagl (name withheld by request), interview with the author in France, February 20, 2020. According to this individual, all of the prison literature was available online and Abu Abdo al-Bagl, one of the most prolific book scanning services, was very active at the beginning of the protests in 2011. By 2011, they had scanned and uploaded more than 5,000 books, which can be found on their Facebook group or their website. This Facebook group allowed book requests; if a book was unavailable, someone would buy, scan, and distribute it to people online. The motivation for offering this service was to protect the books from being confiscated and deleted, to make the books available to people in need, and to raise awareness. The initiative works collectively and depends on a team for scanning, selecting, and publishing the scanned books. The online catalogue of books is available at http://abuabdoalbagl.blogspot.com/, accessed July 18, 2023. The private Facebook group for requesting books is available at https://www.facebook.com/groups/abuab/about/, accessed July 18, 2023.

244 A virtual private network (VPN) allows an individual to access a proxy server, making it appear as if their computer is accessing the internet from another city or country. It can be used to bypass certain internet restrictions imposed by the state or various companies.

245 Jennifer Preston, "Syria Restores Access to Facebook and YouTube," *The New York Times*, February 9, 2011, accessed July 18, 2023, https://www.nytimes.com/2011/02/10/world/middleeast/10syria.html.

246 *Orient*, "*Awwal Riwāya li-Riwāʾī Sūrī ʿan al-Thawra al-Sūriyya*" [The First Novel by a Syrian Novelist on the Syrian Revolution], November 3, 2012, accessed July 18, 2023, http://bit.ly/39KpoSM.

247 Samar Yazbik, *Taqāṭuʿ al-Nīrān: Min Yawmiyyāt al-Intifāḍa al-Sūriyya* [Crossfire: The diary of the Syrian intifada], Arabic (Beirut: *Dār al-Ādāb*, 2012).

from twenty-four Syrian authors, including novelists, poets, and activists.[248] In addition to this series, it published several other books and articles about citizenship.

The liberation of the key visual, audio, and musical spaces and distribution of books in Syria thus paved the way for the emergence of a new revolutionary language. The following chapter dives in detail into the resulting slogans, graffiti, banners, revolutionary performances, and new everyday terms that were used in the country. It also tries to offer a thick description as a synonym for the act of thick translation for the examples discussed by showing the movement, use, thick layers, and histories of this language as illustrated above in relation to a number of songs.

248 Samar Yazbek, *Taqāṭu' al-Nīrān: Min Yawmiyyāt al-Intifāḍa al-Sūriyya* [Crossfire: The Diary of the Syrian Intifada] (Beirut: *Dār al-Ādāb*, 2012).

Chapter 3
Different Revolutions in Language

ماذا يبقى من إنجيل الثورة ،
حين تقرر قتل مغنيها ؟
ماذا يبقى من كلمات الثورة ،
حين ستمضغ أكباد بنيها ؟
ماذا يبقى ؟
حين تخاف الدولة من رائحة الورد ،
فتحرق كل مراعيها ..
ماذا يبقى من فلسفة الثورة ،
حين تخاف طلوع الشمس ،
وتنتف ريش كناريها ؟.
حين تبول الثورة فوق كلام نبييها . . .
نزار قبّاني
كتابات على جدران المنفى

[What would be left intact of the Bible of the Revolution,
When its chanter is murdered?
What would be left of the Revolution slogans,
When demonstrators' livers are devoured?
What would be left? What?
What would be left?
When a state fears the flower fragrance such that it burns its gardens,
What would be left of the Revolution ideology?
When it fears sunrise,
Plucks its canary feathers and desecrates its prophets' catechisms?

Nizār Qabbānī, "*Kitābāt ʿalā Ḥāʾiṭ al-Manfā*" [Writings on the Exile Walls]

A New Symbolic Product: New Everyday Terms

The Syrian public space produced different terms that had not previously been used in Syria. The change of products in the symbolic market led to a change in the nature of competition among individuals, who either created a different type of language that had never been used before, or subverted words and phrases or overloaded them with new layers of meaning. Beyond the changes in the symbolic market, this change in language was the result of the oppressive language that the regime imposed on Syrians being insufficient to express the exceptional, revolutionary events that took place from March 2011. New terms, vocabularies, and expressions thus entered everyday Syrian spoken language, and ultimately affected the revolutionary language. This relationship between what had existed and what

https://doi.org/10.1515/9783110767698-004

existed during the revolution took the shape of a subversive mechanism on both sides which complicated and accumulated more layers of meaning and was characterized by thick contexts and thick histories at different periods of time. Following the logic of the market and competition over symbolic products, the existence of such a language—with its rich new vocabularies and terms—not only resulted in the composition of new linguistic symbolic products on the pro-revolution side, but also affected the language used by supporters of the regime. In this chapter, I will show more specifically how a new type of language started to emerge due to the mutual influence of the regime and the revolution, as seen in visual and musical production. This language is limited to terms, slogans, *bakhkh*/graffiti, and pro-regime and revolutionary performances.

Tansīqiyya

Tansīqiyya [coordination] was one of the first new terms to be used in Syria after the beginning of the protests. I was unable to find any trace of this word in government newspapers, Facebook, YouTube, Syrian dramas or magazines before the protests began in Syria, and none of the people involved in the protests knew where the word originated or why it was used for political gatherings for anti-regime Syrians. While the political meaning of *tansīqiyya* could not have existed during either of the Assad eras, I also looked for similar terms that might have had a meaning related to "political gathering" to trace the social and cultural development of this term in the Syrian context. I was unable to find anything with a similar meaning of coordinating or gathering for a single goal, as *tansīqiyya* implies beyond the context of the protests. I therefore attempted to trace the origin of this word, first asking the founders of *tansīqiyya* in Syria why it was chosen. Here I explain the origin of *tansīqiyya* found in different contexts, activists' own opinion of the term, its revolutionary functionality, lexical meaning, and finally its use by supporters of the regime.

The term *tansīqiyya* was first used during the "Black Spring" protests in Algeria in 2001, and prior to this a similar term was used in the Amazigh context: "The National Council of Coordination," founded in 1994.[249] In this context, *tansīqiyya* was used for political activists who fought for their Amazigh cultural and national rights. When I asked Ahcene Graichi, an Amazigh activist, why the political move-

---

249 Didi Lsawt, "*Qirā'a fī al-Ḫiṭāb al-Amāzīghī al-Maghribī*" [A Reading in the Moroccan Amazigh Discourse], *Al Jazeera*, November 27, 2005, accessed July 18, 2023, http://www.aljazeera.net/special-files/pages/182a3618-ec5f-47c2-b2f1-0563f4b2230b.

ment was called a "coordination," Graichi noted that the word likely "...came from the French 'coordination,' the Arabic version of which was *tansīqiyya.*"[250]

Tracing the use of the term in French shows that such types of unorganized and popular activism are referred to as forms of "coordination," meaning "a map of singularities, composed of a multiplicity of committees, initiatives, places for discussion and development, political and union activists, a multiplicity of trades and professions, networks of friendships, and 'cultural and artistic' affinities that are made and unmade with different speeds and purposes."[251] This definition confirms that such a map goes beyond creating a politically united group; its purpose is to be an umbrella, uniting all of the scattered political activism on the ground. As I will illustrate below, this was the case with the Syrian coordinations: people needed to coordinate because they had agreed on a single ultimate objective but had no leadership or prior experience of organizing themselves, since Hafez al-Assad had dissolved all Syrian syndicates on April 8, 1980.[252]

Apart from the use of *tansīqiyya* in the Maghreb context, a Syrian activist who wished to remain anonymous told me that: "I first heard the term from Egyptians as *mansiqiyya.*"[253] Due to his role in structuring the revolutionary Syrian committees, he suggested a similar term. Ultimately, *tansīqiyya* was selected, which translates into English as "coordination," the same as the Amazigh term and the contextual meaning of the term in French. While it was very difficult for activists to have political gatherings or a political headquarters on the ground, other technologies were available, and *tansīqiyya* thus first came into existence on social media before passing through several phases of development. In the early days of the protests, there was only one well-known, pro-revolution Facebook page: the "Syrian Revolution Against Bashar al-Assad." In 2015, the name of this group was changed to the Syrian Revolution Network.[254] According to a 2012 article by Muḥammad Raḥḥāl, the first coordination "was in Idlib, then activists created another one in Bāb Sabāʿ

---

250 Ahcene Graichi (Amazigh activist from Ayt Gigush Coordination), interview with the author, July 14, 2017.

251 Maurizio Lazzarato, "La forme politique de la coordination," *Multitudes* 3, no. 17 (2004): 105–14.

252 Al-Lajna al-Sūriyya li-Ḥuqūq al-Insān [The Syrian Human Rights Committee], "*al-Ḥuryyāt al-Naqābyya fī Sūriyya*" [The Syndicates' Freedom in Syria], January 26, 2004, accessed July 18, 2023, https://www.shrc.org/?p=7200.

253 Interview with a founder of the Syrian Revolution Coordinators Union (activist), May 20, 2017, who wished to to remain anonymous.

254 Syrian Revolution Network, official Facebook page, accessed July 18, 2023, https://www.facebook.com/Syrian.Revolution/.

in Homs."[255] Another activist believed that the first coordination was in Daraa, though was unsure about this.[256]

Trying to understand the meaning of coordination from the Syrian context reveals the use of *tansīqiyya* in the names of Facebook pages, usually in conjunction with the names of cities in which secret discussions were held about coordinating and cooperating to organize protests between activists in the same geographic location (such as a city, town, suburb, or quarter). All of the coordination Facebook pages are similar to the Syrian Revolution against Bashar al-Assad page, which sets out the role and objectives of the coordination: "1- working on the revolution media campaigns; 2- coordinating with revolutionary committees on the ground; 3- conveying what is happening in Syria to the world; [and] 4- offering media products to be reused in different websites and media."[257] The stated objectives of other coordination pages are almost identical, suggesting both a diffusion of knowledge and/or experience—people were clearly copying from a successful model—and the need for a media presence to fulfil these functions and support protests on the ground that could be filled by these committees. It is further apparent that the coordinations developed from the original Syrian Revolution against Bashar al-Assad Facebook page, having similar objectives but engaged in specific regions of Syria.

The word does not therefore appear to have originated in Syria, but likely developed from the influence of Egyptian or Algerian/Amazigh activism—as well as the French language used in both Algeria and Syria—given that both societies have a longer history of political activism than Syrians, who lived for four decades with almost no civil or political activism against the regime. The meaning of *tansīqiyya* in Arabic is derived from *nassaqa*, which means to put things in order. *Tansīq*, according to the *al-Wasīt* dictionary, means "the regularity of things to be organized and stabilized."[258] The word *tansīqiyya* can be defined as a neologism coined in various contemporary political contexts and obtained by nominalization of the same adjective—*tansīqiyya* constructed with the *nisba* suffix *-iyya*. The word thus means to organize the various revolutionary powers on the ground. The role of the *tansīqiyya* was to secretly organize revolutionary acts via private groups on Facebook. The term, therefore, developed as a result of revolutionary activists' progress

---

255 Muḥammad Raḥḥāl, "*al-Thawra al-Sūriyya: al-Māl al-Siyāsī wa-Atharuhu fī-Ifshāl al-Thawra*" [The Syrian Revolution: Politicized Money and its Impact on the Failure of the Revolution], *Alankabout online*, July 12, 2012, accessed June 1, 2020, http://www.alankabout.com/issues_and_opinions/5985.html (no longer available).

256 Syrian Revolution Coordinators Union activist, interview.

257 Syrian Revolution Network, official Facebook page.

258 *al-Wasīt Dictionary*, s.v. "Tansīq," accessed July 18, 2023, https://bit.ly/35vtVYf.

in gathering large numbers of Syrians on the ground, reflecting the recurrence of the term in the French, Algerian, and Egyptian contexts.

The success of this strategy also led to the launch of fake coordinations by the regime, via the Syrian Electronic Army. These were used to obtain activists' identities, and endangered the function of the coordinations.[259] According to a report by *Russia Today*, the Syrian Electronic Army was founded in 2011.[260] Because of the many fake coordinations—some of which were originally real pages subsequently infiltrated by the regime, while others were specifically created as fake pages—the revolutionary side took the step of creating a new group: the Local Coordination Committees, which later joined together to form the Syrian Revolution Coordinators Union to unify the coordinations and delete redundant pages where multiple ones existed for a single geographic location. This union of the revolutionary groups was designed to "improve [the coordinations'] strategies in reaching the international media."[261] From June 2011, the term Syrian Revolution Coordinators Union thus incorporated 26 coordinations and in the month of Ramadan [August] 2011 added 150 coordinations.[262] Regardless of the origin of these revolutionary committees, the Syrian public space witnessed a huge shift at the level of using new terms. Terms like coordination(s), Coordinators Union, or local coordination committees were all new to Syrians. The rapid change in using the terms is explained by the novelty of the political experience for Syrians: the ability to name what they had founded through their own initiative.

New terms in the Syrian public space were defined by both parties vying to produce terms, in their competition to produce symbolic products in the market. Although the term coordination was a new social and revolutionary phenomenon in Syria established by the revolutionary side, regime supporters also used the same term for the goals of the regime. Due to the weakness of regime-supported news channels and the fluff pieces that they broadcast, regime supporters created new Facebook pages with similar names to revolutionary ones, imitating the revolutionary activists' successful strategy in order to bypass the regime's official media outlet, though for slightly different purposes. Broadcasting news through

---

259 Laura Smith-Spark, "What is the Syrian Electronic Army?" *CNN*, August 28, 2013, accessed July 18, 2023, https://edition.cnn.com/2013/08/28/tech/syrian-electronic-army/index.html.

260 *Russia Today*, "The Syrian Electronic Army: Different Types of Tasks," YouTube, video, uploaded April 9, 2013, accessed June 1, 2020, https://www.youtube.com/watch?v=LZcJGTuxSbM (no longer available).

261 Wael Sawah and Salam Kawakibi, "Activism in Syria: Between Nonviolence and Armed Resistance," in *Taking to the Streets: The Transformation of Arab Activism*, eds. Lina Khatib and Ellen Lust (Baltimore: Johns Hopkins University Press, 2014), 136–71 (142).

262 "*Lamḥa ʿan Taʾsīs Itiḥād Tansīqiyāt al-Thawra al-Sūriyya* [A Brief Summary on Constituting the Syrian Revolution Coordinators Union], accessed July 18, 2023, http://www.syrcu.org/?ID=11566.

the regime media required permission from the secret service and had to pass through a number of steps, delaying its release. Forming a coordination was thus a more effective way for regime supporters to quickly obtain news from the pro-regime side. Regime coordinations were generally named with a combination of the phrases "Coordination of [a geographic location]" and "Assad," such as the "Pro-Assad Coordination of Quṭayfa," the "Assad Coordination of Qāra," the "Assad Coordination of Ṭarṭūs," and so on.[263]

Shabbīḥa

While *tansīqiyya* was a term coined solely for the revolutionary context and then used by regime supporters, *shabbīḥa* was not coined for the revolution but passed through different phases of meaning until 2011, when it acquired different layers of meaning and dimensions of use from those of the prior four decades. The meanings of *shabbīḥa* developed according to the different contexts of the Hafez al-Assad and Bashar al-Assad eras, and then the new dimensions of the protests.[264] The protests created new meanings for some terms, which were, therefore, used differently before and after March 2011. Such terms gained more layers of meaning due to the sequence of events and the political, social, and cultural context of Syria post-2011. Terms here are seen as interactive entities. While words and terms are intangible, they change in accordance with the environment and can gain additional attributes, as happened with *shabbīḥa*. The new expressive space that opened up in the Syrian public sphere allowed these terms to acquire new meanings consistent with the revolutionary context.

After the protests broke out, *shabbīḥa* took on a new existence, and, therefore, new meanings were added to the word. After 2011, *shabbīḥa* took on the new meaning of a group of people used to suppress the protestors. Initially this was the secret police, disguised as either civilian street vendors or outlawed groups who kept their weapons on show in order to intimidate passersby. The first news report to use the term *shabbīḥa* in this way was published in *al-ʿArabiyya* on March 27, 2011, which spoke about the existence of criminal gangs, called *shabbīḥa*, suppress-

---

263 *Tansīqiyyat al-Quṭayfa al-Muʾayda* [The Pro-Assad Coordination of Quṭayfa], Facebook page, accessed July 18, 2023, http://bit.ly/39OncK0; *Tansīqiyyat Qāra al-Assad* [The Assad Qara Coordination], Facebook page, accessed July 18, 2023, https://www.facebook.com/www.qara.net/; the Assad Coordination of Ṭarṭūs page has since been deleted.

264 Please see Chapter 1 on how the term *shabbīḥa* was used during the Hafez al-Assad era.

ing demonstrations.[265] This marked a turning point in the use of the term *shabbīḥa*, because it had a distinctively different meaning than before and was later publicly used on TV.

As the protests spread, *shabbīḥa* groups of disguised secret police were unable to cover all of the protest areas. Government employees were then "recruited" to suppress the protests using a carrot-and-stick approach: the carrot being overtime bonus pay, and the stick the threat of being perceived as insufficiently loyal to the regime. This shift in the meaning of *shabbīḥa* to refer to civilians who worked for the regime changed the term, so that it was used to describe anyone involved in suppressing demonstrations. For example, in the first few months of the protests, "personnel were recruited from the 'Military Installations' in Kafr Sūsa, Damascus to suppress demonstrations, with this work considered to be overtime which guaranteed for them security with certain privileges and earning extra money."[266] Proving one's loyalty to the regime at that time was very important in order for people to keep their jobs and their loved ones safe. Not participating in suppressing demonstrations on a Friday might be seen as a suspicious act that could lead to interrogation by the secret police.

Later, with the huge number of demonstrations in Syria, the term *shabbīḥa* was further extended to encompass anyone defending the regime. A *shabbīḥ* could therefore be a pro-regime media activist, politician, or regime supporter, as well as someone actively involved in suppressing demonstrations. In relation to this, the revolutionary side used the word pejoratively to describe people with no morals, and successfully turned it into an insult. Regime supporters therefore tried to subvert the term, giving it additional meanings and dimensions: *shabbīḥa* thus became people loyal to the regime and its president. This term was changed from a specific group (as explained above) and used to describe all Syrians who were clearly with Assad's supporters, due to the mixture of meanings where Assad represents Syria instead of vice versa. The attempted shift in connotation was supported by discursive products such as the one in Figure 10, claiming "I am a *shabbīḥ*" and

---

265 *Al Arabiya*, *"al-Shabbīḥa 'Iṣābāt Ijrāmiyya Tanshur al-Fawḍā wa-l-Ru'b fī al-Lādhiqiyya bi-Sūriyā"* [*Shabbīḥa* Criminal Gangs Spread Chaos in Lattakia, Syria], March 27, 2011, accessed July 18, 2023, https://www.alarabiya.net/articles/2011/03/27/143208.html.

266 Nadya Turkī, Yūsuf Dīyāb, and Muhammad 'Alī Sāliḥ, "Maṣādir Sūriyya Tattafiq 'alā Wujūd 'Alāqa Wathīqa Bayna al-Shabbīḥa wa-l-Sultāt" [Many Syrian Sources Agree on the Relationship between the Authority and Shabbīḥah], *Souria Houria*, January 28, 2012, accessed July 18, 2023, http://bit.ly/2wxS1nU.

the famous slogan "We are *shabbīḥa* forever for the sake of Assad," which was performed in the presence of Assad during his speech in January 2012.[267]

Figure 10: A poster uploaded to social media, which says: "If loving Syria and its president is Tashbīḥ, [exaggerated form of the verb Shabbaḥa] then I am 'Shabbīḥ'".

---

267 Image from personal research archive; for slogan use, see, for example, Bashar al-Assad, "Ra'īs Sūriyā Dr. Bashar al-Assad fī-Sāḥat al-Umawiyyīn bayna Abnā' Sha'bbihi" [The Syrian President Dr. Bashar al-Assad in Umayyad Square], 00:02:50, *HananNoura* (YouTube channel), uploaded January 11, 2012, accessed July 18, 2023, https://www.youtube.com/watch?v=oLUpQ46bkrQ&t=41s.

Minḥibbak

In a continuation of the famous presidential slogan and song launched by Syriatel (one of the two mobile phone companies owned by cousins of Bashar al-Assad), the phrase *minḥibbak* [we love you] went through new developments and was subverted by the revolutionary side. This phrase was one of the key slogans and terms used by regime supporters on banners and in slogans to express their loyalty to Bashar al-Assad.[268] The revolutionary side called these people *minḥibbakjī*, meaning people who blindly believe in the "We love the president" banner. The addition of *jī* at the end of a word is derived from Turkish, and when it is added to the end of a noun, "it indicates a person associated with a profession: güreşçi 'wrestler', lokantacı."[269] In colloquial Syrian, for example, this suffix is used to designate tradespeople, as in *khuḍarjī* for greengrocer, which is derived from *khuḍār* [vegetable], and *jawāhirjī* for jeweler, which is derived from *jawhara* [jewel]. Unlike in Turkish, the "*jī*" suffix can also function as a depreciative suffix, used in the political lexicon in particular to demean an opponent. For example, *waṭanjī* means a person with an exaggerated or fanatical belief in *waṭan* [homeland], while *qawmajī* is used in a derogatory way to mean someone with a strong belief in nationalism, especially Arabism, referring to *qawmayya* [nationalism].

The word *minḥibbakjī* thus suggests a person who has an exaggerated or outsized love for Bashar al-Assad and holds banners reading *minḥibbak* in pro-regime demonstrations. A YouTube video posted on the Kibrīt pro-revolution channel explains *minḥibbakjī*, saying: "My hypocrisy is lies for your lies. My stupidity is because of your ignorance. My money is [earned by] stealing from you. I am ready to be bribed. My conscience seeks to betray you. My beloved people are *shabbīḥa* for you."[270] This short monologue, which lacks logic and is full of contradictions from the point of view of the protestors, expresses what a *minḥibbakjī* thinks about the people who are supporting the revolution. It presents the immoral acts that the pro-revolution side saw the *minḥibbakjī* as doing to gain everything in their lives—such as being stupid and immoral (stealing and bribery)—as being morally correct. The pro-regime side felt that as long as it meant love for the leader of the country, *minḥibbakjī* was not a negative word. Like the word *shabbīḥa*, *minḥibbakjī* was thus subverted in the Syrian public space, with posters springing up using the

---

268 Please see Chapter 1 for more details about the term in the context of the Bashar al-Assad era.

269 Asli Göksel and Celia Kerslake, *Turkish: A Comprehensive Grammar* (London: Routledge, 2005), 58–9. In Turkish, the letter "c" sounds like "j," and this suffix may be spelled -cı, -ci, -cu, or -cü, according to the rules of vowel harmony.

270 Kibrīt, "*Anā Minnḥibbakji*" [I Am A *Minnḥibbakji*], video blog post, uploaded June 17, 2011, accessed July 18, 2023, https://www.youtube.com/watch?v=0WXsnL3eD4c.

term positively, for example with the famous slogan: "Yes, I am *Minḥibbakjī*. Is there a problem with that?"

When *minḥibbakjī* became a positive term among regime supporters, the revolutionary side produced a more offensive term derived from *minḥibbak—minḥibbak jaḥshī*—which contains both *minḥibbak* [we love you] and *jaḥish* [used in Modern Standard Arabic (MSA) literature to mean a burro or small donkey]. In the Syrian context, being called a *ḥimār* [donkey] is an insult and indicates stupidity, ignorance, apathy, and indifference. The word *jaḥish* [donkey foal], however, is even more insulting, because it suggests a newborn animal without any experience and thus even more stupid than an adult donkey. The newly coined term thus equated a *minḥibbakjī* with a *jaḥish*, essentially stating that a person with an exaggerated love of Assad was a young donkey. The last iteration of this term was not received in the same way as *shabbīḥa* or *minḥibbakjī*. It was considered an insult, and the regime side stopped reacting to it.

***Mundassūn*, *jarāthīm*, and other Terms**

Just as regime supporters attempted to subvert the negative terms coined by the revolutionary side, supporters of the revolution followed the same strategy. It is important to highlight here that most of the regime's words and terms describing demonstrators used MSA. This indicates that the regime had not learned that MSA was not as effective as colloquial Syrian for criticizing protestors. As discussed in Chapter 1, colloquial Syrian was used more than MSA after Bashar al-Assad came to power, and the use of MSA to reproach people in Syria is typically seen as funny or even silly. This gave the supporters of the revolution a way to invert or subvert the regime's terms.

Supporters of the regime created numerous terms in the public space to insult protestors, the most famous being *mundassūn* [infiltrators] and *jarāthīm* [germs]. These two terms were used by Bashar al-Assad in his first three speeches following the outbreak of the protests (the third one was on June 20, 2011) to describe those who were protesting.[271] While *mundassūn* indicates an individual who secretly enters a group, organization, or territory for the purpose of sabotaging it, demonstrators subverted the word to mean a brave person, in order to ridicule the regime's conspiracy discourse. As discussed in the previous chapter, the most famous revolutionary forum and blog was, for example, called *al-Mundassa al-Sūri-*

271 *Al-Jadeed News*, "*Khiṭāb Bashar al-Assad 20-06-2011*" [Bashar al-Assad's Speech, June 20, 2011], uploaded June 20, 2011, accessed July 18, 2023, https://www.youtube.com/watch?v=f3dNMienjX8.

yya [the Syrian [female] Infiltrator]. Adding "Syrian" here to the name of the blog served to invert the term, because it is not possible to be both a Syrian loyal to one's homeland and to infiltrate the country at the same time, since an infiltrator must come from outside the country or be disloyal to it. The use of "Syrian" in this title thus presented infiltrators as being Syrians but against the regime. The same strategy for subverting regime terms was used when Bashar al-Assad described demonstrators as *jarāthīm*, which was taken up by protestors in ironic slogans and banners.[272] After Assad used this word in his speech, a banner was created in Barza, reading "this speech was sponsored by Dettol," a well-known disinfectant liquid in Syria used to kill bacteria and germs.[273] This term was then subverted to become the slogan "hey Syria. . . Assad is a germ in Syria."[274]

Conversely, the regime sought to replace new revolutionary terms in order to change their positive dimensions into negative ones, especially when the terms were abstract and had generally positive meanings, such as *thawra* [revolution] and *al-Jaysh al-Ḥurr* [the Free Army]. In addition to the generally positive meaning of *thawra*, which implies change and hope, prior use of it by the regime had been positive, for example in describing the *coup d'état* of March 8, 1963, regarding which Patrick Seale, for example, says "their revolution had succeeded."[275] For this reason, it was not possible to contest the term "revolution" in the Syrian public space. Instead, *thawra* was replaced by *fawra* [surge or spree]. In Arabic, the word *fawra* is based on the pattern of *fāʿil*, known as *Ism al-Marra*, characterized by a single occurrence of an action; the action happens only once and then is finished.[276] Pro-regime supporters thus used *fawra* to describe the revolution, in order to insult it and scorn it, using a rhyming word.

Fawra was not the only negative term used to describe the revolution; regime supporters also sought to replace *al-Jaysh al-Ḥurr* [the Free Army]. This term capitalized on the highly positive dimensions of pro-revolution supporters, because the army was primarily founded by soldiers who had refused to kill Syrians and defected. The Free Army was seen as credible in terms of its patriotism by pro-rev-

---

272 Jamāl Al-Qaṣṣāṣ, "*Zuʿamāʾ Yashtimūn Shuʿūbahum: min ʿal-Jirdhānʾ ilā ʿal-Jarāthīm*" [Leaders Insult Their Peoples: From Germs to Rats], *Asharq al-Awsat*, July 1, 2011, accessed July 18, 2023, http://archive.aawsat.com/details.asp?section=45&article=629141&issueno=11903#.WWyzZ4SGNaQ.

273 *Ugarit News, Syria*, "Ugharit, Dimashq, *Lāfitāt Barza*" [*Ugharit*, Damascus, the Banners of Barzeh], *Ugarit News – Syria*, uploaded June 24, 2011, accessed July 18, 2023, https://www.youtube.com/watch?v=JR6dWyblNv8.

274 "*al-Lādhiqiyya al-Ṭābiyāt Hayy Sūriyā wa-l-Asad Jirthūma fīhā*" [Lattakia al-Tabyat, This is Syria and Assad is a Germ in it], *Tansīqiyyat al-Lādiqiyyah* [Latakia Coordination], uploaded October 22, 2011, accessed June 1, 2020, https://www.youtube.com/watch?v=3BynAboWTI8 (no longer available).

275 Seale, *Asad*, 78.

276 Roughly equivalent English phrases include "a flash in the pan" or "a tempest in a teapot."

olution Syrians and was also considered by many to be an icon of the revolution. The existence of such a term threatened the regime's term *al-Jaysh al-ʿArabī al-Surrī* [the Syrian Arab Army]. This is why regime supporters created a ridiculing rhyming alternative, *al-Jaysh al-Kurr* [the Donkey Army]. In one example, a pro-regime news agency used this term, saying "some ridiculed General Riyāḍ al-Asʿad, the leader of the Free Army (the Donkey Army according to [regime] supporters)."[277]

In addition to the clear pro- or anti-regime/revolution terms that appeared after 2011, other terms were coined by people who did not want to be involved with the everyday politics of what happened in Syria but cared only for their safety, regardless of what happened in the country. This group of Syrians were called *al-ramādiyyūn* [the Gray People]. They frustrated both the pro- and anti-revolution factions because they were unreliable, and their attitudes and beliefs varied depending on what was happening to them and where they were. In order to adapt to the situation in Syria and ensure they did not show support to either side, the Gray People used the pronoun "it" to talk about what was happening. In this way, they were able to avoid calling events in Syria a "revolution" or "conspiracy," and thus support neither the regime nor the revolution. A particularly well-known phrase was coined by this group as a response to any daily story: "*Allāh yaṭaffīhā bi-nūru*" [May God put IT [the revolution/crisis/conspiracy] off with his light].

A final group of terms that became widespread after 2011 was the coded language used in the closely monitored and censored electronic lives of Syrians. Telephone calls, SMS, and messages on social media or mobile phone apps were all monitored by the regime. Regardless of their stance regarding the regime, Syrians thus used coded language to elude the secret police. For example, instead of using the word "demonstration" on the phone for something that had happened or was yet to occur, the caller would say there had been (or was going to be) a big wedding. Wedding guests were divided into two categories: the family of the bride were protestors, while the family of the groom were *shabbīḥa*. The number of *shabbīḥa* was indicated by the number of buses. Based on a pre-determined code, the question that had to be asked in order to obtain this information was how many bags of bread they brought or the number of loaves in a bag. Places controlled by the Free Syrian Army were referred to by the stars on the flags of each side: since the stars on the revolutionary flag are red, but those on the official flag are green, people used to say that a particular city was controlled either by the "red eyes," referring to the Free Syrian Army, or the "green eyes," referring to the regime.

277 ʿĪsā ʿAwwām, "*Al-Insiḥāb al-Taktīkī Yushʿil Sukhriyyat al-Muʾayyidīn wa-l-Muʿāriḍīn ʿalā al-Fāysbūk*" [Tactical Withdrawal Viewed with Sarcasm by Anti- and Pro-Regime Supporters on Facebook], *Syria Now News*, March 7, 2012, accessed July 20, 2017, http://syrianownews.com/index.php?p=53&id=2985 (no longer available).

The users of this coded language can be described as a "speech community," defined as "a group of people who share a set of norms and expectations regarding the use of language."[278] Speech communities consist of groups that have "at least two members but there is really no upper limit."[279] Shared norms and expectations are sufficient grounds for creating a community, as seen in the above examples where these three groups can be considered speech communities in the context of the Syrian protests: the first group shared the set of norms of the regime presented in the symbolic market and its products; the second group, because of its subjugation by the first group, created their own coded language and chose to defy the first; while the third group—the Gray People—chose to create its own linguistic patterns for security reasons, rather than due to pressure from the anti- or pro-regime speech communities. The existence of these three groups in Syria affected the symbolic products and language used by Syrians in the post-protest context of 2011. The impact of these groups on the symbolic market was not random; their shared norms, expectations, and experiences were represented by new symbolic products that functioned to change the value of, and provide an alternative to, the existing products of the regime. This not only affected musical and visual products and the language and terms used in everyday life, but also created a new space for a new revolutionary language (limited in this study, by necessity, to slogans, banners, *bakhkh*/graffiti, and performance). This changed Syria from the single oppressed speech community of Assad to three distinct but interacting communities.

Revolutionary Language

The interaction of the indicators discussed above—the four decades of compulsory discursive performance that the regime required of Syrians, the beginning of the protests, and the emergence of new speech communities—created a new milieu for Syrians to shape new discursive spaces in the Syrian public space, represented by new symbolic products. In March 2011, the new resistance to the regime paved the way for new discursive practices. These new practices were apparent in all of the elements studied here: the Syrian public space acquired new terms, and new visual and music spaces, and these practices were characterized by performance. They contained slogans, banners, *bakhkh*/graffiti, and revolutionary performative elements. All of these elements were the main components of the revolutionary

278 George Yule, *The Study of Language*, 6th ed. (Cambridge: Cambridge University Press, 2016), 256.
279 Ronald Wardhaugh, *An Introduction to Sociolinguistics*, 6th ed. (Chichester: Wiley-Blackwell, 2010), 118.

language, and also symbolic products, like the musical ones. What distinguished them is that they were performed in the revolutionary field such as streets and squares. It is very difficult to separate one element from the others, since all of these elements of the revolutionary language were performed at once. This meant that when there was a demonstration, slogans, banners, and *bakhkh*/graffiti were present at the same time. The location of the demonstration was the setting for these acts. It was a performance with a spontaneous text, performed in the street, with protestors as actors. Demonstrators sprayed the walls with *bakhkh*/graffiti as they passed through the area, chanted slogans, and held up banners. While none of these acts were performed separately, here I will try to present and analyze them in individual categories and treat them as symbolic products that simultaneously influenced and were influenced by the language of the regime due to the competition to produce more convincing symbolic products.

Slogans

The slogans presented here were not chosen randomly, since the purpose of this section is not only to examine the revolutionary slogans, but also to illustrate the mechanism by which they were created through interaction with the political, cultural, social, and economic circumstances of Syrians and, above all, the slogans of the regime. I will therefore compare revolutionary slogans to those of the regime I presented in Chapter 1 to show the importance of both sides' influence in producing slogans. This method of taking equivalent slogans from both sides helps to narrow down a selection of the thousands of slogans produced in Syria in 2011, and also helps foster greater understanding, as they can be compared from the perspective of the protestors and regime supporters. This approach also continues to underline the competition between symbolic products in the Syrian space, which shows through its selection here that the products produced and recreated on both sides were the most important in the symbolic struggle to gain more of the Syrian public space.

According to a study by Jamāl Shuḥayyid, "the number of slogans in the first year of the revolution exceeded 10,000."[280] Given this huge number, I decided to narrow down my choices and present slogans categorized into three groups, which emphasize them as symbolic products that were produced and re-imagined to compete in the symbolic market of the Syrian public space. These categories show the revolutionary language through the language of the regime and vice versa. The first group consists of revolutionary slogans that reused old regime slogans—those

280 Jamāl Shuḥayyid, "Slogans of the Supporters."

used under Hafez al-Assad, which had nearly been forgotten but were revived and reshaped after March 2011—and can be seen as a continuation of them. The second group consists of revolutionary slogans that were (re)created by regime supporters, using the same form but with different content in order to reimagine the revolutionary slogans for use by the regime. Finally, the third group contains new slogans that were produced by the regime after March 2011. The classification of slogans into these groups does not encompass all of the slogans that were used, but clearly shows the interactions between both sides in producing symbolic products and the competition involved. Moreover, it shows how difficult it is to produce a thick description of a revolutionary or pro-regime slogan without digging deep into the layers of meanings and histories on both sides. A pro-revolution slogan is unclear in isolation without reference to the equivalent one on the pro-regime side due to the mutual dynamics of the interplay of meaning, which fluctuates between both spheres: the opposing and the dominant ones.

The First Group of Slogans

As described in Chapter 1, the cult of Assad went through a series of phases in which Hafez al-Assad was presented as a president, a leader, the only leader, a prophet, and finally the god of Syria. Similar slogans were also used under Bashar al-Assad, with "Hafez" simply replaced by "Bashar" and the continued use of slogans that only mentioned "Assad." This group of slogans, written in MSA, was characterized by new revolutionary slogans to take the place of those supporting Hafez: they were originally produced by the regime, but then reworked by revolutionary activists to subvert the old regime slogans still in use or present in the public space under Bashar al-Assad. This act of rewording old slogans was not solely an act of replacement but also one of deconstructing and destroying the cult and legend of Assad. It can be seen as releasing new products in the Syrian public space. Looking at this in conjunction with the timeline of the pro-Hafez slogans shown in Table 6 below illustrates a new network of meaning for slogans that were first chanted in 2011, but whose root or original first performance had taken place decades before.

This first group contains three sub-groups that explain the development of slogans that were created based on the slogan's themes. The first subgroup consists of the "forever" and "immortality" slogans as shown in Table 6 below.

The column on the left shows the gradual development of the slogans for Hafez al-Assad that ultimately portrayed him as a god. These slogans were generally not reused under Bashar al-Assad, other than those from the final development of the godlike phase, with slogans such as "there is no God but Assad" and "Oh, God! It is time to give your place to Bashar." This is an indication that Bashar al-Assad's slogans' repertoire and development continued where his father's ended, rather

Table 6: The "forever" and "immortality" slogans of the Assad regimes.

Hafez al-Assad	Bashar al-Assad
إلى الأبد يا حافظ الأسد Assad Forever قائدنا إلى الأبد الأمين حافظ الأسد Our leader forever, the honest [trusted, faithful] Hafez al-Assad لم يتبق لنا إلا أنت No one left for us but you حلك يالله حلك يطلع حافظ محلك Oh, God! It is time to give your place to Hafez.	
	لم يتبق لنا إلا أنت No one left for us but you حلك يالله حلك يطلع بشار محلك Oh, God! It is time to give your place to Bashar.

than starting over, thereby indicating Bashar had inherited his father's godlike status, as shown in the right-hand column.

The group of slogans that subverted the "immortality" and "forever" slogans of the regime after 2011 were based primarily on insulting Assad's family, proving the falsity of the immortality slogans, and pointing to Assad as the source of Syrian corruption. Syrians knew that this oppressive regime was not only caused by Bashar but also by his father, who had created it, and demonstrators thus chanted:

يلعن روحك يا حافظ[281]

[Hafez, curse your soul]

This curse targets the holiness and sanctity built up through Hafez al-Assad's godlike status. While logically it should have targeted Bashar—the one currently oppressing demonstrators—rather than his father, this slogan targeted the halo surrounding Hafez al-Assad. It would have had more impact if it had been performed under Hafez al-Assad's rule, but the goal was to "curse the whole political era of the Assad regimes."[282] This slogan broke the most dangerous taboo in Syria—insulting the god of Syria—and thus represented the beginning of revenge for the oppression

---

281 *Sham News Network*, "*Ḥamā Ughniyat Yalʿan Rūḥak Yā Ḥāfiẓ*" [Hamah, the Song of Hafez, Curse Your Soul], uploaded July 22, 2011, accessed July 18, 2023, https://www.youtube.com/watch?v=D-K7qSP6J-xo.
282 Māhir Sharaf al-Dīn, "*al-Maḍmūn al-Siyāsī li-Shiʿār "Yilʿan Rūḥak yā Ḥāfiẓ*" [The Political Content of "Hafez, curse your soul"], *Mouatana Magazine*, August 10, 2013, accessed June 1, 2020, http://www.mouatana.org/archives/7834 (no longer available).

suffered under Hafez. Cursing the latter's soul not only affirmed that he was the main figure to blame for constituting the regime but also emphasized his mortality. This was further stressed by slogans such as:

يا حافظ طلاع [سماع] وشوف، صرنا نسبك عالمكشوف[283]

[Hafez, [hear] come back to life and see, we curse you explicitly]

This slogan was a public challenge to the deceased leader, calling on him to come back from his grave and see what was happening in Syria. The second part highlights how people could now publicly curse him, in contrast to previous decades when a single word spoken against the Assad family would put a Syrian's life at risk. The curse-invoking slogans were a direct message to regime supporters that the cult around Hafez al-Assad had reached its end. In addition to this, calling Hafez al-Assad solely by his first name, and the tone of performance, was a threat to Hafez al-Assad (deceased, but immortal in Baath symbolism) and a clear sign of disrespect, since autocrats like Assad typically have dozens of titles that must be used before the first name. The use of the *yā* style in Arabic also denotes vilification; it was a confirmation that he had passed away, and the only part left was his soul.[284] This is why invoking a curse was the only, and final, act available to the protestors. It is important to note that from a social and religious perspective the concept of "cursing" people is unacceptable in Syria, since a "curse" in the Islamic context means "the dismissal of God's mercy."[285] The other reason that "cursing" is counted as a deadly sin is that the only one in Islam to be cursed by God is the Devil, which implies that cursing Hafez al-Assad turns him from the god of Syria to the devil of Syria. Due to the sensitivity of the topic of cursing Hafez, numerous articles were written about whether this was permissible, most of which allowed (pro-revolution) Syrians to curse his soul.[286] "Cursing" someone in Arabic does not necessarily

---

283 "*Yā Hafez Ismāʿ wa-Shūf Ṣurnā Nisibbak ʿāl-Makshūf, al-Inshāʾāt,*" YouTube, uploaded June 30, 2011, accessed July 18, 2023, https://www.youtube.com/watch?v=MwF1fMreLUg.

284 *ʿUmar ʿAbdilhādī ʿAtīq, ʿIlm al-Balāgha bayn al-ʾAṣāla wa al-Muʿāṣara* [Rhetoric Between Authenticity and the Contemporary] (Amman: Dār Usāma li-l-Nashir, 2012), 198. The "yā" style in Arabic can be used for many purposes, including vilification or glorification: "the distinguishing criteria between vilification and glorification is the relationship between the addressed and the addressee. If the relationship is one of respect and honor, then the purpose is glorification. If the relationship is a hostile and hateful one, then the purpose is vilification."

285 Muhammad ʿAli al-Ṣābūnī, *Rawāʾiʿ al-Bayān: Tafsīr Ayāt al-Aḥkām* [The Beauty of Eloquence: Interpretation of the Commandment Verses], 3rd ed. (Damascus: Al-Ghazali Publishing House, 1980), 114.

286 Numerous Facebook posts and YouTube religious channels put out a Fatwa that permitted cursing Hafez al-Assad, see, for example: *Sheikh Abdilʿazīz al-Ṭuraifī, "Ḥukm Laʿn al-Ẓalimīn ka-Bashar al-Assad: al-Sheikh Abdilʿazīz al-Ṭuraifī"* [Cursing Bashar al-Assad: Sheikh Abdilʿazīz al-mean excluding them from God's mercy, but does mean wishing divine suffering on them now and/or in the afterlife. The success of this slogan in provoking Assad supporters led to it being extended and developed into different forms and versions, such as:

يلعن روحك يا حافظ على هالجحش يلي خلفتو[287]

[Hafez, curse your soul for this donkey [Bashar] you gave birth to]

This slogan was made specific to Bashar, as the son of Assad, by describing him as *jaḥish* [colloquial pronunciation of MSA *jaḥsh*], which implies that Hafez al-Assad is a donkey, since the primary meaning of *jaḥsh* is a donkey's foal. Other forms of the same slogan replaced Hafez with Anisa, the wife of Hafez, assigning her responsibility for giving birth to the *jaḥish* Bashar, or used other concluding phrases such as:

جنو جنو البعثية، لما طلبنا الحرية ، يلعن روحك يا حافظ يا ابن الحرامية[288]

[Poor Baathists, they became crazy when we asked for freedom. Hafez, curse your soul, you are a thief like your father]

By vilifying the "Baathists," this slogan attacked the cornerstone of the regime based on the Baath ideology, and had the same function of going back to the root of regime corruption and exploitation, by stating that it was the Baath Party that had robbed the country for decades. This development is highlighted in Table 7. The concluding part of this slogan was seen as highly provocative by regime supporters and was, therefore, performed in different forms and remixed into a number of songs to highlight the vilification of Assad as Syria's leader. These remixes were produced as a

---

Ṭuraifī], *Zidnī ʿIlman* [Increase Knowledge in Me] (YouTube channel), uploaded September 9, 2012, accessed July 18, 2023, https://www.youtube.com/watch?v=7RIyKHYyEqc; *al-Sheikh al-Ṣayāṣnih, "Raʾī al-Sheikh al-Ṣayāṣnih fī-Laʿn Rūḥ al-Maqbūr Hafez al-Assad"* [The Opinion of al-Sheikh al-Ṣayāṣnih on Cursing the Kicked Away Hafez al-Assad], *Syria* (YouTube channel), uploaded February 2, 2013, accessed July 18, 2023, https://www.youtube.com/watch?v=vDJBaarafbs; *al-Sheikh ʿUthmān al-Khamīs, "Ḥukm Laʿn al-Ẓalimīn ka-Bashar al-Assad: al-Sheikh ʿUthmān al-Khamīs"* [Cursing Bashar al-Assad: Sheikh ʿUthmān al-Khamīs], *Dr. Othman Alkamees* (YouTube channel), uploaded January 3, 2017, accessed July 18, 2023, https://www.youtube.com/watch?v=0wLTPNEYcSw; *Islamweb*, "Laʿn al-Ẓālimīn" [Cursing Oppressors], Fatwa Number 221458, September 25, 2013, accessed July 18, 2023, https://bit.ly/2P1qz7Q.

287 "Yilʿan Rūḥak Yā Hafez ʿalā hal-Jaḥish illī Khallaftū" [Hafez, curse your soul for this burro [Bashar] you gave birth to], *Abu Fedāʾ al-ḥamwwī* (YouTube channel), uploaded August 24, 2011, accessed June 1, 2020, https://www.youtube.com/watch?v=iBQoeUb8OvU.

288 *"Bāb Qiblī Jannū al-Baʿthiyya"* [Bāb Qiblī The Baathists Went Crazy], *CrAzY-MaN* (YouTube channel), uploaded January 18, 2012, accessed July 18, 2023, https://www.youtube.com/watch?v=I_is6qVx7os.

development of the slogans originally used in various mediums as songs or ringtones. They also drew on well-known characters to show the popularity of cursing Assad. For example, pro-revolution activists created a "Curse Your Soul, Hafez, the Smurfs" remix aimed at children showing the Smurfs dancing, taking the original work and employing its function for the political purpose of the slogan, and another "Curse Your Soul, Hafez, with Timon, Pumbaa and Simba."[289] The same slogan was also used in remixes of a Cheb Khaled song and a Michael Jackson song,[290] and a ringtone was created for mobile phones entitled the "Ringtone of Hafez, Curse your Soul; Bashar, Curse your Soul" to encourage people to use it publicly.[291] Such symbolic products were important because they enhanced the visibility of the main slogan cursing Assad and proved that it had a high value in the symbolic market. This was seen not only through the products from the revolutionary side but also by those from the regime.

Table 7: Curses as slogans in opposition to the "immortality" slogans of the regime.

Revolution	Development
يلعن روحك يا حافظ Hafez, curse your Soul يلعن روحك أنيسة Anisa, curse your Soul يلعن روحك يا حافظ على هالجش يلي خلفتو Hafez, curse your soul for this donkey [Bashar] you fathered	يا حافظ طلاع [سماع] وشوف، صرنا نسبك عالمكشوف Hafez, [hear] come back to life and see, we curse you explicitly[292] جنو جنو البعثية، لما طلبنا الحرية ، يلعن روحك يا حافظ يا ابن الحرامية" Poor Baathists, they became crazy when we asked for freedom. Hafez, curse your soul, you are a thief like your father

289 *"Yil'an Rūḥak yā Hafez al-Sanāfir," Mustafa Alabdan* (YouTube channel), uploaded December 13, 2012, accessed June 1, 2020, https://www.youtube.com/watch?v=7kCIfcy0pz8 (no longer available); *"Yil'an Rūḥak Yā Hafez Tīmūn wa-Būmbā wa-Sīmbā," syrebeltv* (YouTube channel), uploaded August 28, 2012, accessed July 18, 2023, https://www.youtube.com/watch?v=O1OxO4pT3ZA.
290 *"Yil'an Rūḥak yā Hafez – Shāb Khālid," sourianow* (YouTube channel), uploaded September 28, 2012, accessed July 18, 2023, https://www.youtube.com/watch?v=efr4vu-KEL4; *"Yil'an Rūḥak Yā Hafez Māykil Jāksūn," sourianow* (YouTube channel), uploaded March 24, 2012, accessed July 18, 2023, https://www.youtube.com/watch?v=SGfPMd1iH9Q.
291 *"Naghmit Yil'an Rūḥak Yā Hafez Yil'an Rūḥak Yā Bashar," shabeeha2011* (YouTube channel), uploaded January 30, 2012, accessed July 18, 2023, https://www.youtube.com/watch?v=vQ2FtdkpKfI.
292 Personal research archive. This slogan was used in different cities with slight differences. Sometimes it used *Ismā'* [hear] and others *Iṭlā'* [come back to life].

Once a symbolic product was released, the other side rushed to produce an alternative product. The pro-regime response to the curse slogans consisted of slogans that resembled the rhythm of "Hafez, mercy on your soul" or "Hafez, God bless you" (see Table 8). The regime frequently used the slogan "Hafez al-Assad. . . The symbol of the Arab Nation/Revolution." This slogan was replaced on the revolutionary side by "Hafez al-Assad, the dog of the Arab Nation." Instead of "Forever Hafez al-Assad," demonstrators chanted "Freedom forever, over your will, Assad."

Table 8: Slogans of regime supporters in response to the curse slogans.

Pro-regime responses	
Abu Hafez [three claps]	أبو حافظ (ثلاث صفقات)
Hafez al-Assad. . . The symbol of the Arab nation/ revolution	حافظ أسد رمز الأمة\الثورة العربية
Hafez, rest in peace / Mercy [blessing] for your soul	يرحم روحك يا حافظ

The second subgroup can be termed the "sacrifice slogan" group, since it consisted of subverting slogans originally created to praise the Assads and urge Syrians to sacrifice their souls for them.

Table 9: Sacrifice slogans from the Assad regimes post-March 2011.

Hafez al-Assad	Bashar al-Assad	Revolution	Development	Pro-regime response
بالروح بالدم نفديك يا حافظ	بالروح بالدم نفديك يا بشار	بالروح بالدم نفديك سورية (اسم مدينة)	يادرعا (اسم مدينة) نحن معاكي للموت	بالروح بالدم نفديك يابشار
Hafez, we redeem you with our souls and blood	Bashar, we redeem you with our souls and blood	Syria [or name of a city], we redeem you with our souls and blood[293]	Daraa [or name of a city], we are with you until death[294]	Bashar, we redeem you with our souls and blood

For Syrians, it would be considered a normal sign of people's love and gratitude toward their leader to give up everything they held precious for him, especially their blood and souls. Almost identical slogans were thus used in both the Bashar and Hafez eras:

293 This slogan either appeared in the way I have presented it here, or with "Syria" replaced by the name of a city, for example, "Homs, we redeem you with our souls and blood."

294 Just as "Syria" could be replaced by the name of a city, the name of a specific location could also be substituted for "Daraa."

بالروح والدم نفديك يا بشار\حافظ.

[Hafez/Bashar, we redeem you with our souls and blood]

In contrast, however, the protestors argued that it was not Hafez or Bashar who were worthy of sacrifice, but rather the country itself. They thus used the same slogan, as seen in Table 10, but replaced Bashar or Hafez with Syria or the name of a city instead. Regime supporters did not respond to these slogans with a new one, but instead continued to repeat the sacrifice slogan.

Table 10: An example of subverting Bashar al-Assad's slogans.

Hafez al-Assad	Bashar al-Assad	Revolution	Development	Pro-regime response
------------------	الله سورية بشار وبس God, Syria, and only Bashar	الله سورية حرية وبس God, Syria, and only freedom	------------------	الله سورية بشار وبس God, Syria, and only Bashar

The last subgroup of slogans are those that subverted the combination of Assad and Syria into a single unit. As explained in Chapter 1 regarding the visual products of Hafez al-Assad, there was a violence inherent to breaking the signification of meanings in such a way that limited the meaning of Syria to Assad alone, while simultaneously opening up and increasing the meanings of Assad, as demonstrated, for example, by calling Syria "Assad's Syria."[295] The slogan in Table 10 above represents this concept, and was one of the first colloquial Arabic slogans to be used during the era of Bashar al-Assad. The element changed in this slogan by protestors was the substitution of "Assad" with "freedom," implying that having freedom necessarily meant dropping Assad from Syria and, therefore, emancipating the slogan from Assad. Assad supporters did not respond to the new revolutionary slogan, but instead kept repeating the original version.

As we can see from this group of slogans, the aim was to subvert the old regime's legacy of rhetoric that had controlled Syrians for decades. This shows how the old slogans that controlled the public space of the regime's marches were threatened by new ones. These are only a few examples of slogans that the regime had produced throughout the previous four decades being subverted or replaced with other ones by the revolutionary power to compete in the Syrian public space.

295 For more information, please see Chapter 1.

The Second Group of Slogans

Table 11: Pro-regime imitations and subversions of revolutionary slogans.

Revolution	Regime
سلمية... سلمية Peaceful . . . peaceful [we are demonstrating peacefully]	سنية. . . سنية Sunni . . . Sunni [we demonstrate as, or for, Sunni people]
الموت ولا المذلة We would rather die than be humiliated	الفول ولا البازيلا Beans or green beans
زنقة زنقة دار دار بدنا نشيلك يابشار From every alley and house, we will eradicate Bashar	زنقة زنقة دار دار نحن رجالك يابشار In every alley and house, we are your men, Bashar.
الشعب يريد اسقاط النظام People want to overthrow the regime الشعب يريد اعدامك بشار People want to execute you, Bashar	الشعب يريدك بشار People want you, Bashar
يلا أرحل يابشار Bashar, get lost.	نحن رجالك يا بشار Bashar, we are your men.
هيبي يلا مامنركع إلا ل الله We kneel only to God.	هيبي يلا بشار من بعد الله[296] Hey, Bashar is after God

This group of slogans were originally produced by the revolutionary side but then inverted by the pro-regime supporters as a response to the revolutionary discourse, and to cast doubt on the goals and aims of the demonstrators. Contrary to the revolutionary subversion of products that made fun of regime icons, the regime's act of subversion was to (deliberately) misinterpret the aims and intentions of demonstrators. For example, one of the first and most commonly used slogans was "silmiyya, silmiyya" [Peaceful, peaceful], meaning "we are demonstrating peacefully." The regime showed videos in which protestors appeared to be saying "sinniyya, sinniyya" [Sunni, Sunni], meaning "we are demonstrating as, or for, Sunni people," suggesting that the protests sought to establish a Sunni regime or state.[297] Pacifism was a symbol of the revolution, and making the public doubt the pacifism of pro-

296 "Jāmiʿat Ḥalab Ashbāl al-Assad fī Kuliyyat al-ʿUlūm."

297 "Silmiyyat al-Thawra Bidnā Naḥkī ʿal-Makshūf ʿAlawiyya mā Bidnā Nashūf" [The Peacefulness of the Revolution, We Want to Talk Explicitly, We Do Not Want to Meet Alawites Anymore], *alawiyoun14* (YouTube channel), uploaded July 19, 2011, accessed July 18, 2023, https://www.youtube.com/watch?v=ve-b2qtEd1o.

testors thus served to invert and subvert their intentions. The regime also claimed that the demonstrators sought to create an Islamic emirate in Syria, and supported the idea through dozens of articles falsely describing what had happened in many cities and fabricating videos.[298] The regime's goal was to discredit the revolutionary discourse in general, countering the same strategy from the other side. A slogan such as "al-mawt walā al-madhalla " [We would rather die than be humiliated] was sarcastically performed by Assad supporters as "al-mūz Walā al-bazilla" [Beans or green beans]."[299] This slogan, which replaces "death" with "beans" and humiliation with "green beans," can be interpreted at various levels: first, as a direct distortion of the revolutionary slogans; second, as a comic close phonetic parallel of the original slogan, designed to ridicule it; and finally to suggest that demonstrators were talking about ideas and concepts they did not understand. It also suggested that social class and poverty were the drivers behind such slogans. One famous slogan used in all of the Arab protests, "People want to overthrow the regime," was changed to become "People want Bashar al-Assad/Abu Hafez in particular" or "People want you, Bashar."[300] Another famous slogan influenced by Ghaddafi's speech went:

زنقة زنقة دار دار بدنا نشيلك يابشار

[From every alley and house, we will eradicate Bashar]

298 See, for example, Ibrāhīm Ghaybūr, "'An Qatanā: al-Fusayfisā' al-Suriyya wa-'Ā'ilat al-Qādirī" [About Katana: The Syrian Mosaic and the Family of Qadri], *Syrian Days Online Magazine*, July 28, 2011, accessed July 18, 2023, https://bit.ly/3c101Ot; *Syria Line*, "Hakadhā Takallama al-Jaysh al-Sūrī fī Bāniyās" [This is How the Syrian Army Spoke in Banyas], May 14, 2011, accessed July 18, 2023, https://bit.ly/37Qc3a3; *Al-Mijhar*, "al-Salafiyyūn Yu'linūn Talkalakh Imāra Islāmiyya" [Salafists Declare Talkalakh an Islamist Emirate], May 13, 2011, accessed July 18, 2023, https://goo.gl/7A7S4U; *Syriarose Online Magazine*, "Muṣādarat Mu'iddāt wa-Makhṭūṭāt li-Inshā' Imāra Islāmiyya fī Bāniyās" [Confiscating Instruments and Plans for Founding an Islamic Emirate in Banyas], May 11, 2011, accessed July 23, 2018, https://bit.ly/2T819Hp (no longer available). These are just a few of the articles published by the regime in the first two months after protests broke out to support the (false) claim that the protestors sought to establish an Islamic emirate.

299 "*Jāmi'at Ḥalab Ashbāl al-Assad fī-Kulliyyat al-'Ulūm*" [Aleppo University, the Sons of Assad in the Biology Faculty], *Ashbal Asad* (You Tube Channel), uploaded June 19, 2012, accessed July 1, 2018, https://www.youtube.com/watch?v=OLcFjjufDSo (YouTube channel since deleted).

300 "*al-ShSha'bb Yurīd Bashar al-Assad*" [The People Want Exactly Bashar al-Assad], *Mrabogabi* (YouTube channel), uploaded October 12, 2011, accessed July 18, 2023, https://www.youtube.com/watch?v=yxf2gi8nKac.

This slogan became:

زنقة زنقة دار دار نحن رجالك يا بشار

[In every alley and house, we are your men, Bashar]

Another type of slogan was performed as a street revolutionary song, generally with few or no musical instruments. This genre consisted of neither a song nor a single slogan, but of a long slogan-song performed in demonstrations in the streets or in squares. The short slogans discussed above each consisted of a single theme, topic, and rhyme, and changing the slogans illustrated a different theme, rhyme, and performance. The performance of these slogan-songs was closely related to Syrian folk performances or songs performed by soccer fans at a match. Each one lasted for several minutes and presented a dialogue between the regime and the protestors. They stood out in their use of musical instruments like drums and resembled some folk dances, such as *Dabka*. The context of their use was however totally different, giving their performance a very different feel to supporting a soccer club or dancing at a wedding. As this section discusses the second group of slogans—those that the regime reformulated from revolutionary slogans—I will provide only one example of this type of slogan-song.

The song "yallā irḥal yā Bashsār" [Bashar, get lost] was written and performed by Ibrahim al-Qashoush. It became very famous and was performed in many different cities during demonstrations after a rumor spread that al-Qashoush's throat had been cut by the secret police.[301] After his supposed death, al-Qashoush became a revolutionary icon. Every demonstration singer became Qashoush, and started to call themselves the al-Qashoush of Homs, Aleppo, etc.[302] The song was also produced as a symphony by Malek Jandali, called the "Freedom Qashoush Sympho-

301 *"Hamā, Nafī Istishhād Munshid al-Thawrah fī-Hamā"* [Hamah Denying the Death of Revolution Singer in Hamah], YouTube, uploaded July 10, 2011, accessed July 18, 2023, http://bit.ly/3bZ7Y6K; James Harkin, "The Incredible Story behind the Syrian Protest Singer Everyone Thought Was Dead," *GQ Magazine*, December 7, 2016, https://www.gq-magazine.co.uk/article/syria-civil-war. In 2015, people were surprised to learn that Ibrahim al-Qashoush was alive and had not been murdered, contrary to reports. Despite al-Qashoush's alleged death, this song had a powerful impact on revolutionary activism and became a source of inspiration for many artistic and revolutionary songs. Here, it is important to note that belief in his "death" persisted in spite of the number of activists who said he had not been killed, and the fact that al-Qashoush filmed a short video on July 10, 2011 declaring that he was still alive. Because of the brutality of the Assad regime, people did not believe he was still alive until this was explicitly confirmed in an article published outside Syria.

302 Simon Dubois, "Les Chants se Révoltent," in *Pas de Printemps pour la Syrie: Les Clés pour Comprendre les Acteurs et les Défis de la Crise (2011–2013)*, eds. François Burgat and Bruno Paoli (Paris: La Découverte, 2013), 197.

ny."[303] Later, many revolutionary activities and acts were named after al-Qashoush, such as the al-Qashoush speakers that were hidden in public places and loudly played the al-Qashoush song "Yallā irḥal yā Bashsār."[304]

The lyrics of the song are based on Bashar's speeches over the previous few months and his reaction to demonstrations. It has three main parts. The first begins by portraying Bashar and his brother Maher as strangers to Syrian society, asking him to leave the country and step down, and then describing the lies in his speeches; the second part ridicules Bashar and Maher, calling them traitors and describing them as agents for the Americans; and the final part describes the economic corruption of Syria, naming the members of the Assad family who are consuming Syrian resources. As such, it provides a clear summary of why Syrians were protesting against the regime.

Due to the powerful impact of the song, and similarly to "Yā Ḥīf," regime supporters reworked it and called it "Bashar, We Are Your Men." This was a song dedicated to Bashar to prove the loyalty of Syrians. There are many versions of the song available on YouTube, but unfortunately all but one have poor sound quality. One of the differences between the original song and the regime version is the length, as Table 12 illustrates. Some of the lines are broken and cannot be sung, because they do not fit the rhythm, or the musical phrases are too long for the performer to have enough breath to finish the line. This is noticeable in performance, when the performer cannot continue some lines or sings them quickly without rhythm. In terms of its content, the song added nothing more than the regime agenda of national unity with Bashar al-Assad, propaganda about his sacrifices and struggles to remain in power, and praise for the power of Bashar and Maher Assad and the weaponry of the regime army.

---

303 Malek Jandali, composer and pianist, "Freedom Qashoush Symphony," with the Russian Philharmonic Orchestra, conducted by Seregey Kondrashev, track 8 on *Emessa – Homs, Malek Jandali* (YouTube channel), uploaded February 11, 2012, accessed July 18, 2023, https://www.youtube.com/watch?v=ax5ck0fzyaU.

304 Videos of the "freedom" or Qashoush speakers:

a) "*Sbīkarāt al-Ḥuriyya fī-al-Jāmiʿa al-ʿArabiyya*" [Freedom Speaker at European University], *Abood ha* (YouTube channel), uploaded April 30, 2012, accessed July 18, 2023, https://www.youtube.com/watch?v=mqzxw8tXhjY;

b) *Sham News Network*, "*Shām, Dimashq, Mabnā Mudīriyyat al-Māliyya Sbīkarāt al-Qāshūsh Taṣdaḥ wa-l-Amn Yuḥāwil Iskātahā*" [Damascus, Qashoush Speakers Sing and the Secret Police Try to Silence Them], Finance Center Building, Damascus, uploaded October 31, 2011, accessed July 18, 2023, https://www.youtube.com/watch?v=hpw9Zsv4q4k.

c) "*Sbīkarāt al-Ḥuriyya fī-Kuliyyat al-ʿUlūm*" [Freedom Speakers in Science Faculty], *Sam alhorani* (YouTube channel), uploaded November 18, 2011, accessed July 18, 2023, https://www.youtube.com/watch?v=2UCnTmSqjYg.

Table 12: "Bashar, Get Lost" and the inverted version "Bashar, We Are Your Men".

يلا أرحل يابشار[305] **"Bashar, get lost"** **(Revolution)**	نحن رجالك يابشار[306] **"Bashar, we are your men"** **(Regime)**
يابشار مانك منا، خود ماهر وارحل عنا وهي شرعيتك سقطت عنا Bashar, you are not one of us. Take Maher [Bashar's brother] and get lost. Your legitimacy is gone away.	عنا لحمة وطنية، اسلام ومسيحية، هيدي ام الحرية، برئاسة الدكتور بشار We have national unity, Muslims, Christians. This is the best form of freedom, under Dr. Bashar.
يابشار ويا كذاب، تضرب أنت وهالخطاب، الحرية صارت عالباب Oh, Bashar, you are a liar. Do you really believe your speeches? Freedom is getting very close.	رح نحكي بكل اللغات، رح نبقى نحن الثبات حب الوطن للمات برئاسة الدكتور بشار We will say it in all languages. We will love our homeland until death, under Dr. Bashar.
ياماهر وياجبان، وياعميل الأمريكان، الشعب السوري مابينهان Oh, Maher, you are a coward and an agent of the Americans [traitor]. Do not forget that the Syrian people shall not be humiliated.	يابشار وياغالي، بفديك بدمي وبمالي، ترئيسك والله بيحلالي وبالدم منفديك يابشار Oh, Bashar, you are very dear to us. We sacrifice for you our blood and money. You being a president to us is what I like.
يابشار طز فيك، وطز بلي بيحيك، والله مامنطلع فيك Bashar, screw you and your supporters. You have lost our respect.	من دون ماهر الجيش بينذل، والله لأجلك منشرب دم، واليوم عرسنا أحلى عرس بوجود الدكتور بشار Without Maher, the army will be humiliated. For you, we will drink blood. Today our wedding [the march] is the best wedding with the presence of Dr. Bashar.
لسا كلّ فترة حرامي، شاليش وماهر ورامي، سرقوا اخواتي وأعمامي We have serial thieves, like Shālīsh, Maher and Rami who stole from us.	نحن نزلنا بالساحات، ورح ندعم الاصلاحات، ونحن رجالك للمات We are in all squares to support the reforms, and we are your [Bashar's] men until death.
يابشار ويامندس، تضرب أنت وحزب البعث، وروح صلح حرف ال"س" Oh, Bashar, it is you who are an infiltrator, screw you and the Baath Party. Can you pronounce "S" properly? [a reference to his lisp]	يا بشار يا حبيب، والله أنت الطبيب،ويلي معك مابيخيب Oh, Bashar, you are very dear to us and you are our doctor. Who follows you will never be deceived.
	لعيونك يا أبو حافظ، على سورية بدنا نحافظ، اهلينا نصروا حافظ Oh Abu Hafez [Hafez's father, referring to his father and also Bashar's son (Hafez)], we will protect Syria because our families supported Hafez [the deceased Hafez al-Assad].

305 Ibrahim al-Qashoush [Abū al-Fidā' al-ḥamawī], "Ughniyyat Yallā Irḥal yā Bashshār," uploaded August, 14, 2012, accessed July 18, 2023, https://www.youtube.com/watch?v=KBiWEqz712U.
306 *"al-Sabiʿ Baḥrāt Bisnādā Bayt Niḥnā Rijālak yā Bashar,"* [at the Seven Fountain Square, We are your Men Bashar] unknown artist, *Assadsyria* (YouTube Channel), uploaded October 12, 2012, accessed June 20, 2018, https://www.youtube.com/watch?v=8fFYHwRNaRU (no longer available).

Table 12 (continued)

يلا أرحل يابشار **"Bashar, get lost"** **(Revolution)**	نحن رجالك يابشار **"Bashar, we are your men"** **(Regime)**
	بالجنة عنا شهداء، ماماتو بعدون أحياء، ولو قطعونا أجزاء بنبقى رجالك يابشار In heaven, we have martyrs. They are not still alive. Even if we are cut into pieces, we will stay loyal and be your men, Bashar.
	معلمنا عندو تاريخ، يمكن شافوا الصواريخ، ولوبتصيروا عالمريخ Our master has a history. Maybe they saw our missiles. Even if you are on Mars.

The regime supporters' imitation of revolutionary slogans provides evidence of the successful impact that the revolutionary slogans had at that time, as well as the slogans or other symbolic products generated by the regime and imitated by the revolution. All of these symbolic products were important due to the influence they had on the Syrian public space in consuming, exchanging, and competing to produce more convincing products.

The Third Group of Slogans

As can be observed from the examples above, the regime side was mostly dedicated to responding to revolutionary slogans in the Syrian public space rather than releasing new ones, with the exception of old slogans and those specifically composed after 2011. From my personal observations, participation in marches in 2011, and analysis of YouTube videos of regime marches, I have determined that the main slogan themes were Hafez al-Assad and his sons, Bashar and Maher. The main slogan themes in this group did not, therefore, change from the 1970s through to 2012. Some focused on the name Hafez, used as a suggestion that the next regime president would be Hafez, Bashar's son. The following slogans were used after 2011 because the context in which they were used did not previously exist. For example:

يا بشار ويا حنون. . . الشعب السوري ما بيخون[307]

[Oh, affectionate Bashar, Syrians will never betray you]

307 *"Min Amām al-Qunṣuliyya al-Briṭāniyya fī-al-Lādhiqiyya"* [In Front of the British Consulate in Lattakia], *Hanan Asad* (YouTube channel), uploaded July 12, 2011, accessed August, 10, 2017, https://www.youtube.com/watch?v=_kACnwSvg1E (no longer available).

What is interesting here is the thought-provoking use of the adjective *ḥanūn* [affectionate], which is not appropriate to the situation because of the violent response to the protests. It is used simply to improve the rhyme of the slogan, and has the connotation of begging the favor of Bashar al-Assad, as if he were in a strong position in 2011 and able to show mercy rather than grappling with rising tension.

The new regime slogans used after 2011 were slightly different. Before 2011, a slogan like the one above, used by regime supporters to prove their loyalty to their leader, would have been absurd—for who would have dared to be disloyal? This shift led to many other new slogans. For example:

وحياة ترابك يا حافظ، على بشار لنحافظ.[308]

[Swearing by the ashes of Hafez, we will protect Bashar]

Here, to express the holiness of Hafez al-Assad, his grave is presented as a divine symbol to be sworn by, similarly to God and saints. The context of this slogan and the one before it was the beginning of the protests, and the expression of solidarity with the regime. When the political and economic situation began to impact on the Syrian lira and its value decreased, regime supporters said:

لو حطونا عالحديد، غيرك أسد مامنريد"[309]

[Even if we become bankrupt, we do not want any other Assad but you]

When the *shabbīḥa* and the army massacred demonstrators, supporters said:

ليش ما عم تحكو ليش.. ماحمانا إلا الجيش.[310]

[Why are you not saying that only the army protected us]

When Turkish intervention escalated in Syria, a new slogan emerged:

كيفك فيا.. كيفك فيا.. جاية دورك تركيا[311]

[Hey Turkey. . . It is your turn now]

This slogan reflects the regime propaganda that foreign countries were controlling the demonstrations on the ground, and was also intended to further the conspiracy theory that the regime had promoted since the beginning of the protests: that the West would spread chaos in the Middle East. Other slogans specifically targeted the media, or what they called the "channels of sedition," for example:

308 Personal research archive.
309 Personal research archive.
310 Personal research archive.
311 Personal research archive.

يا جزيرة ياخنزيرة.. تعي شوفي هالمسيرة.[312]

[Hey, piggy Jazeera. Come and see this march]

These examples are merely some of the regime slogans that were produced and performed for the first time in 2011. If we try to connect the two periods (before and after March 2011), we find that the regime side activated its tools and instruments to produce new slogans just after March 2011. Before this, all slogans were used to support the leader's cult and glory (whether Bashar or Hafez). Slogans addressing Syria were rarely used, and when they were used they referred to Assad's Syria. Both periods also show the development of slogan language, and the flexibility of regime slogans, competing with the same revolutionary vocabulary after subverting it. After March 2011, the regime used the same themes that glorified the Baath regime and its leader, but in a more flexible way that was able to respond to, or produce similar, symbolic products to those of the revolution.

The consumption and use of slogans on both sides confirms the mutual dependency of the pro-regime and anti-regime factions in imitating each other. This imitation was seen in terms of form, but not content, with the content proving that the aim of this replacement was not only to reproduce a product but also to subvert it.

Banners

While banners were an essential part of the revolutionary language of Syrians, regime supporters were largely inactive in holding up banners other than images of Bashar al-Assad or the Syrian flag centered on a photo of him. This section does not, therefore, follow the pattern of the previous one by juxtaposing revolutionary banners with their regime counterparts, but instead presents a short summary of the general characteristics of revolutionary banners.

Many cities became a focus for the media because of their innovative banners. Binnish, for example, was famous for what were called "moving banners," where dozens of banners were held up by demonstrators to form a painting or single banner, while Kafr Nable was well-known for the most outspoken banners, using sarcastic cartoons and paintings. Zabadani was known for its creative use of language and the smartest messages for the international community, and in Deir Ezzor the "Kartoneh" initiative was founded, displaying banners made from cardboard.

An analysis of my personal research archive, which contains photos taken by me personally, photos from activists' archives, and from digital media, reveals

312 Personal research archive.

several elements shared by the revolutionary banners. First, most of them were written in MSA at a time when colloquial Syrian was being used more frequently, contrary to the Arabism drive over the previous decades of Baath reign. Second, banners functioned as messages to the international community. Demonstrators were aware of this essential function, and wrote messages on banners asking about the world's responsibility for what was happening, because they were able to broadcast their demonstrations worldwide. These addressed three groups: Arab countries, Muslim countries, and the international community. Banners were also therefore frequently written in English as well as in Arabic.

Third, there was an informal, internal competition for cities to write or draw the most outspoken or creative banners and, as a consequence, some cities copied banners from others. There was also an ongoing dialogue between cities, with banners written in response to those displayed in other locations. Fourth, banners had different topics and themes, including political sarcasm, jokes, political messages, and interpretations of events from the point of view of demonstrators.

Fifth, another aspect of the exceptional performance was holding up banners for photos taken from behind. Demonstrators risked their identities when broadcasting their demonstrations via videos or photos because the regime was able to recognize and arrest people. They thus had two options to maintain their safety: first, to blur the videos; or second and easiest, to film demonstrators from behind. This required demonstrators to hold their banners backwards in order to keep them readable, and camera operators to film from the back of the demonstration. Sixth, banners were used unintentionally to document the events. During the early weeks of the protests, banners did not display any dates or places, enabling the regime to tell the story that these protests were old ones, or that the photos had been taken in another country. Camera operators then took responsibility for showing the date and place on a small piece of paper, mentioning it at the beginning of the video. Months later, the way of writing dates was changed by adding the year the demonstration took place. If banners were collected in chronological order, they could provide a narrative of the events. In addition to this, opposition politicians accepted the banners as a way to deliver the opinions of and messages from protestors, such as accepting the proposals of the National Council.

Bakhkh

Similarly to other symbolic products, *bakhkh* [graffiti] was used to control the public space in Syria. Before the protests, walls were used either to hang Assad's portraits or to praise him in poetry, especially on the walls of military establishments. Despite this dedicated purpose of walls in Assad's Syria, *bakhkh* passed through different phases to take shape in the first phase of protests. Anything sprayed on a wall post-March 2011 was called *bakhkh*, raising the question of which term should be used. Across the Arab revolution countries, messages sprayed on walls were explicitly called *grāfītī* in Arabic, using the word borrowed from European languages, but in Syria people favored the terms *bakhkh* [spray] and *ḥīṭān* [walls].

To investigate the reasons for this, it is important to understand whether the practice of graffiti was already present in Arabic culture or if there had been any similar type of art prior to current use of the term in Arabic. The oldest source in which it can be found is *Adab al-Ghurabā'* [The Literature of Strangers] by Abū Faraj al-'Aṣbahānī (897–967).[313] This book is a compilation of the poetry and sayings written by unknown people on walls in different regions seen by the author in the course of his travels. His book is an old document that illustrates that the act of *bakhkh* was not a new practice, for in it the author claims to have recorded what he heard and saw of what people "wrote on the walls of taverns, fields and houses in different lands."[314] Other books describe the act of writing on walls, including *Nafḥ al-Ṭīb Min Ghuṣūn al-Andalus* [The Pleasant Fragrance from the Tree Branch of al-Andalus] by al-Muqarrī al-Tilmisānī (1577–1632) and *Ḥayāt al-Ḥayawān al-Kubrā* [The Great Life of the Animal] by Kamāl al-Dīn al-Dumayrī (1344–1405).[315] Each of these indirectly mention the practice of writing on walls as a tool for repressed people, especially for strangers and those who feel alienated from their homelands.[316] All of these sources describe writing on walls in different terms, but notably focus on the performativity of the act, its performance, and the quite specific verbs written on the wall as a result. This is why there was no special term for the writing people put on walls but merely a description of the act of writing itself: similarly to the Syrian revolutionary context in which *bakhkh* described the act, rather than the product.

In the first few months of the protests, demonstrators did not call their work "graffiti." Spraying messages on walls was referred to as *bakhkh*, until the *Isbū' Grāfītī al-Ḥuryya* [Freedom Graffiti Week] event, which took place from April 14 to 21, 2012.[317] Before and after this event, Syrians called their sprayed walls either

---

313 Abū Faraj al-'Aṣbahānī, *Adab al-Ghurabā'* (Beirut: Dār al-Kitāb al-Jadīd, 1972).

314 Abū Faraj al-'Aṣbahānī, *Adab al-Ghurabā'*, 22.

315 Aḥmad Ibn Muhammad al-Muqarrī al-Tilmisānī, *Nafḥ al-Ṭīb Min Ghuṣūn al-Andalus*, vol. 1 (Beirut: Dār ṣāder, 1988); Kamāl al-Dīn Muhammad Ibn Mūsā al-Dumayrī, *Ḥayāt al-Ḥaywān al-Kubrā* (Damascus: Tlas, 1992).

316 al-'Aṣbahānī, *Adab al-Ghurabā*, 22.

317 Donatella Della Ratta, "Syria: Art, Creative Resistance and Active Citizenship," *Freemust*, October 2012, accessed June 1, 2020, https://bit.ly/2IedhE.o (no longer available).
"Freedom Graffiti Week, Syria" official event Facebook page, accessed June 1, 2020, http://bit.ly/3adZcAj (no longer available).

ḥīṭān or *bakhkh*. This has to do with the fact that, as a general rule, colloquial Syrian tends to use fewer foreign loan-words than other Arabic dialects. The prevalence of the word might also be linked with its appearance before 2011 in an episode of the famous series *Buq'at Daw'* called "al-Rajul al-Bakhkhākh" [The Spray Man].[318] When the writer of this sketch was later arrested, newspaper headlines read "The Spray Man Was Arrested."[319]

Describing *bakhkh* as murals or graffiti does not adequately reflect the specifics of the Syrian context, since these definitions are too narrow for post-2011 Syria. A mural is "a large picture that has been painted on the wall of a room or building," while graffiti is defined as "words or drawings, especially humorous, rude, or political, on walls, doors, etc. in public places."[320] "Graffiti" is closer to describing Syrian *bakhkh* only when it is seen as an act of resistance and an "act of vandalism."[321] In cities like Sarāqeb, for example, regime troops were present, and paintings on walls were called *ḥīṭān*, not murals. In order to reflect the specifics of the Syrian context, and in reference to the TV series *Spotlight* that gave this word to Syrian graffiti, *bakhkh* is the more culturally appropriate term for this act as performed in Syria. One of the spray-women told me that: "you can consider it as graffiti and an artistic act of what we do. For us, we see it only as *bakhkh*, not graffiti."[322] In addition to this, *bakhkh* does not describe the final product on the wall. It illustrates the act of doing it rather than the writing on the wall itself as the purpose, giving greater importance to the act of spraying rather than what is actually sprayed.

As a symbolic product, *bakhkh* resembled others designed to gain the most visual space possible on walls by changing them. Since *bakhkh* and graffiti are acts of resistance, the regime did not start creating this product in its own controlled visual public spaces since these areas already displayed the regime's visual materials. For this reason, the most apparent visual spaces were the regime's signs. It intervened in the production of symbolic products of *bakhkh* only in two ways: first, by

---

318 *Al-'Arabī al-Jadīd*, "*al-Rajul al-Bakhkhākh: Ḥikāyāt al-Thawra al-Sūriyya*" [The Spray Man: Stories of the Syrian Revolution], March 17, 2018, accessed July 18, 2023, http://bit.ly/32pvPrI.

319 *Zamān al-Waṣl*, "*al-Rajul al-Bakhkhākh 'Adnān Zarā'ī Kharaj wa-lam Ya'ud mundhu Thalāth Sanawāt*" [The Spray Man *'Adnān Zarā'ī* Left and Has Not Come Back for Three Years], July 19, 2015, accessed July 18, 2023, https://www.zamanalwsl.net/news/article/62607/.

320 Kate Woodford, ed., s.v. "mural," *Cambridge Advanced Learner's Dictionary online* (Cambridge: Cambridge University Press, 2003); Kate Woodford, ed., s.v. "graffiti," *Cambridge Advanced Learner's Dictionary online* (Cambridge: Cambridge University Press, 2003).

321 Michael Walsh, *Graffito* (Berkeley: North Atlantic Books, 1996), 12.

322 The first spray-woman in Damascus (name withheld by request), interview with the author, Istanbul, October 15, 2015.

responding to revolutionary *bakhkh* with alternative products or subverting them; and the second after invading a city that was actively demonstrating. Due to the intensity of *bakhkh* in the countryside rather than cities, especially during the first few months of the protests, I have divided this section on *bakhkh* into two parts: *bakhkh* in towns and the countryside, and *bakhkh* in the cities.[323]

Towns and the CountrySide

In the early months of the protests, as the number of demonstrations increased, demonstrators changed the visual public space of their surroundings, especially in places that witnessed many demonstrations. If a city was crowded by demonstrations, *shabbīḥa* and secret police were unable to regain control over it. They retreated to their posts without intervening in revolutionary acts until they had backup, giving demonstrators the opportunity to change the public space visually. Usually, *bakhkh* on the walls did not last for long. At most, it was three days before the *shabbīḥa* intervened and arrested demonstrators. This short period of time was an opportunity to change all the names of public places, including squares, streets, and schools, to include the word "freedom," such as Freedom Square, Freedom School, Freedom Hospital, and so on. Walls were repainted and new banners made to redefine the places. "Freedom," "revolution," and "peaceful" were the words most frequently written on walls. Once the army regained the city, walls, banners, and *bakhkh* were either destroyed or erased. As with other symbolic products, *bakhkh* was the scene of a metaphorical war between the regime and the demonstrators. After all the traces were removed by the regime, activists would sneak out at night to change the words written by the regime. Once a place was visited by Assad's army, they left messages on the walls to activists and protestors, as shown in the photos below.

---

323 In Syria, "countryside" and "city" areas do not refer to rural and modern urbanized places respectively, but are administrative divisions. The countryside can therefore be a big, very crowded city. The only difference is that in the countryside, the secret police have less of a presence than in big, administrative cities due to the larger size and importance of regime institutions in cities. However, this does not mean that all countryside areas are urban and crowded areas like cities.

Figure 11: "Assad or we Burn the Country, If you come back here, we are back Greetings from Assad's men".[324]

Figure 11 above and below display the *bakhkh* the regime's *shabbīḥa* used to write on shops and walls when backup arrived in the city. Checkpoints were created in the city or the town, and as they passed through the streets, the *shabbīḥa* sprayed or painted many messages on shops and walls. The message in Figure 11 reads "Either Assad or we burn the country," showing that the only reason the Assad regime started the act of spraying on walls was to signal that the space was still controlled by the regime, at least in visual terms. The right side of the photo reads "If you are back, we come back, Greeting from Assad's men". "You" here refers to protestors or as the regime called them, terrorists. The second photo (Figure 12), taken at Aleppo University, shows how these acts of spraying were subverted by changing the *bakhkh* itself with the addition of a word or two: here with the addition of the verb *naḥruq* [burn] before the sentence and the word *nabnī* [build], so the message becomes "We will burn Assad and build the country." Additionally, the word *ṭuzz* [Fuck] is written under Assad in green which might illustrate that there were two people involved in the subversion of this *bakhkh*. Another example of exchanging symbolic products in Syria is seen in the third photo (Figure 13). When the regime army passed through a city to "purge" it of "terrorists," it wrote "the Assad army has passed through here"—an indication of the importance of visual spaces for the regime (Figure 13). It was normal for activists to distort the phrases written on the

324 Yallair7al (pseud.), "Assad Graffiti. Translated: '[Either submit to] Assad or We Burn The Country,'" location and date unknown, *The Revolting Syrian* (*Tumblr* blog), posted in 2012, accessed June 1, 2020, http://therevoltingsyrian.com/image/18230221834 (no longer available).

walls. By putting a single dot over *mīm* in Arabic to become *fā'*, the meaning was changed from "passed through" to "ran away from."

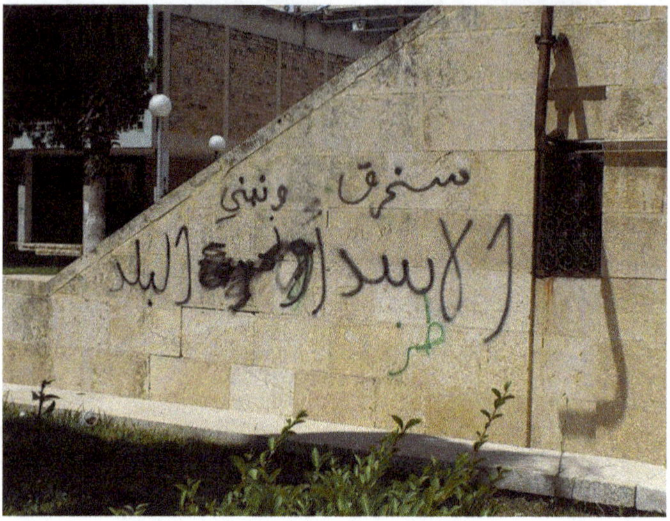

Figure 12: Subverting bakhkh between the pro-revolution and regime supporters. The regime bakhkh says "Assad or we burn the country." This was changed to "We burn Assad and we build the country."[325]

Figure 13: Subverting bakhkh between the pro-revolution and regime supporters. The regime bakhkh says "Assad's soldiers passed through here," which was changed to "Assad's soldiers ran away from here."[326]

---

325 "University of Sciences – Aleppo. The Specific Business Squad of the College of Science scolded the college squares in revolutionary terms and erased the writings left by the Shabiha," Aleppo University, Freedom Graffiti Week Syria official Facebook page, July 9, 2012, accessed July 18, 2023, http://bit.ly/39YJWa7.

326 Bdama, Idlib, October 9, 2012, saved to my personal research archive and verified by Tamer Turikmane, the Syrian Revolution Archive.

Cities

In cities, the iron fist of the regime was firmly present and the situation was different. The act of spraying had to be done quickly wherever it took place, but here people had even less opportunity to spray compared to the countryside and towns. They did however see the spread of a new revolutionary technique called a *muẓāhara ṭayyāra* [flying demonstration].[327] This was a type of demonstration that lasted only three to five minutes and worked well due to the crowdedness of the streets. When the demonstration ended, demonstrators would scatter and sneak out with the passersby. During those few minutes, however, demonstrators changed the visual public space of the street by throwing pamphlets and quickly spraying on the wall. The act of spraying was initially performed spontaneously, but later became organized. New terms were then coined, including *al-rajul al-bakhkhākh* [spray man], *al-mar'a al-bakhkhākha* [spray woman] and *katībat bakhkh* [spray battalions] and "brigades," referring to groups of these protestors and known according to the geographic area where they sprayed, such as the spray battalion of Muhājirīn or Mīdān. In my interview with the first spray woman in Damascus, who refused to give her name, she told me: "my work was spontaneous and never planned in advance. For example, on the birthday of Bashar al-Assad, I wrote in the Muhājirīn [Presidential] neighborhood, 'today is the birthday of the Duck' and 'Have a black day, Duck'."[328] With the development of the revolutionary spraying techniques, demonstrators tried to organize themselves more by moving *bakhkh* away from a spontaneous act toward unity in messaging and spray techniques. This included organizing numerous spray campaigns, the best known of which was the "Freedom

---

327 See Yanna Vogiazou, *Design for Emergence: Collaborative Social Play with Online and Location-Based Media* (Amsterdam: IOS Press, 2007), 23–6. *Muẓāhara ṭayyāra* literally translates as "flying demonstration" because it lasts only a few minutes. While superficially similar to a flash mob, it is a totally different phenomena. Flash mobs are social, apolitical experiences that do not put their participants at risk and consist of a performance, like the first flash mob in London in 2003. Usually, flash mobs last for half an hour. In contrast, flying demonstrations are political performances by nature and do not last more than five minutes. Longer performances are risky, as demonstrators may easily be arrested (some are even arrested as they scatter).

328 First spray woman, interview. Some You Tube videos of the Spray campaign in Damascus: "Dimashq, *Arwaʿ Maqṭaʿ li-l-Rajul al-Bakhkhākh fī Minṭaqat al-Mazza*" [Damascus, the Most Wonderful Footage for the Spray Man in Mezzeh], YouTube, uploaded December 3, 2011, accessed July 18, 2023, https://www.youtube.com/watch?v=-aVuGlFnMNA; "*Faʿāliyyāt al-Rajul al-Bakhkhākh Fi al-Mazza al-Qadīmah*" [The Events of the Spray Man in Old Mezzeh], YouTube, uploaded November 9, 2012, accessed July 18, 2023, https://www.youtube.com/watch?v=-aVuGlFnMNA, Personal research archive, Mazza, Damascus, January 6, 2012; "*al-Muhājirīn, Ṭalʿat al-Shūrā, 100 mitr min al-Maktab al-Riʾāsī*" [Al-Muhājrīn, ṭalʿat al-Shūrá [a neighborhood in Damascus] 100 meters from the presidential office], uploaded June 28, 2012, accessed July 18, 2023, https://www.youtube.com/watch?v=L-xcZ2IcQmI&feature=youtu.be.

Graffiti Week" that took place from April 14–21, 2012. What was different about this campaign was its international dimension, because it spread into several Arabic countries, in addition to those who participated in this campaign in Syria. A close reading of the context and images shows that the use of stencil graffiti provided a new way of expressing the revolutionary act. Activists uploaded all of the possible stencils to Facebook, enabling everyone to print them out for spraying later. This meant that the themes and sprayed phrases were rather different from the spontaneous ones seen at the beginning of the protests. The most famous graffiti was "Stop the killing," accompanied by an image of Bashar, and then written below it "stomp here," "break the chains," "freedom for [name of an arrested person]," or photos of known Syrian activists or historical figures. For the first time, Syrian symbols were revived in this campaign: Sultan al-Atrash (1891–1982), Ibrahim Hananu (1869–1935), and Fares al-Khury (1877–1962) were mentioned in reference to the democratic Syria of the 1950s, while new icons were seen in Bāsil Shiḥāda, Ḥamza al-Khaṭīb and others. Prior to this campaign, graffiti had not drawn on Syrian icons, figures of the great revolution, or from Syrian history.[329]

As explained above, the revolutionary side was generally the first to start the competition for distributing such symbolic products. *Bakhkh* was a sensitive issue for the regime, in spite of the fact that it did not last very long before the secret police came to erase or paint over it. It is important to note that, unintentionally and with the help of the political context of the Arab Spring, the first symbolic product released in the Syrian public space was *bakhkh* by the revolutionary side, when a group of elementary school pupils in Daraa wrote on the wall of their school: اجاك الدور يا دكتور، الشعب يريد اسقاط النظام "It is your turn, doctor," and "People want to overthrow the regime." The children were tortured and had their nails pulled out, causing people to demonstrate for the first time in Daraa for the freedom of their children.

329 Some photos of the Graffiti Week, "The Liberated Countryside of Aleppo, City of Manbij," Manbij, Aleppo, Freedom Graffiti Week Syria official Facebook page, September 24, 2012, accessed July 18, 2023, http://bit.ly/2Plutch; Ḥay Kāshīf, Darʿā [Daraa, Al-Kashef Neighborhood], Freedom Graffiti Week Syria official Facebook page, June 22, 2012, accessed July 18, 2023, http://bit.ly/2Pm6bip. Daraa, Freedom Graffiti Week Syria Official Facebook page, June 15, 2012, accessed July 18, 2023, http://bit.ly/2SXpAbP; "*Munthir Iskān Al-ḥayy al-Gharbī Qāmishlī* [Munthir Iskān, the West District], Graffiti Week Syria official Facebook page, June 5, 2012, accessed July 18, 2023, http://bit.ly/2STjx7P; "*Kulyyat al-ʿAmāra – Jāmiʿat Dimashq – Bāsil – Raḥ Yishtaqlak al-Ṭarīq*" [Damascus University, Faculty of Architecture – Bāsil, Our Path Will Miss You], Damascus University Faculty of Architecture, Freedom Graffiti Week Syria official Facebook page, uploaded June 1, 2012, accessed July 18, 2023, http://bit.ly/2SUIMXJ.

Performance

The symbolic products discussed above can be properly understood when combined with performance. The importance of performance is that it is shared by many symbolic products. For example, a written slogan is incomplete if its performance is ignored since all of its performative aspects are eliminated, reducing the slogan to merely its textual elements. In this section, therefore, I turn to studying and offering a translation of the revolutionary performances as the final component of the revolutionary language. While the aim of this study is not to analyze pro-regime street performances, I refer to and compare them to revolutionary street performances in order to support the understanding and analysis of the latter.

What gives these performances meaning is not only the linguistic content of these products but also their speech acts. Competing performances can be distinguished by the level of organization. Revolutionary demonstrations were more ordered in terms of their form and organization, particularly after the formation of coordinations designed to organize such demonstrations. Pro-regime performances, on the other hand, were chaotic; people were gathered to enact a performance but did not know what to do and were directed by disguised secret police officers, who showed that they were part of this spontaneous chaotic performance. They sometimes gave instructions about what to chant, what to carry, and where to go in case the crowd lost its way in chanting. In addition to this, a significant number of participants were obliged to attend because they worked for the regime or had otherwise been forced to do so. Pupils were unexpectedly taken from schools to join marches, and loyalist marches were organized prior to and following all of Assad's speeches.[330] The "million people marches" of the regime were also made possible by "forcing the government employees from Damascus and other cities to join the march."[331] These acts of praising Assad and his regime were not, therefore, the spontaneous events they were declared to be by regime outlets but instead forced performances. Some people were forcibly gathered and acting out of fear. This is why the regime performances were soulless and simply an act of self-glorification. This does not, however, exclude the existence of some people who were convinced by what they did in such performances and did them while believing in the regime.

The revolutionary performance of demonstrations was substantially different for many reasons. Joining a demonstration against the regime was not as safe as pro-regime marches and many were killed; demonstrators were thus free to protest

---

330 Bishāra, *Syria: The Path of Suffering*, 253.
331 *Al Jazeera*, *"Al-Assad Yaḥshud Muwālīn wa-l-Muʿāraḍa Tarudd"* [Assad's Regime Deploys its Loyalists and the Opposition Answers], October 12, 2011, accessed July 18, 2023, http://bit.ly/39d-DdJi.

or stay safely at home. The decision to take part in a demonstration, despite the danger, was also itself an act of emancipation from the regime, defying it, and liberating bodies in the public space. As argued in this chapter, both sides sought to fill the maximum amount of space—physically, visually and symbolically. The act of filling public streets with bodies was thus an act of emancipation for the geographic space, first of all, as well as for the bodies risking themselves in the streets. Finally, the revolutionary performances were better organized than those of the regime. People might not have known exactly how to demonstrate, but they allowed their bodies to be in tune with the momentarily liberated street or square. This gave protestors control over the area through unprecedented cultural practices in a way that was unnecessary in pro-Assad marches since the regime normally controlled the public space. If the regime organized a march, it served only to confirm ownership of a space, while protestors were making their own claim on the public space. The act of demonstrating was one of emancipation from the regime itself, even if only for a short period of time.

The cultural practices used in the revolutionary demonstrations varied from one city to another, based on geographic location. This regional variation in the performances was not a negative point, but was rather perceived as reflecting Syria's rich cultural identity. In some cases, this encouraged demonstrators to imitate another city's performative demonstration or to borrow its slogans. One of the first slogans was "min ḥūrān hallat al-bashāyir" [From Hūrān comes the good omen], which was performed differently in Daraa compared to in Damascus. [332] Because of their pride in their cultural differences, demonstrators elsewhere went on to perform it as it had been performed in Daraa. The rhythm of the performance was close to that of a popular type of song performance called *Shurūqī*.[333] There are many types of *Shurūqī*, including *al-Ḥadā, al-ʿArḍa, al-Sāmirī, al-Marbūʿ*, and others.[334] These types of folksongs are usually performed in the Ḥūrān valley, not in the cities, and are frequently seen at social occasions like weddings. The distance from Hūrān was not why these slogans were not performed the same way in other Syrian cities that lacked this colorful folk tradition in their culture; instead, the differences were due to the feeling of being united with all of the folk and cultural aspects of other regions as proud representations of all Syrians. Folklore traditional singing moved to the big cities like Damascus or other cities performing some of

332 *"Al-Thawra al-Sūriyya: Ṭafis min Ḥūrān Hallat al-Bashāyir"* [The Syrian Revolution: Tafess—from Hūrān Comes the Good Omen], *Anas Al Aswad* (YouTube channel), uploaded April 28, 2011, accessed July 18, 2023, https://www.youtube.com/watch?v=EZF2Gh_pIwk.

333 *Shurūqī* is a type of singing that Hūrān is known for. Historically it is a type of song used in wars and has many variations.

334 Choukair, interview.

the Hūrānī pieces or other traditional practices of other cities to reflect the union with the cities that witnessed the protests. Another type of cultural practice that was seen in other geographic places was the folk dance *Dabka*, seen in cities like Damascus, or *Dabka Jūfiyya*, a special type of *Dabka* that characterized other revolutionary performances in Daraa, and which begins with two rows of dancers, standing face to face.

Al-Harīqa Demonstration

An example of the cultural practices of demonstrators and their differences from regime marches can be seen in the first demonstration in the al-Harīqa quarter of old Damascus on February 17, 2011,[335] when a group of civilians protested against the police beating a young man in Harīqa Souk.[336] Suddenly people felt humiliated by what had happened to this man, since such a practice was normal in Syria, and unexpectedly gathered and started chanting "thieves," in reference to the police. What characterizes this event is the impossibility of analyzing any single detail in isolation, especially its performance. The main components of the performance were four slogans:

حرامية، الشعب السوري مابينذل، لا إله إلا الله، بالروح بالدم نفديك يابشار.

> [Thieves; Syrians will not be humiliated; There is no god but God; We redeem you, Bashar, with our blood and souls][337]

It is impossible to read these slogans separately from the geographic context and the performance. Despite the fact that the last slogan praised Bashar, it is understandable only within the chain of the four slogans together. What is more important is the performance of the demonstrators, which makes it possible to interpret the slogans themselves and indicates the presence of additional dimensions. According to witnesses, activists, and news reports, "this demonstration was not a prepared one," but was spontaneous.[338] The first slogan, "thieves," described the police and

---

335 Bishāra, *Syria: The Path of Suffering*, 75.
336 *France 24*, "*Muẓāhara Iḥtijājiyya fī-Dimashq baʿda Iʿtidāʾ al-Shurṭa ʿalā Shāb*" [A Protest in Damascus After the Police Beat a Young Man], February 18, 2011, accessed July 18, 2023, https://bit.ly/3c0U43X.
337 "*Muẓāhara fī-Dimashq – Ḥayy al-Ḥarīqa al-Tijārī 17–2–2011*" [A Demonstration in Damascus in al-Ḥarīqa, the Commercial Quarter, February 17, 2011], *Souriana4all* (YouTube channel), uploaded February 17, 2011, accessed July 18, 2023, https://www.youtube.com/watch?v=NykGjfKn3TU&t=42s. The slogans were extracted from the only existing video of the demonstration.
338 Interview with a demonstration participant, October 10, 2015.

the routine bribery and corruption that Syria was suffering under. The second slogan, "Syrians will not be humiliated," was an honest reaction to the beating of a man, representing not only the moment of the demonstration but also the tragedy of Syria under the Assad regimes, in which humiliation was only one facet of the symbolic violence inflicted on the people. The slogan after that was "there is no god but God," which at first glance appears to be out of context, since the relationship between this slogan, the context and the event itself is not immediately apparent. The final slogan, "We redeem you, Bashar, with our blood and souls," also seems inappropriate for the situation: why would an anti-regime protest end with a slogan praising Bashar?

This brief description of the slogans provides only limited information about them. The act of interpretation can be misleading if the performance does not accompany these slogans. Performativity is "a reiteration of a norm or a set of norms, and [. . .] it acquires an act-like status in the present."[339] By looking at the repeated acts, gestures, and movements of demonstrators, we can piece together a rhythm that might provide additional information about the performativity of the demonstration, as long as performativity is "a repeated act" in the first place.[340] This repetition is then what signifies the performance itself and the agent(s) when *doing*, not *being*, demonstrators.

The footage of the demonstration shows certain repeated norms, acts, and gestures.[341] For example, we see all of the demonstrators in different scenes are facing in the same direction, and all of their hands are positioned upward. Importantly, a man who appears to be leading the crowd is standing on top of a car or elevated object, using it as his platform. All of these performative aspects and repetitive acts illustrate the Syrian cultural practices inherent in the demonstrators' identity. This identity is informed by *ʿArāḍa*, a traditional Syrian folk performance done with the help of a crowd to celebrate the happiest events, such as weddings, parties, the return from a pilgrimage to Mecca, or the birth of a baby. In such folk performances, a leader is usually carried on the shoulders of the crowd to get the crowd excited and to cheer them on by reciting *Shiddiyyāt* [a special type of poetry] that they repeat after him. The performance that is shown in the YouTube video, depending on the performative aspects of the demonstrators, can thus be described as an adjusted *ʿArāḍa*, used not to celebrate but to protest.

339 Judith Butler, *Bodies that Matter: On the Discursive Limits of "Sex"* (New York: Routledge, 1993), 12.
340 Butler, *Bodies that Matter*, 107.
341 *"Muẓāhara fī Dimashq Ḥayy al-Ḥarīqa al-Tijārī 17-2-2011."*

Throughout the video, the slogans are chanted in different sequences. Each slogan is about a theme or issue that is not closely related to the theme or issue of the others. The first slogan attacks the police as thieves, and is repeated more than once for emphasis. After this, the demonstrators in the video do not appear to know what to do. They start screaming and whistling in a way that suggests they may have begun to feel afraid of the dangers associated with protesting. Later, during this chaotic noise, one person says, "Syrians will not be humiliated," and all of them seem happy to repeat the slogan. Then the whistling and shouting suggesting fear start again. To fill the chaos of this gap, someone says, "There is no god but God," and the demonstration becomes quiet and humble as they repeat it respectfully, in line with the practice typical in Syrian funerals.

After this slogan, another is shouted that seems to contradict the essence of the demonstration itself and its aims: "We redeem you, Bashar, with our blood and souls." The slogan does not seem to fit with the rest and appears to be out of context, particularly since the demonstration is against the regime. Prior to this moment, the demonstrators had refused to leave and remained in a crowd until the minister of the interior came to reassure them that they would not be arrested afterward. When he arrived, he reproached them through the window of his car for protesting, saying "Shame on you! This is a demonstration." One of the demonstrators, justifying their demonstration and framing it as not anti-regime, says: "I swear to God, they beat us." Then, in order to be able to leave in peace, demonstrators chant the pro-regime slogan to prove that they are not against the regime.

As this example highlights, incorporating the performative aspects of the demonstration steers its interpretation as occurring under the only permissible occasions for Syrians to demonstrate under Assad. This can be seen in the repeated gestures that are very similar to those of *ʿArāḍa* and the humbly voiced repetition of "there is no god but God" as if at a funeral, during which this phrase is repeated after the funeral prayers are finished and while the deceased is being carried to the cemetery. These two cultural sources were the sole ones that helped demonstrators in Damascus to perform a dissident act, likely for the first time in their lives. This is why the act of protesting in this context is not only an act of objecting but one of emancipating the body itself.

The performances of demonstrators developed over time and primarily drew on Syrian cultural identity and folk culture. The geographic location of a demonstration could be deduced from the revolutionary performance, for example by the inclusion of *ʿArāḍa, Dabka* in Damascus and *Dabka Jūfiyya* in cities like Daraa. Demonstrations in Daraa, for example, took a different form to those in other cities. The rows of demonstrators were circular, almost resembling a wedding more than a protest, and activities like *Zajal* and *ʿAtābā* were practiced, in addition to reciting poetry and clapping in a way typical of *Ḥūrānī* culture.[342] Another source of inspiration for demonstrators was visible in Homs, where demonstrators used soccer stadium techniques in demonstrations, such as performing a wave, and certain soccer slogans were reused in demonstrations.[343] Although the revolutionary demonstrations had highly innovative and culturally based aspects, the revolutionary performance was shaped over weeks and months. The cultural characteristics noted above were present during the later months of demonstrations against the regime, once people had had the time to think about the best way(s) to demonstrate.

Comparing the Regime and Revolutionary Performances

These cultural practices were less noticeable in pro-regime street performances. To trace the nature of the performances of regime marches in the streets and squares, I examined three events that caused the regime to call for large marches in Damascus. These marches centered on, first, burning the Israeli flag on December 2, 2011; second, receiving the Arab League delegation on October 26, 2011; and third, the Million People March on October 12, 2011. These three examples highlight the practices used by the regime to form its marches during protests. Visually, all the videos of these marches show photos of Bashar al-Assad everywhere—on shirts, flags, banners, streets and anything that could hold a visual product (see photos below; Figures 14 to 17). In terms of music, despite the fact that the slogans were taken from different occasions, they are almost the same, to the extent that it is impossible to differentiate between the marches or state when they occurred. Such marches appear to be rituals in praise of Assad. The aim of such performances was not to deliver a message, but to control the people and demand their obedience. This is why it seemed to be unimportant to the regime for the performance to have any meaning other than to send a message to international observers that the regime still controlled Syria.

342 *Zajal* and *ʿAtābā* are types of traditional art that are constructed on colloquial poetry mostly known in Syria, Lebanon, and Palestine but also performed in other Arab countries.
343 *Al Jazeera*, "*Sūriyā Laḥn al-Amal: al-Qiṣṣa al-Kāmila*" [Syria, the Hope Melody: The Complete Story], uploaded April 21, 2014, accessed July 18, 2023, https://www.youtube.com/watch?v=wKuQA-nrSnDY.

Figure 14: Pro-regime march in Damascus.[344]

The slogans used in these marches included:

> الله سورية بشار وبس"، "بالروح بالدم نفديك يا بشار"، "يا بشار يا حبيب دمّر دمّر تل أبيب"، "الجيش السوري يا عيوني ما تخلي ولا صهيوني
>
> [God, Syria and only Bashar; We redeem you, Bashar, with blood and soul; Oh, Bashar, destroy Tel Aviv; Oh, Syrian army, do not let any Zionist live]

These slogans add nothing to the context of the protests and provide nothing but praise for Bashar al-Assad, extolling him as Syria's only possible leader, despite the specific titles and events instigating the marches. This meant that the regime marches might change slightly from one to the next but generally had the same form and content even over a long period of time. The language of the regime is not a flexible one, and it rarely adapts to aspects of new situations and new political contexts. The only slogan seen is "we love you," meaning support for Assad as the leader in Syria forever.

---

344 *"Fīdyū Khaṭīr min Qalb Masīra Mu'ayyida fi-al-Sabi' Baḥrāt"* [An Important Video from the Heart of the Pro-Regime March in Saba' Baḥrāt], *Ali Alkadri* (YouTube channel), uploaded October 12, 2011, accessed June 1, 2020, https://www.youtube.com/watch?v=ibSBM5PAuVU (no longer available).

Figure 15: Pro-regime march in Damascus.[345]

The photos below (Figures 16 and 17) show visually how the revolutionary performance in the streets developed, beginning with normal gatherings and protests against the regime during the first few months of protests. The movement and actions of people resembled the way that regime supporters walked during marches. The act of simply walking and the presence of Syrian flags at the beginning were the essential elements of the early protests in Syria. Very quickly, however, the protests began to include speakers, banners, and even special types of dances and new body movements.

Figure 16 shows the similarities between the practices of regime supporters and the pro-revolutionaries in their way of walking and the presence of Syrian flags. The visual signs are also very similar to those of pro-regime marches, other than the images of Assad displayed in the latter. Figure 17, meanwhile, shows a different revolutionary performance with protestors in rows, putting their hands on each other's shoulders and jumping up and down as though they were in a soccer stadium supporting their favorite club. This arrangement of people is also similar to how people are arranged when they pray in a mosque, giving the impression of solidarity against the regime. The banners are also hung as if for a live broadcast for soccer fans. In addition to this, the protestors interweave their arms and shoulders in a very similar way to the folk dance of *Dabka Jūfiyyah* in Daraa where

345 "*Masīra Malyūniyya Jadīda Tastaqbil Wafd al-Jamiʿa al-ʿArabiyya*" [A Million People March for Receiving the Arab League Delegation], *Talal zena* (YouTube channel), uploaded October 27, 2011, accessed July 18, 2023, https://www.youtube.com/watch?v=ZgQhEyPchGk.

Figure 16: An early anti-regime demonstration in Daraa.[346]

the demonstration took place. Another element is the way in which the footage is taken: from behind, in order to reduce the chances of anyone being identified by showing their faces.

Figure 17: An anti-regime demonstration in Daraa in December 2012.[347]

---

346 *"al-Thawra al-Sūriyya Muẓāharāt Madīnat Dāriyā bi-Jawda ʿĀliyya"* [The Syrian Revolution, the Demonstrations of Daryah in High Quality], *Nezar Darayya* (YouTube channel), uploaded June 1, 2020, accessed July 18, 2023, https://www.youtube.com/watch?v=5Nad7rETYLA.

347 *Sham News Network, "Shām, Darʿā, Naṣīb, Muẓāharāt al-Aḥrār Nuṣratan li-Arḍ al-Buṭūla Rāʾiʿa 25 12 2011"* [Shām News Network, Daraa, Nasseb, Demonstrations for Supporting the Land of

The Revolutionary Language: A Break or Continuation?

The Syrian public space post-March 2011 was shaken up as new powers emerged to change the general scene of the consumption and distribution of symbolic products. This was apparent across all of the types of products discussed previously, including musical and visual materials, and extended to the language used in the streets, including slogans, *bakhkh*, banners, and the performance of each one. Chapter 1 traces the elements that helped the Assads control Syria symbolically, from the *coup d'état* of 1970 through to Bashar al-Assad's inheritance of power in 2000. Chapters 2 and 3 look at the same elements that helped the Assad regimes control Syria symbolically but from the revolutionary side. The first to be attacked by the revolutionary side was the symbolic domination, as seen in the subversion of the entire symbolic heritage of the regime. For example, while the regime fought for forty years to prevent free newspapers, magazines and alternative media outlets, the revolutionary side immersed the public space with many of these in just the first few months. Instead of the Baath popular organizations, there were many political movements, possibly briefed by *tansīqiyya*. The visual domination that the Assads had worked on for decades was destroyed, and the public space took on an alternative appearance. In addition, an active wave of ebooks helped to fill the gap left by publishers in Syria, represented by new publishers and the easier circulation of ebooks among Syrians, which affected everyday Syrian language through the creation of new terms.

All of these elements helped the regime and pro-revolutionaries change the dynamics and distribution of symbolic products to win over supporters from the other side. These products are represented by music, slogans, *bakhkh*, and performance. The metaphorical map drawn in these last two chapters reveals similarities and differences between the dominant and opposing discourses. The similar and different products and aspects of performance ultimately leads the research to consider one inquiry regarding this matter. After demonstrating different types of the regime language and that of the revolution, the question of whether the revolutionary language constitutes a break from or a continuation of the regime language can be considered. The importance of this question runs through both chapters and indicates that, in addition to attempts to subvert the products of each side through imitation and recreation, there is more to be said about the imitation and learning strategies and practices of each side. I argue that the revolutionary discourse was born decades ago from that of the regime and, over time, became the

Champions, Wonderful], uploaded December 25, 2011, accessed July 18, 2023, https://www.youtube.com/watch?v=0VXK7Jb_5I8&t=1s.

opposing one. This act of splitting from the dominant discourse started after 2011, moving it from an internal opposing discourse to a separate one. In saying that the opposing discourse was part of the dominant one, I do not mean that the opposing discourse used identical instruments, but rather that it had similarities that prove it was internal to the dominant one until clearly separating from it. This period of splitting would not have happened without imitating the performance practices of the regime, including in music, slogans, and performance.

The first aspect was seen in (re)productions of similar songs on both sides with the aim of subverting them. This practice led the pro-revolutionaries to produce covers of decades-old regime songs like "Syria My Beloved," and the revolutionary song "Yā Ḥīf," which were subsequently re-released several times by regime supporters. The similarity can be seen not only in producing the songs and covers but also in how the pro-revolutionary language followed the same path as that used by the regime to produce its songs.

The second musical aspect was the quasi- and would-be national anthems. In addition to the gradual use of colloquial language in producing songs, on the part of both pro-revolution and pro-regime factions, the Assad regime had released three other quasi-national anthems in addition to the official one since the Baath Party came to power: the anthems of the Baath Party, the Youth Union, and the Baath Vanguard.[348] Similarly, over a two-year period, the revolution produced several national anthems.

The third aspect is seen in the way in which the revolutionary songs imitated the regime songs at the level of language. A graph of all the compiled songs released in the period from March 23, 2011 to March 23, 2012 categorizing them as sung in MSA or colloquial Syrian,[349] shows that over the course of this year, the revolutionary songs developed in the same way as the nationalist songs of the regime over the previous decades of Assad's rule: in the first few months, MSA was used more than colloquial Syrian, but gradually the latter took over from MSA in the revolutionary discourse.

The fourth similar aspect was the regime and revolutionary performances, including their visual appearance. Although the revolutionary performance of demonstrations was unique and creative in most cases, it initially resembled that of the regime. Figures 8 and 9, for example, compare a regime and revolutionary

348 For more information on Baath national anthems, please see Chapter 1.

349 This graph was created using a small database of 315 existing revolutionary songs on YouTube and ordered according to the date uploaded and interviews with some of the singers to ascertain the release date of the songs. See appendix for checking the songs and to listen to them, please visit the following link on Creative Memory of the Syrian Revolution: https://creativememory.org/ar/collections/eylaf-bader-eddin-songs.

activity: both were engaged in an unspoken competition to display the longest Syrian flag. Without the photo caption, it would be impossible to tell which photo was of the regime or revolutionary activity. Therefore, it is clear that the early revolutionary performances originated in those of the regime. In addition to this, both sides used the same techniques (in this case, releasing balloons) to send a similar but different messages.

The fifth aspect was slogans. Both sides co-opted, imitated, or recreated slogans from the other, but only the most influential symbolic products were recreated. This is shown in the three groups of slogans that I describe in this chapter, with the tables showing the mutually influential relations between the old regime and new revolutionary slogans. The most influential slogans were the ones that protestors copied from the regime. It is important to note that copying the other side's slogan not only meant imitating its words or form, but also its performance as well.

The final aspect was the other revolutionary terms that came into sudden, widespread use after their meanings were subverted or changed. Despite the presence of newly coined terms, the ones most commonly used were taken from the regime discourse or Assad's speeches, even though the aim of reusing these terms was to ridicule and subvert the regime language.

Examining these six aspects confirms that the relationship between the revolutionary language and that of the regime was neither one of total rupture nor of total continuation, but represents the long journey of development of the opposing discourse, which took four decades to take the shape that was seen in 2011–2. It was not a language composed in a day, but rather the result of long-term social, political, cultural, and economic processes with contradictory goals: domination (for the regime) and emancipation (for the revolution). The revolutionary language began as an internal discourse to the dominant discourse of the regime and, over time, separated and created itself using the linguistic realm and practices of the regime on which it was based.

Chapter 4
Are They Being Heard?

.لا تستهن بالكلام، تحت غطائه، تدور في العالم كله، أقدار الدول والبشر
فوّاز حدّاد
من رواية "المترجم الخائن"

"Don't underestimate speech, for beneath it simmers the fates of states and peoples."

Fawwāz Ḥaddād, *al-Mutarjim al-Khā'in* [*The Treacherous Translator*]

The previous chapters illustrate how the language of the revolution was a thick language, the understanding of which is dependent on different layers of meanings, connected to multiple contexts, histories, discourses, and narratives. While a simple slogan like "Hafez, curse your soul" might seem very clear in 2011, a close reading of it takes us into the dense history of Syria, and it is given another layer of meaning by the 2011 protests. What happens to this type of language—including slogans, *bakhkh*, and banners—when it is translated into English? How has it been translated and by whom, and is it possible to convey all of these complicated layers in translation? This chapter looks at translation(s) through different approaches and two different layers or acts of translation. The first of these is the translation of revolutionary products by local and external agents, produced for and received by Western audiences, and represented by three books: *A Woman in the Crossfire: Diaries of the Syrian Revolution* by Samar Yazbek; *Syria Speaks: Art and Culture from the Frontline*, edited by Malu Halasa, Zaher Omareen, and Nawara Mahfoud; and *The Story of a Place: The Story of a People*, by Creative Memory of the Syrian Revolution.[350] These are the only three books to have been produced as a translation of the artistic and revolutionary languages from a published Arabic source language text. The second act of translation I will consider is by local agents, showing how Syrians (activists, bloggers, and academics) perceive Syria and the language of the revolution and present it in English. This is represented by initiatives such as the Free Syrian Translators, *Mundassa*, and other individual projects. This approach of examining different layers of translation sheds light on the sociology of translation by considering the translation market, which is dependent on many factors for its existence and distribution. These factors, as Johan Heilbron and Gisèle Sapiro

350 Creative Memory of the Syrian Revolution does not use a definite article "the" in its name in order to show it is not *the* only archive or collection of the revolution; it is merely *an* archive or just one collection. Their insistence on this is part of their strong belief in the pluralism of archives and interpretations of the Syrian revolution and its memory/memories.

https://doi.org/10.1515/9783110767698-005

explain, are "the international field and its structuring in cultural exchange, the economic and political constraints that influence these exchanges, and the agents of intermediation and the processes of importing and receiving in the recipient country."[351] The materials and translations used in this chapter are open-ended in that they necessarily begin with the revolutionary events of 2011 and extend through 2019. It is impossible to limit the timeframe of the act of translation, since new translations may be produced during, or after, the writing of this chapter.

Discourse(s) and Narrative(s)

The existence of different levels of translation, different agents, and different audiences obliges this study to draw on the concept of dominant and dominated discourses in translation, combined as I will show with narrative theory. This chapter draws on the concept of discourse to interpret the translations and books. Michel Foucault defines discourse as:

> a group of statements in so far as they belong to the same discursive formation. . . [Discourse] is made up of a limited number of statements for which a group of conditions of existence can be defined. Discourse in this sense is not an ideal, timeless form. It is, from beginning to end, historical—a fragment of history [. . .] posing its own limits, its divisions, its transformations, the specific modes of its temporality.[352]

A collection of statements thus forms a discourse by sharing a description or act of telling or by providing information on the same subject. Because the described act of telling creates shared attributes for the subject, these statements then create a discursive formation that can exist in multiple media, including speeches, narratives, research providing background information or analysis and, in our case, translations—as the act of importation or exportation.[353] The relationship between these elements (translation as importation and translation as exportation) is a dialectical one: each element resists and opposes the others because each is written from within a specific discourse, for a specific audience, and by different agents. One characteristic of discourse is pre-existing repertoires; we do not use our own words but instead reproduce what is already present in our competences, using

351 Johan Heilbron and Gisèle Sapiro, "Outline for a Sociology of Translation: Current Issues and Future Prospects," in *Constructing a Sociology of Translation*, eds. Michaela Wolf and Alexandra Fukari (Amsterdam: John Benjamins, 2007), 93–107 (93).

352 Michel Foucault. *The Archeology of Knowledge*, trans. A. M. Sheridan Smith (New York: Pantheon Books, 1972), 117.

353 The act of translation as importation or exportation depends on whether the language is dominant or dominated, and will be further explored in this chapter.

words, stories, terms, and expressions from existent repertoires. In other words, what we say about and how we describe Syria has been "already said."[354] For that reason, we are not free to write about this subject without using the existing competences on Syria and the narratives—Arabic and English—that shape our discourse. In this sense, we are using one another's statements when we discuss and create our perceptions of this topic and subject. This requires us to consider the narratives of Syria that have been created by both activists and translators. Are narratives translated by activist translators and translators from Arabic into English similar or different, given that the presence of two languages means (at least) two discourses? To provide a counterpoint to Foucault's "already-said," I use Roland Barthes' perception of text and combine it with narratives and discourse. In Barthes' words, "the text is a fabric of quotations, resulting from a thousand sources of culture [. . .] the writer can only imitate an ever anterior, never original gesture, his sole power is to mingle writings, to counter some by others."[355] In what ways can translation represent a discourse and fill in the gaps of what is translated and how? Moreover, as Ngugi Wa Thiong'o notes:

> Over the years, I have come to realise more and more that work, any work, even literary creative work, is not the result of any individual genius but the result of a collective effort. There are so many inputs in the actual formation of an image, an idea, a line of argument and even sometimes the formal arrangement. The very words we use are a product of a collective history.[356]

It is thus important to consider narrative theory in combination with the translational act and discourse, since narratives are what shape our perceptions of the reality of our world. Narratives are defined as "stories that we come to subscribe to—believe in or at least contemplate as potentially valid—and that therefore shape our behavior towards other people and the events in which we are embedded."[357] These groups of stories structure a particular way of perceiving the world around us that consists of stories embedded in ourselves. This narrative system accordingly indirectly urges us to reproduce similar narratives in much the same way as discourse functions. Drawing on the theory of narratives and the influence of daily narratives in translation shows us not only two different narratives from different discourses, but also illustrates the translator's choices in selecting between alterna-

354 Foucault, *Archeology of Knowledge*, 25
355 Roland Barthes, *The Rustle of Language*, trans. Richard Howard (Berkeley: University of California, 1989), 53.
356 Ngugi Wa Thiong'o, *Decolonizing the Mind: The Politics of Language in African Literature* (London: Heinemann, 1986), x–xi.
357 Mona Baker, "Reframing Conflict in Translation," in *Critical Readings in Translation Studies*, ed. Mona Baker (London: Routledge, 2010), 117.

tive translations of this narrative into a new culture and language. When a translator chooses a specific term for another in the source language text, they are aware that this choice is shaped by discourses that use different repertoires. One example of this, somewhat removed from the events of 2011 but relevant to Syria, is how to translate the phrases "Ḥarb Tishrīn al-Taḥrīriyya" [October War of Liberation], "Ḥarb al-ʿĀshir min Ramaḍān" [10th of Ramadan War], or "Milḥimit Yum Kipur" [Yom Kippur War] from Arabic discourse into English. The translator will be well aware that these are three different terms each representing the narrative of a particular country or state. The first, for example, has been used by the Syrian regime to indicate that it liberated the governate of Qunayīṭerah; the second emphasizes the holiness of the month of Ramaḍān in the Egyptian context; and the third is the Israeli name for this war, which highlights that it happened on this religious day. Translating this war into English, regardless of which term is used, illustrates the narrative(s) in which a translator has chosen to inscribe themself. By unpicking the cultural and political dimensions of the term in three different countries, this example highlights how the act of translation goes beyond questions of domestication and foreignization.[358] No matter what the translator chooses, their choice will reflect the narratives and discourse of the target language and the narratives and discourse that dominate the translator. With the example above, my aim is not to prove which label is "right," but rather to show that in the context of such political phenomena, the act of naming is impossible without drawing on our narratives as translators. In this sense, translation works hand-in-hand to either confirm the previously constructed discourse of a domesticated translation, or to provide an opposing view through a foreignized translation that seeks to provide an additional cultural dimension and external narratives.

Hence, in the Arabic discourse, choosing to describe what happened in Syria in 2011 as a "revolution" represents the revolutionary discourse and narratives, whereas labeling it a "crisis," "sedition," or "conspiracy" refers to the discourse and narratives of the regime. In this case, the well-informed translator will be fully aware that their choice of term tips the balance toward a specific discourse in Arabic. Some translators might strive for neutrality by choosing different labels, creating a new group of terms not used in Arabic such as "civil war," "clashes," "unrest," "rebellion," and so on. This third way (neither labeling Syria as a revolution nor a crisis) is not necessarily as neutral as it claims to be, but it does create a linguistic space for new narratives and stories about the same political phenomena in Arabic and is also able to reuse part of the same repertoire of the English discourse. Here I will not make a case for the "right" label for Syria in 2011, but I will

358 Lawrence Venuti, *The Translator's Invisibility: A History of Translation* (London: Routledge. 1995).

argue that different names and labels activate different narratives and discourses in treating Syria as a topic, subject, event, and a place.

Translation as Importation and Exportation

Given the importance of narratives and discourses, the act of translation is done from various perspectives depending on other elements related to the sociology of translation and its market. As Heilbron and Sapiro explain, there are three dimensions of "the transnational circulation of goods; firstly, the structure of the field of international cultural exchanges; secondly, the type of constraints—political and economic—that influence these exchanges; and thirdly, the agents of intermediation and the processes of importing and receiving in the recipient country."[359] In this chapter, these three elements will be analyzed as follows: first, by broadly mapping the translation market—what is translated and what books are available; second, by focusing on translation as a good or commodity with economic value, in the shape of translational projects funded by external organizations and foreign agents (i.e., *Syria Speaks*, *A Woman in the Crossfire*, and *The Story of a Place*); and third, by looking at participants in the translation process. The latter may include public organizations, such as cultural centers and universities, but translations can also be produced without state or institutional funding, as will be seen in the examples of the Free Syrian Translators, *Mundassa*, and individual initiatives. In line with these three elements, this chapter is divided into three main sections. The first section, entitled "Translating through absence," provides a general overview of the translation market from Arabic into English, presenting recent statistics for translations and new data gathered for this study on what is translated (topics and themes), when, and for what audiences.

The second and third sections are dedicated to translation as importation, and translation as exportation respectively. These two acts of translation are based on the characteristics of each language, specifically its classification as either a dominant language (English) or a dominated one (Arabic). To put it differently, "dominant countries export their cultural products [translations] widely and translate little into their languages, the dominated countries 'export' little and 'import' a lot of foreign books, principally by translation."[360] The second section thus looks at how revolutionary works have been translated into English (a dominant language) as represented by its agents, experts, and publishers. In the context of my research,

359 Heilbron and Sapiro, "Sociology of Translation," 93.
360 Heilbron and Sapiro, "Sociology of Translation," 96.

I refer to this as importation: the act of translation is done by foreign agents, organizations, and translators with the goal of importing the translation into the foreign market, and the specific work to be translated is decided based on the needs of the English book market. In this case, English—as a language—has imported translations without any Syrian government funding. Thus, the importation of a translation into English is produced by English agents for Syrian texts.[361] The second type of translation, termed translation as exportation, is done either by Arabic agents who may be defined as activists, bloggers, translators, or through local, unfunded initiatives.

In addition to those mentioned above, the analysis of sample translations will be accompanied by a "thick translation" that illustrates additional possibilities for the translated material. Thick translation, as a concept and tool, is used to dissect translations and shed light on the plurality of meanings in the source language text, as for example in Kwame Appiah's thick translation of the proverbs.[362] The cultural dimensions and meanings that are rarely transferred from the source in translation foster the use of this theoretical concept. Thick translation empowers the source language text by localizing and structuring translation, as Martha Pui You Cheung shows with "*xin*, *da*, and *ya*."[363] Localizing and structuring helps a translation to "bring depth and breadth to the representation of culture, even if the representation can never be total, never complete."[364] This process makes a reader a "traveler not a tourist"[365] while reading the translated text. In this study, I have sought to undertake a thick translation by paraphrasing the language of the revolution within offering a parallel English text and a description of it in Arabic in the previous chapters.

361 Regardless of whether or not the translators are native speakers, here I use the term "English translators" to refer to translators working from within the English discourse.
362 Kwame Anthony Appiah, "Thick Translation," *Callaloo* 16, no. 4 (1993): 808–19.
363 Martha Pui Yiu Cheung, "On Thick Translation as a Mode of Cultural Representation," in *Across Boundaries: International Perspectives on Translation Studies*, eds. Dorothy Kenny and Kyongjoo Ryou (Newcastle: Cambridge Scholars Publishing, 2007), 22–37 (23). In this article, Cheung shows how the translations of *xin*, *da*, and *ya* [literally translated as: faithful, clear, and graceful] do not serve the source language culture of these words but are a translation of English concepts in English, meaning that the English translation of these three Chinese words cuts off the root of the Chinese words and represents English concepts instead. Cheung uses thick translation to activate the links between the Chinese and English.
364 Cheung, "On Thick Translation," 32.
365 Cheung, "On Thick Translation," 32.

Translating Through Absence: Translating Literature

While it might seem unusual to describe the state of translation as one of absence in a time in which cultures and nations meet through translation, this is nevertheless the most appropriate description of the current translation market into English. Numerous studies and articles have been written about the (absence of) translations of Arabic fiction into European markets and languages. In their study of translation from Arabic to Spanish, Maria Luz and Gonzalo Fernández Parrilla deplore the lack of books translated from Arabic in the Spanish translation market.[366] In a 2000 study on the German market, Hartmut Fändrich lamented that "the sale of any well-written and highly interesting novel of contemporary Arabic literature may not be sufficient to even cover the costs of production," and found that public interest in translation essentially ends with the *Arabian Nights*.[367] Two decades after Fändrich's work, little seems to have changed: in her recent 2021 study of translations of Arabic into German, Sandra Hetzel states that "Arabic literature in translation continues to play a marginal role in the German-speaking literary market."[368] In France, meanwhile, the proportion of books translated into French from Arabic in 2018 was "less than 1% of all translations, disregarding the continuous immigration of perhaps five or six millions from the Maghreb countries in a total population of 68 millions," as shown by a *Diversity Report*.[369] In his major case study of the translation of Arabic literature into French, Richard Jacquemond writes: "Arabic production appears in the French market as the most translated non-European literary production, with an average of 10 to 20 translations published each year in the 1980s (out of 2,000 to 3,000 translated books)."[370] Focusing on the Italian market, Monica Ruocco's survey of books translated from Arabic into Italian shows that

---

366 Maria Luz Comendador, Luis Miguel Canada, and Miguel Hernando de Larramendi, "The Translation of Contemporary Arabic Literature into Spanish," *Yearbook of Comparative and General Literature* 48, (2000): 115–25; Gonzalo Fernández Parrilla, "Translating Modern Arabic Literature into Spanish," *Middle Eastern Literature* 16, no. 1 (2013).

367 Hartmut Fändrich, "Viewing 'the Orient' and Translating its Literature in the Shadow of the *Arabian Nights*," *Yearbook of Comparative and General Literature* 48 (2000): 95–106.

368 Sandra Hetzl, "Translation of Arabic Literature in German-Speaking Countries 2010–2020," ed. Alexandra Büchler, 5, https://www.lit-across-frontiers.org/wp-content/uploads/2013/03/GERMANY-Arabic-Lit-Translation-2021.pdf.

369 Rüdiger Wischenbart et al., *Diversity Report 2018: Trends in Literary Translation in Europe* (Vienna: CulturalTransfers.org, 2019), accessed July 18, 2023, https://bit.ly/36kF2nr.

370 Richard Jacquemond, "Translation and Cultural Hegemony: The Case of French-Arabic Translation," in *Rethinking Translation: Discourse, Subjectivity, Ideology*, ed. Lawrence Venuti (London: Routledge, 1992), 139–58 (146).

only 33 such translations were published between 1987 and 1997.[371] The same issue is apparent in Sweden, as Marina Stagh has shown.[372]

In the English translation market, things are much the same. There have been a number of studies—both old and new—discussing English translations of Arabic literature. In 1990, Edward Said noted that: "Arabic remains relatively unknown and unread in the West, for reasons that are unique, and remarkable, at a time when tastes here [in the United States] for the non-European are more developed than ever before."[373] Ten years later, Peter Clark added that "Arabic literature is still largely the preserve of the Middle Eastern specialists. It has not come out of the ghetto."[374] Hosam Aboul-Ela agreed, writing eleven years after Said's article that: "[t]he American literary marketplace may be more disinterested in and ignorant of contemporary Arabic literature today than it was eleven years ago."[375]

Salih J. Altoma's *Modern Arabic Literature in Translation: A Companion* sheds further light on the lack of Arabic translations into English.[376] Altoma divides contemporary translation from Arabic into English into three periods.[377] The first period saw the translation of ten literary works, and lasted from 1947 to 1967. The second period was characterized by giving higher importance to modern Arabic literature, and lasted from 1968 to 1988. During this time, most of the translated books were by Egyptian writers, with the exceptions of Tawfiq Awwad (Palestine), Halim Barakat (Lebanon), Abd al-Raḥmān Munif (Jordan/Saudi Arabia), Tayyib Salih (Sudan), and Muhammad Shukri (Morocco).[378] The period resulted in "some sixty novels and forty anthologies of short stories" being translated.[379] The third and final period ran from 1988 to 2003, during which time there was a higher number of translations into English, especially after Naguib Mahfouz was awarded the Nobel Prize in Literature.[380]

---

371 Monica Ruocco, "A Survey of Translation and Studies on Arabic Literature Published in Italy (1987–1997)," *Arabic & Middle Eastern Literatures* 3 no. 1 (2000): 63–75, https://doi.org/10.1080/13666160008718230.

372 Marina Stagh, "The Translation of Contemporary Arabic Literature into Swedish," *Yearbook of Comparative and General Literature* 48 (2000): 107–14.

373 Edward W. Said, "Embargoed Literature," *The Nation* 251 no. 8, September 17, 1990.

374 Peter Clark, *Arabic Literature Unveiled: Challenges of Translation* (Durham: University of Durham, Centre for Middle Eastern and Islamic Studies, 2000), 13.

375 Hosam Aboul-Ela, "Challenging the Embargo: Arabic Literature in the US Market," *Middle East Report* 219 (2001): 42–4.

376 Salih J. Altoma, *Modern Arabic Literature in Translation: A Companion* (London: Saqi, 2005), 54.

377 Altoma, *Modern Arabic Literature*, 54.

378 Altoma, *Modern Arabic Literature*, 55.

379 Altoma, *Modern Arabic Literature*, 55.

380 Altoma, *Modern Arabic Literature*, 55.

Bringing us closer to the present day, a *Literature Across Frontiers* report on Arabic translations shows that "close to 596 titles in the category of literature (fiction, short stories, essays, memoirs, poetry) were published in the period of 2010 to 2020, which are distributed in the UK and Ireland."[381] Despite the fact that publications of translations have increased over time, the numbers indicate a lack of publishing interest in translating books from Arabic, since less than 1% of the total number of books published annually in Arabic—an estimated 17,000 titles per year—are translated into English.[382] This relationship between languages—the source language of a text and target language of a text—can be explained at two levels.[383] The first depends on the value or "literariness" of the language in "the world republic of letters," according to Pascale Casanova, while the second is dependent on what Heilbron calls the "world system of translation."

The World Republic of Letters

Language has a value that imposes itself in what Casanova calls "the world republic of letters."[384] The value that Casanova terms "literariness" is responsible for assigning importance to a specific language based on "the political sociology of language [that] studies the usage and relative 'value' of languages only in political and economic terms."[385] In this sense, some types of languages, by "virtue of the prestige of the texts written in them, are reputed to be more literary than others, to embody literature."[386] This positions Arabic—as a language and literature—as having less merit and does not show its value, not only because of the unfair law of competition and struggle between languages and their literatures, but also because of the need for "the existence of a more or less extensive professional milieu."[387] In general, Arab countries do not have a professional milieu providing literary salons

381 Alexandra Büchler and Abdel-Wahab Khalifa, "Translation of Arabic Literature into English in the United Kingdom and Ireland 2010–2020," *Literature Across Frontiers*, September 2011, accessed July 18, 2023, https://www.lit-across-frontiers.org/wp-content/uploads/2013/03/UK-IRELAND-Arabic-Lit-Translation-2021-1.pdf.
382 *Al-Bayān*, "Al-Nashr fī al-Waṭan al-ʿArabī. Waqiʿ Mutaraddī wa-Mustaqbal Ghāmiḍ" [Publishing in the Arab World: A Deteriorating Situation and an Ambiguous Future], December 17, 2010, accessed July 18, 2023, https://www.albayan.ae/paths/art/2010-12-17-1.3417.
383 By language here I mean not only the literal language but also the whole cultural complex behind it.
384 Pascale Casanova, *The World Republic of Letters* (Cambridge: Harvard University Press, 2004), 4.
385 Casanova, *World Republic.*
386 Casanova, *World Republic*, 17.
387 Casanova, *World Republic*, 15.

or ensuring freedom of speech—a natural consequence of authoritarian regimes. Some of the Gulf states have established initiatives and study centers for translation, such as the Kalima Project, the Doha Institute for Graduate Studies, and the Hamad Bin Khalifa University Press (initially managed by Bloomsbury), as well as organizations that provide funding to translate highly-awarded Arabic literature into English.[388] These have, however, had very little effect because of the nature of the political and economic core of countries that is required for the freedom and creativity of their citizens. It is also impossible to exclude the impact of the Western view of Arabic literature, particularly with regard to publishers that are highly selective when dealing with Arabic culture. Peter Clark, for example, describes how one publisher refused to accept short stories by Syrian writer ʿAbd al-Salām al-ʿUjaylī for translation, saying: "There are three things wrong with the idea. He's male. He's old. And he writes short stories. Can you find a young female novelist?"[389] Clark's account may provide an explanation for the policies of certain publishing houses, which select works to be translated from Arabic not based on their literariness but rather on the basis of clearly ageist and/or sexist demands. In addition to this, financial considerations might put pressure on publishers to find an Arab writer to write directly in one of the European languages without the need for translation, which is cheaper and quicker

The World System of Translation

Another explanation for the problem, complementing Casanova's analyses, can be found in Heilbron's "world system of translation"; an institution that is "hierarchical and [. . .] comprises central, semi-peripheral and peripheral languages."[390] Arabic is a peripheral language in this system as it, along with other languages like Chinese, Japanese and Portuguese, comprises "less than one per cent of the world

388 Kalima Project, "What is Kalima," official website, accessed July 18, 2023, http://kalima.ae/en/default.aspx; Doha Institute for Graduate Studies, "Overview," official website, accessed July 18, 2023, https://www.dohainstitute.edu.qa/EN/About/DohaInstitute/Pages/Default.aspx; Hamad Bin Khalifa University and Press, "Homepage," accessed July 18, 2023, https://hbkupress.com/en. The Kalima project is a translation project located in the UAE for Arabic/English–English/Arabic translation; the Doha Institute for Graduate Studies is a private, nonprofit institution for educational, research, and public service purposes in Qatar; and the Hamad Bin Khalifa University and Press, also based in Qatar, seeks to promote "literature, literacy, scholarship, discovery and learning."
389 Clark, *Arabic Literature Unveiled*, 63.
390 Johan Heilbron, "Translation as a Cultural World System," *Perspectives: Studies in Translatology* 14 (2000): 9–26.

market."[391] This tiny percentage is quite shocking, given the huge number of speakers of so-called "peripheral languages" regionally and globally, not to mention the global impact of these countries, cultures, and migrants. There are, for example, 423 million Arabic speakers in the Arab world alone,[392] and according to the International Organization for Migration, approximately 6 million Arabs living in Europe, excluding Arab descendants.[393] Millions more live in Canada and the United States. And yet the rankings of central, semi-peripheral, and peripheral languages rarely change. When they do, it is typically because of political reasons, such as the rapid decrease in translations from Russian as a direct effect of the division of the Soviet Union and its impact on Eastern Europe.[394]

While I have thus far described the situation and context of the market for Arabic translations to English, this has largely been in regard to translations of literature. While no one can deny the importance of translating literature, this study does not concern literary works and translation, but rather the translation of a political event into English. To the best to my knowledge, there are no statistics for translations of nonfiction works from Arabic into English. The available statistical data only show the number of translated fictional works from Arabic, with little to no coverage of translations of nonfiction.

Before moving on to consider nonfiction translations from Arabic into English, it is important to observe a new attitude to selecting literary works to translate from Arabic into English. This focus can be seen in the translation of literary works that reflect the reality of the Arab social and political world through literature. One type of literature that flourished after 2011 is testimonial literature, which is not a diary or novel, but rather a mixture of the two. In the case of Syria, the selection of such translations serves as a replacement for importing knowledge in Arabic, though it is important to note that imported testimonial literature that reflects the Syrian situation through literature is not synonymous with how Arab or Syrian intellectuals analyze the situation. This strategic act of publishing literature specifically on Syria after 2011 came about not only because of the needs of the book market but also as a result of the increased number of displaced Syrians in exile. This relationship created and affected a new Syrian cultural field in exile, in par-

391 Heilbron, "Translation as a Cultural World System," 14.

392 World Population Review, "Arab Countries 2020," accessed July 18, 2023, http://worldpopulationreview.com/countries/arab-countries/.

393 International Organization for Migration, *World Migration Report 2010 – The Future of Migration: Building Capacities for Change* (Geneva: International Organization for Migration, 2010), accessed July 18, 2023, https://bit.ly/2Y4JhQT.

394 Heilbron, "Translation as a Cultural World System," 15.

ticular through the increased demand for translating Syrian works.[395] In France, for example, "after 2011, Syrian literary works were translated more than before 2011."[396] This increase observed by Richard Jacquemond is really due to the specific setting of the Syrian cultural field. In contrast to France and Germany, where many Syrians fled, English-speaking countries like the UK and USA did not receive many Syrian refugees—due to a combination of geographic location and asylum policies—and this meant that they had no real impact on the artistic field in English and, with the exception of some individual initiatives, especially on translation into English.[397] Even in Germany, Syrian writer Yassin Haj Saleh has argued that the landscape of Syrian works translated into German "isn't always very interesting. It's survival literature," and has called for more translations of "wider things like philosophy, religion studies, culture, and sociology."[398]

To identify the precise number of translations of nonfiction works, or as Saleh calls it, "survival literature," within the framework of my study, nonfiction works translated from Arabic into English are the most salient to the understanding of knowledge production from Syria in Arabic and also on Syria in English. Due to the lack of any statistics for this field, I compiled information from a variety of sources in order to produce a dataset of nonfiction books translated from Arabic into English. I first used the website ArabLit to research any translations published from 2009 to March 2019. This provided me with the name of a few dozen publishers translating works from Arabic. I then examined their online catalogs and asked them for any official statistics for their translated works. I also used the catalog information to search the Amazon website for other publishers offering translations from Arabic into English, enabling me to locate 104 publishing houses producing translations

---

395 For more information about the new Syrian artistic field in exile, please see Simon Dubois, "Négocier son identité artistique dans l'exil. Les recompositions d'un paysage créatif syrien à Berlin," in *Migrations Société* 174, no. 4 (2018): 45–57.

396 Richard Jacquemond, *"al-Adab al-Filisṭīnī Mutarjaman ilā al-Firansiyya: Tārīkhuhu wa-Atharuhu"* [Palestinian Literature Translated into French: History and Impact], *Alif: Journal of Comparative Poetics*, no. 38 (2018): 94–119, https://www.jstor.org/stable/26496387?seq=1.

397 *Refugee Action*, "Facts About Refugees," accessed July 18, 2023, https://www.refugee-action.org.uk/about/facts-about-refugees/; *Statista*, "Syrian Refugee Arrivals in the United States from 2011 to 2019," accessed July 18, 2023, https://www.statista.com/statistics/742553/syrian-refugee-arrivals-us/; Marion MacGregor, "Germany: Refugee Numbers in Context," *Info Migrants*, June 20, 2019, accessed July 18, 2023, https://www.infomigrants.net/en/post/17641/germany-refugee-numbers-in-context. It is expected that by June 2020, the UK will have received 17,051 Syrians. The total number of Syrian refugees in the USA is 21,725 (for the entire period of 2011 to 2019). In contrast, Germany has 1.06 million refugees, half of whom are Syrian.

398 Mari Odoy and Yassin al-Haj Saleh: 'We, as Syrians, are Allowed to be 'Witnesses,'" in *Arablit & Arablit Quarterly*, series on Syrian writers living in Berlin, September 30, 2020, accessed July 18, 2023, https://arablit.org/2020/09/30/yassin-al-haj-saleh-we-as-syrians-are-allowed-to-be-witnesses/.

from Arabic into English.[399] After browsing publishing catalogs and emailing publishers to inquire if they had any official statistics or reports about their translation market, I created a list of all translated nonfiction works. Table 13 shows the number of nonfiction works published on an annual basis. Using these methods I identified a total of 643 fiction and nonfiction books translated from Arabic into English between 2009 and 2020. The year 2009 was chosen as a starting point as this was when the ArabLit website was launched, and beginning two years before the Arab revolution made it possible to exclude a sudden interest in revolutions and translations.

Table 13: Number of nonfiction and fiction books translated from Arabic into English by year.

Nonfiction books translated from Arabic into English by year		Fiction books/diaries/testimonies translated from Arabic into English by year
Year	**Number**	**Number**
2009	5	37
2010	5	29
2011	16	26
2012	6	25
2013	9	35
2014	9	38
2015	6	35
2016	11	42
2017	8	31
2018	2	40
2019	4	25

The table above shows the dearth of translations from Arabic into English for nonfiction books, with far fewer nonfiction works translated than literary works. The numbers are very small and illustrate an almost total lack of interest in translating nonfiction works for the English-speaking market. A closer reading of the titles indicates that these translations are primarily religious books or classics. For example, four of the five books translated in 2009 were written by Ibn Battuta (1304–69), Isḥāq Ibn Sulaymān al-Isrāʾīlī (c. 832– c. 932), Ibn Ṭufayl (1105–85), Aḥmad Ibn ʿAjība (1747–1809), and the final one was about Imam Aḥmad Sibaʿī (1905–84) in Mecca, followed by critical essays on his autobiography. The types of books translated did not change in 2010, when the five books translated were about Ibn Ḥanbal (780–855), al-Ghazālī (1058–1111), Ibn ʿAbd al-Ḥakam (800–871),

399 See Appendix 3.

Ibn Khallīkān (1211–82), and ʿAbd al-Jabbār (935–1024). Almost the same content and types of translated books can be seen for the years that followed. These figures and the themes of the nonfiction works translated from Arabic into English indicate that Western interest in Arabic translations is mostly in classical or religious books, with the almost total exclusion of contemporary nonfiction Arabic books. The almost complete absence of translated nonfiction books raises the question of how English books on Syria are written and explains the existence of just three books including a translation of the language of the Syrian revolution. How do these books translate such a thick and rich language, and on what level are they able to convey the complexities of the context of revolution? Below I examine translations of the revolutionary language by looking at two different acts of translation: translation of imports and exports of/to the dominant language, first through books translated by agents from the center, and second through translations done by activists working on the periphery.

Translation as An Act of Importation into the Dominant Language

In this section, I will show the marginal space—reflected by just three books—that has been assigned to translation as a direct act of transferring Syria in 2011 from Arabic into English. A number of other books that were published as translations but not published in Arabic were excluded from the study due to the absence of a published Arabic source text. What distinguishes this act is that it is done by external initiatives, and translations are published by English-language publishing houses with the cooperation of Syrian authors. The three translations I will examine are *A Woman in the Crossfire: Diaries of the Syrian Revolution* by Samar Yazbek; *Syria Speaks: Art and Culture from the Frontline*, edited by Malu Halasa, Zaher Omareen, and Nawara Mahfoud; and *The Story of a Place: The Story of a People*, by Creative Memory of the Syrian Revolution (CMSR).[400] Each book will be analyzed

400 Samar Yazbek, *A Woman in the Crossfire: Diaries of the Syrian Revolution*, trans. Max Weiss (London: Haus Publishing 2012); Samar Yazbek, *Taqātuʿ al-Nīrān: Yawmiyāt al-Intifāḍa al-Sūsriyya* [Crossfire: The Diary of the Syrian Revolution] (Beirut: Dar al-Adab, 2012); Malu Halasa, Zaher Omareen, and Nawara Mahfoud, eds., *Syria Speaks: Art and Culture from the Frontline* (London: Saqi Books, 2014); Zāhir ʿUmarīn, Mālū Hālāsā, and Nawwāra Mahfūḍ, eds., *Sūryā Tataḥaddath: al-Thaqāfa wa-l-Fann min ajl al-Ḥurriyya* [Syria Speaks: Culture and Art for Freedom] (Beirut: Dar El Saqi, 2014); Creative Memory of the Syrian Revolution (CMSR), *The Story of a Place, The Story of a People: The Beginnings of the Syrian Revolution (2011–2015)*, eds. Sana Yazaji, trans. Rana Mitri (Beirut: Friedrich-Ebert-Stiftung, 2017); CMSR, *Qiṣṣat Makān Qiṣṣat Insān: Bidāyāt al-Thawra al-Sūriyya 2011–2015*, ed. Sana Yazaji (Beirut: Friederich-Ebert-Stiftung, 2017).

using its translations, comparing the similarities and differences of both the Arabic and English texts. I will also combine the analysis with an act of thick description, highlighting where the existing translation fails or succeeds to show the multilayered text and interpret the materials.

A Woman in the Crossfire: Diaries of the Syrian Revolution

The first of these translations on the Syrian revolution, published in 2012, was written by Samar Yazbek and entitled *A Woman in the Crossfire: Diaries of the Syrian Revolution.* The Arabic title cautiously describes the events as an "uprising" or "Intifada," rather than a "revolution" as in English, showing that the publisher and translator were taking a stand regarding Syria in 2011.[401] Moreover, unlike the English, the Arabic title does not mention that the owner of these diaries is a woman. The original edition of the book in Arabic was released in December 2012, the English version in August 2012, the German one in February 2012, and the French one in April 2012.[402] The timing of these releases sheds light on the translation market and audience, and the strategy of garnering more attention to the writer by first releasing the book in foreign languages (English and German) before the original Arabic. This has a number of implications, the most important one being that although the book was originally written in Arabic, its audience is a Western one. This can be seen by comparing the reception of the Arabic and English versions on Goodreads: while the Arabic version has 231 ratings, with an average score of 3.57 of 5 stars,[403] the English one has 415 ratings and an average of 3.71 of 5 stars.[404] This suggests

---

401 The title in Arabic, *Taqāṭuʿ al-Nīrān: Min Yawmiyāt al-Intifāḍa al-Sūriyya,* literally translates as "Crossfire: From the Diaries of the Syrian Intifada."

402 According to the ISBN Search website, accessed July 18, 2023, https://isbnsearch.org/isbn/9789953892368; Samar Yazbek, *A Woman in the Crossfire: Diaries of the Syrian Revolution,* trans. Max Weis (London: Haus Publishing, 2012), Amazon page, accessed July 18, 2023, https://www.amazon.de/Woman-Crossfire-Diaries-Revolution-English-ebook/dp/B008VLJSK8/ref=sr_1_1?keywords=a+woman+in+crossfire&qid=1559217663&s=digital-text&sr=1-1; Samar Yazbek, *Schrei nach Freiheit. Bericht aus dem Inneren der syrischen Revolution,* trans. Larissa Bender (Germany: Verlag Nagel & Kimche AG, 2012), Amazon page, accessed July 18, 2023, https://www.amazon.de/Freiheit-Bericht-Inneren-syrischen-Revolution/dp/3312005310; Samar Yazbek, *Feux croisés: Journal de la revolution syrienne,* trans. Rania Samara (Paris: Buchet-Chastel, 2012).

403 *Goodreads,* "Editions—*Taqāṭu' al-Nīrān: Min Yawmiyyāt al-Intifāḍa al-Sūriyya*" [Editions—Crossfire: The diary of the Syrian intifada], by Samar Yazbek (Beirut: Dār al-Ādāb, 2012), accessed June 1, 2020, https://www.goodreads.com/work/editions/19243091?expanded=true.

404 *Goodreads,* "A Woman in the Crossfire: Diaries of the Syrian Revolution," by Samar Yazbek, trans. Max Weis (London: Haus Publishing, 2012), accessed June 1, 2020, https://www.goodreads.com/book/show/13591934-a-woman-in-the-crossfire.

that the English-speaking readership of the book is greater than its Arabic readership, an indication that the book—and especially its translation—are intended for English-speaking readers (though one limitation of this conclusion is the fact that the accessibility of Goodreads may vary more across the Arab-speaking world).

In an interview with English PEN, the organization that provided a grant to fund the translation of Yazbek's book, translator Max Weiss said that he selects books for translation "to make an impact [on] how the Anglophone readerships will encounter and comprehend the Arabic-Speaking World."[405] This provides support for the idea that the translator's choices are made in such a way as to improve the understanding of others. Weiss' translation is fascinating in its choice of words and expressions, and a high level of cooperation with the author herself was required to produce the work in English. Weiss explains in the interview that "the particularity of her story remained relatively obscure to me until I delved into the diaries at the level of literary translation and began to correspond with Samar on a personal level."[406]

In general, Weiss tries to offer a literal translation for the text. Most of the time, he provides eloquent alternative equivalences for the Arabic text. In some instances, however, the English wording could lead to a different understanding or interpretation compared to the original. This is of particular significance with regard to the revolutionary language, which cannot be translated word-for-word despite its simple structure. For example, Weiss provides notes to explain his translational choices. He may be the first to borrow the term *shabbīḥa* into English, which has generally been translated as "thugs," diminishing the long history of this term in Syria.[407] The way in which a work is translated does not affect the English text, provided the translation is understandable and the English reader does not have sufficient skill in Arabic to compare the two. However, as Lawrence Venuti argues, a transparent translation hides the translator but also hinders recognition of the cultural aspects of the source language text.[408] An act of foreignization, on the other hand, can call back the source language's text with all of its context. A readable, transparent translation might serve the monolingual reader, but distances them from the source language text. At the same time, the act of foreignization can operate in reverse, recalling a com-

---

405 Polly Roberts, "A Word from the Translator- Max Weiss," *English Pen*, July 30, 2013, accessed June 1, 2020, https://www.englishpen.org/translation/a-word-from-the-translator-max-weiss/ (no longer available). Max Weiss is an associate professor of History and Near Eastern Studies at Princeton University and an Arabic translator. He has translated many novels, including several from Syria, including Nihad Sirees's *The Silence and the Roar* (London: Pushkin Press, 2013), and *States of Passion* (London: Pushkin Press, 2018).

406 Roberts, "A Word from the Translator."

407 For more information on *shabbīḥa*, please see Chapters 1 and 3.

408 Lawrence Venuti, *The Translator's Invisibility: A History of Translation* (London: Routledge. 1995), 5.

petence that the reader knows and has negative associations with, as we will see in the examples of slogans from Yazbek's book that were translated into English. These samples were selected because they are slogans from Syrian demonstrations; I will trace their thick meanings to see how effectively the translation is able to transfer the multilayered interpretations of the revolutionary language.

Table 14: Selected patterns of revolutionary language in A Woman in the Crossfire.

Selected patterns of revolutionary language in *A Woman in the Crossfire*			
	Page (English, Arabic)	**English**	**Arabic**
1-	8, 16	Anyone who kills his own people is a traitor	خاين يلي بيقتل شعبو
2-	153, 181	Anyone who kills his own people is a traitor	يلي بيقتل شعبه خاين
3-	119, 144	Freedom, freedom! The Syrian people won't be insulted	حرية حرية الشعب السوري مابينهان
4-	173, 203	Allahu Akbar	الله اكبر [without any explanation]
5-	174, 203	Prince of Jableh	Not in the Arabic text
6-	190, 222	The Free Virgin Women of the Syrian Coast	حرائر الساحل السوري
7-	180, 210	No legitimacy for anyone who kills the people	لا شرعية لمن يقتل الشعب
8-	133, 160	(Clapping), Abu Hafez . . .Abu Hafez	أبو حافظ أبو حافظ (من دون وضع التصفيق في العربي)

Although word-for-word translation is often one of the safest methods of capturing meaning in both the source and target languages, it is not as effective with regard to slogans because of their multilayered nature. Slogans are characterized by their brevity, high level of symbolism, musicality, rhythm, and eloquence, making translation of this performative language a challenge. Revolutionary products such as slogans are multilayered and writerly texts. According to Samia Mehrez, these are the kinds of works "where the reader and translator is confronted with multiple undetermined signs and codes that challenge their expectations of narrative unity and call upon both [the reader and translator] to renounce the passive receptivity of the consumer and to embrace an engaged effort as producers of a text that continues to be written on several levels."[409] This challenge forces translators to negotiate between both cultures and linguistic units and improvise by creating a new translation for these thick contextual expressions and terms.

409 Samia Mehrez, "Introduction: Translating Revolution: An Open Text," in *Translating Egypt's Revolution: The Language of Tahrir*, 1–2, ed. Samia Mehrez (Cairo: AUC Press, 2012).

The translation of slogan 1 in Table 14 above, which generalizes the attribute of treason to *anyone* who kills a fellow Syrian, makes the English translation particularly open to multiple interpretations by English-speaking readers for whom this book provides their only window onto Syria in 2011. Anyone reading the book after 2013, when Syria became a proxy war country, might think that this slogan referred to all fighters—both pro- and anti-regime—but according to Yazbek's account, this slogan was chanted in April 2011, at a time when only one side of the conflict was killing Syrians: the regime. This is one reason why the actual meaning of this slogan is that the *regime* is a traitor because it kills people. A second reason for this interpretation is that there is a possessive pronoun in Arabic, "*sha'bū*," which means one's people; the only side that has a people or nation is the authority—the regime or Bashar al-Assad. A literal translation of this thicker meaning could be "anyone who belongs to a people and kills their people is a traitor." But further confirmation that the perpetrator of the act is the regime is provided by the position of "traitor" at the beginning of the slogan. It could thus be translated to confirm the regime is the murderer ("the regime is a traitor for killing Syrians"). Another possible translation might assume that "*yillī*" [which/who] refers to Bashar al-Assad, and read as "He who kills his people is a traitor." The use of "anyone" in Weiss' translation marginalizes and obscures the identity of the perpetrator, since "*yillī*" literally translates as "the one who." Another suggested translation might generalize the perpetrator of the act of killing, and read: "killing Syrians is betrayal." None of these possibilities are meant to diminish the translation presented in the table, but instead to show that such types of text are open to different meanings and interpretations. This range of possible translations also highlights that the same slogan, at a different time, can be translated differently in its different (new) context. For example, comparing slogans 1 and 2 from Table 14—which have the same English translation, but where the Arabic words are in a different order—shows the importance of putting "traitor" at the end. The translation of two slogans from two different dates should not be the same, suggesting that the translator missed or misread the difference. The first slogan accuses the regime of murder, which is why the emphasis is on the act itself, while the second one emphasizes the perpetrator(s) of the act, including the regime. The different order of the words within the slogan originates, as I read it, from the massacre at Jisr al-Shugūr on June 6, 2011, "where 120 soldiers were killed. The regime accused the terrorists and the opposition accused the regime because they were trying to defect."[410] Here, it does not matter which story

410 *Al Jazeera*, "*Tashkīk bi-Riwāyat Dimashq 'an Jisr al-Shughūr*" [Doubting Damascus' Story Regarding Jisr Al-Shughūr], June 7, 2011, accessed July 18, 2023, https://bit.ly/2QDiDtJ.

was true; with the change to the word order, the slogan is addressing any killers in Syria at that time as a possible reading to the slogan.

Slogan 3 in Table 14 is different in its use of "won't," which indicates Syrians will no longer stand for being insulted. The verb "insult" is not, however, a strong indicator of the Syrian experience of suppression and coercion by the regime. Feelings of insult were provoked among Syrians for decades by the requirement of needing *mukhabarat* [secret service] approval for numerous activities, including registering a marriage with the state (to be recognized on documents and for it to be valid officially), having a wedding party in public, renting an apartment, and selling or buying real estate. Many such normal activities in Syria are connected to feelings of insult and humiliation because of the need to have *mukhabarat* approval for basic administrative procedures, in addition to the exposure of everyday violence and humiliations. The word "insult" is thus insufficient to describe the actual humiliation experienced by Syrians. I would suggest two translations for this slogan. The first one might reflect some of the slogan's musicality in Arabic, as "Not subjugated, not humiliated." The second possibility is similar to this suggestion but would replace "humiliated" with "subjugated and/or humiliated," placing a stronger emphasis on the act of suppressing and repressing Syrians.

Small details can provide additional information for the translated text (though these details are typically omitted in translations). For example, regarding slogan 4, the translator chose to foreignize the expression "*allāhu akbar*" and replicate it without providing any indication of its background in the Syrian and revolutionary context. This is an archetypal act of foreignization on the level of shape but not content. One characteristic of foreignization is to "conform to the dominant cultural values."[411] In this example, we see that the foreignized expression does not defy or challenge the dominant values of the language, but reflects some dimension of the same expression in the target language. "*Allāhu akbar*" is the expression that Muslims use multiple times a day while praying, but when it is foreignized into English, it mirrors the primary meaning that exists in the dominant language: a connection to terrorist attacks.[412] Foreignization, in this case, functions as it is

411 Venuti, *Translator's Invisibility*, 18.

412 See, for example, Steve Almasy and Eva Tapiero, "Attacker Yells 'Allahu Akbar,' Stabs Five in Paris before Police Take Him Down," *CNN*, May 13, 2018, https://edition.cnn.com/2018/05/12/europe/paris-stabbing-attack/index.html; Fabrice Coffrini, "Swiss Muslim Fined £178 for Saying 'Allahu Akbar,'" *The Week*, January 10, 2019, https://www.theweek.co.uk/98878/swiss-muslim-fined-178-for-saying-allahu-akbar; Khaled Beydoun, "The Perils of Saying 'Allahu Akbar' in Public," *The Washington Post*, August 25, 2018, https://www.washingtonpost.com/news/global-opinions/wp/2018/08/25/the-perils-of-saying-allahu-akbar-in-public/. Nearly all (Islamic) terrorist attacks reported in the media have begun with the shouted phrase "Allāhu akbar." This has almost inextricably linked the phrase with Islamic terror in the minds of Western observers.

meant to—showing other dimensions or other foreign cultural resistance—but reifies the accumulated knowledge of the reader. In the Syrian context, "*allāhu akbar*" takes on dimensions that derive from the revolutionary context rather than the social Islamic one. One meaning or interpretation of the term is the Islamic one, "God is the greatest." While the Islamic meaning cannot be divorced from the expression itself, it is used in different contexts and even by non-Muslims in Syria. Socially, "*allāhu akbar*" is used as an exclamation, expressing surprise or regret. For example, if someone has been waiting for something for a long time, when it finally happens, the normal reaction of this person would be to say "*allāhu akbar*," without having any Islamic meaning attached. This can be equated to when an atheist says "Thank God," without actually meaning to show gratitude to a God in whom they do not actually believe. This is commonly seen with all Arabic idiomatic expressions, including the terms *allāh* [God] and *al-Nabī* [the Prophet]. For example, people say "*ṣallī ʿalā n-Nabī*" [Peace be upon Muḥammad] when they mean do not worry, or calm down. Such expressions are too often literally translated into English, thus connoting a religiosity not meant by the Arabic speakers, as seen in the example from Weiss' translation.

Another meaning for "*allāhu akbar*" in the revolutionary context is the Islamic meaning that "God is greater than Bashar," but it is not limited only to this interpretation. A revolutionary meaning, based on the context of protest, requires the knowledge that "*allāhu akbar*" was used as a key expression in the early demonstrations. The official beginning of the demonstration was signaled by a leader saying "*takbīr*," to which demonstrators would answer "*allāhu akbar*." After this, the demonstration started, with songs, music, slogans and so on. "*takbīr*" is an expression that essentially invites people to reply "*allāhu akbar*." As mentioned above, "takbir" is undeniably rooted in Islam, even if demonstrators do not use it for Islamic reasons. Some groups of activists, who wanted to ensure they were not participating in anything religious or Islamic while saying "*takbīr*," interpreted it humorously as "take a beer." The regime used this expression as proof that *takfīrī* and Islamist groups were taking part in the demonstrations, which served its purposes of Islamizing the entire revolutionary act. Thus, foreignizing the term "*allāhu akbar*" in translation does not successfully achieve its purpose but rather does the opposite, bringing negative associations of the foreign culture to the forefront. I would suggest the translation of "Hey! Let's demonstrate" as an alternative domesticated translation. The above example enhances the Islamic interpretation of "*allāhu akbar*," which is also seen in another instance where the English text tries to provide additional information to the reader. A second example that focuses on inserting more information than the Arabic text is:

الاقتحام الثالث لمدينة جبلة 5-6-2011: عندما أشيع نبأ اعتقال "ح" (203)

The English text translates it as "the third incursion into the city was 5 June 2011. When news started to spread about the arrest of H., who had been wanted for some time by the security forces on the charge of murder and who went by the name of 'The Prince of Jableh.'"[413] The English translation provides more detail than the Arabic one and inserts "Prince" as a keyword, which serves the purposes of the regime by strengthening claims that demonstrators were calling for an Islamic emirate to replace the current secular one. The translation does not explicitly state this, which is why I would argue that the use of "Prince" does not serve the Arabic text since it is not present in the Arabic and does not convey any Islamic dimension negatively or positively to the reader in relation to the arrest of this person. In addition to this, and at the level of meaning, adding "Prince" loads the translation with new information that might be (mis)understood as Islamic. It may be that the translator used "prince" instead of "emir" (more frequently used in the English-speaking media) to disconnect the phrase from the Islamic connotations of the latter, but my point here is not about translation choice but rather the insertion of a "Prince" that does not exist in the Arabic text.

A third example that supports an Islamic interpretation of the Arabic text is the decision to translate "*ḥarāʾir al-sāḥil al-sūrī*" as "The Free Virgin Women of the Syrian Coast." Weiss justifies his translation choice as follows:

> Ibn Maja (d.887) in chapter 8 of his Kitab al-Nikah [Book of Marriage] states: "Whoever wishes to meet God in Purity should marry a free virgin woman." However the term appears commonly in Islamic legal and juristic writing... Since the outbreak of the Syrian uprising a women's group loosely affiliated with the Muslim Brotherhood has emerged, calling themselves al-Hara'ir.[414]

An examination of the Arabic book the translation was based on shows the hadith in the translator's note is incorrect. In the Arabic text, the hadith actually reads:

مَنْ أَرَادَ أَنْ يَلْقَى اللهَ طَاهِرًا مُطَهَّرًا فَلْيَتَزَوَّجِ الْحَرَائِرَ.[415]

> Whoever wishes to meet God in Purity should marry free women.

The original hadith does not therefore say "virgin women." I also referred to online translations of the "Book of Marriage" by Ibn Māja, thinking that perhaps the translator had used a translation instead of the Arabic hadith and that maybe the translator of Ibn Maja had changed it into "free virgin women." But translations of the same hadith found at Muflihun.com, Sunnah.com, Ahadith.co.uk, and Islamicfinder.org

---

413 Yazbek, *Woman in the Crossfire*, 174.

414 Yazbek, *Woman in the Crossfire*, 220.

415 Ibn Māja, "*Bāb al-Nikāḥ*" [The Book of Marriage], in *Sunan Ibn Māja* (Riyadh: Maktabat al-Maʿārif, N.D), 324.

all say "free women" rather "free virgin women."[416] While it is possible that Weiss took the phrase from another translation, his translation notes quote the hadith in Arabic, not in translation. His addition of the word "virgin" thus adds a new dimension that combines Islam with the virginity of women and their freedom.

One element that might have affected the translator's decision to translate it this way is that Samar Yazbek herself indicates that *ḥarā'ir* has an Islamic connotation, as demonstrated by her Arabic text.[417] However, in a search of Arabic dictionaries and Islamic sources I was unable to find any relationship between the virginity of a woman and her freedom. Instead, "free woman" means a decent, dignified woman, or *al-mar'a al-karīma.* The term was used in the pre-Islamic era to distinguish such a woman from *al-ama,* which means an enslaved woman. This meaning confirms the term "free" as meaning the opposite of being enslaved, and unrelated to virginity. Trying to interpret the meaning of "free women" in the revolutionary context shows that in the period after August 2011, the Muslim Brotherhood opposition—who had mostly fled Syria—imbued this term with Islamic connotations by claiming that "our sisters are being dishonored by the *shabbīḥa.*" This was not the mainstream position, despite the highly sensitive nature of this topic in Middle Eastern societies in general. For example, Sham News Network, part of the early revolutionary media, used the term *ḥarā'ir* in a neutral way in May 2011[418,419]. The same neutrality is seen on the Facebook page of the "Syrian Revolution Network," which also uses the term. However this use of *ḥarā'ir* in the mentioned Facebook pages does not standardize that the meaning of the term did not change later or took a different path later. Many secular female activists were called *ḥarā'ir* in 2011 before the term got Islamized later in the revolutionary discourse. Examples include Fadwa Suleiman, Mai Skaf, and many others.

The above posts show that there was no Islamic connotation in the use of "free women" in Arabic. These posts were dated after Yazbek's testimony regarding the use of the phrase, which confirms that, as of that date, there was no inextricable Islamic dimension to the term itself.

---

416 Ibn Māja, "Hadith 1862: 9," in Muflihun, accessed July 18, 2023, https://muflihun.com/ibnmajah/9/1862; Ibn Māja, "The Chapters on Marriage," in *Sunan Ibn Majah,* accessed July 18, 2023, https://sunnah.com/ibnmajah/9; Ibn Māja "Sunan Ibn Majah: Marriage Chapter," in *Ahadith,* accessed July 18, 2023, https://ahadith.co.uk/chapter.php?page=2&cid=167&rows=10; IslamicFinder, "Read all Ahadith in the Chapters on Marriage by Sunan Ibn Majah," accessed July 18, 2023, islamicfinder.org/hadith/ibn-majah/marriage/?page=2.

417 Yazbek, *Taqāṭu' al-Nīrān,* 220.

418 Syrian Revolution Network official Facebook page, posted June 28, 2011, accessed July 18, 2023, https://bit.ly/2WPYipB.

419 Sham News Network official Facebook page, posted May 24, 2011, accessed July 18, 2023, https://bit.ly/2ZpqUHO.

In addition to this, Weiss generalizes the women who were protesting to groups affiliated with the Muslim Brotherhood. This interpretation and the translations of the revolutionary language reveal the translator's beliefs and narratives about the women who were demonstrating. Additionally, it offers a simple explanation for women who were wearing the veil and engaging in protests. Despite Weiss' description of women with the veil demonstrating, this phenomenon was seen as a clear challenge to the religious institutions of the regime itself. Women wearing a bright white scarf represented a permitted religious initiative organized by the regime called al-Qubaysiyyāt,[420] and their participation was thus a clear sign of protest against the regime's religious order created in Syria. It was not related to the Muslim Brotherhood phenomena, as claimed by the translator, since his translation generalizes all women wearing hijab as being affiliated to the Muslim Brotherhood in Syria. However, it is important to note that in other places in the book Weiss translates *ḥarā'ir* as "free women" when he did not want to highlight an Islamic background for the female protestors, indicating that he connects virginity as a concept only to link with Islam and Islamization of the context to which he refers.

To sum up, the translation of Yazbek's book shows that the act of translation in this context is a negotiation between representing the revolutionary discourse and inserting additional values that are not present within it. These elements can be clearly seen in the Islamization of the revolutionary discourse and providing simplistic explanations for issues unclear to the translator, as in the example of the women protestors. In the above examples, we see that the translator's choices when producing the translation from the center (English) about the periphery (Arabic) are more influenced by the dominant center rather than how the periphery might see itself, as clearly apparent in the differing and multiple possible interpretations of the examples. What is important to mention here is that—despite his clear desire to voice the Syrian experience—Weiss is unable to avoid being trapped by the narratives embedded in the dominant discourse because it is difficult for a translator to refrain from using the statements of the dominant discourse already present in their native knowledge and language. However, a conclusion from these different translations cannot be reached alone without a more extensive landscape of other

420 Jusūr Center for Studies, "*Jamā'at al-Qubaysiyyāt: al-Nash'a wa-l-Takwīn*" [al-Qubaysiyyāt community: The emergence and formation], April 3, 2018, accessed July 18, 2023, https://jusoor.co/content_images/users/1/contents/666.pdf. Al-Qubaysiyyāt is an Islamic movement for women, the name of which is derived from that of its founder, Mounira al-Qubaisy. It was originally a secret movement for moderate Islam, but after Bashar al-Assad inherited control of Syria from his father, he gave it permission to be a public movement for teaching Islam. The movement is known for its statements before and after 2011, especially that Syrians should not protest against the current political system because it constituted sedition.

translations of the revolutionary language, and as I illustrate in relation to the two following books, it is impossible to generalize in relation to the whole translation if it is not combined with other translations of the revolutionary language.

Syria Speaks: Art and Culture from the Frontline

The second book translated from Arabic I will examine here is *Syria Speaks,* "an anthology of the revolutionary language."[421] Its title alone highlights the importance and goal of speaking and voicing the Syrian revolutionary language and activists. On Goodreads, the English version has 141 ratings, with an average 4.46 of 5 stars, demonstrating that it is visible and read by the English-speaking readership,[422] but the Arabic version of the book has only seven ratings, averaging 3.86 of 5 stars.[423] The Arabic version is not listed on Google Scholar, but the English version has been cited 88 times by different studies.[424] This highlights how the book was created and translated almost solely for the English-speaking market, with the Arabic version remaining almost unread thus far.

Syria Speaks provides a typical example of doubled discourse in different sections of the book: while each version of the book (Arabic and English) presents the same textual content, it does so in line with its own policies of censorship, erasing, in many instances, the cultural and social traces of the source language text. Moreover, both the Arabic and English versions illustrate the values and narratives of that discourse. In other words, the Arabic version glorifies the revolutionary values while the English one tones some of them down and adds others to them. The English book moves closer to the English reader, and there is no negotiation between the translation and original as an act of resistance to the foreign or target culture in some examples. In this sense, the act of translation cannot be seen through the perspectives of domestication and foreignization, because it moves further away from the source language to suit the culture of the target language reader. In an interview, Zaher Omareen explained how the book started:

---

421 Malu Halasa (co-editor of *Syria Speaks*), interview with Eylaf Bader Eddin, April 11, 2019.
422 *Goodreads*, "Syria Speaks: Art and Culture from the Frontline," eds. Malu Halasa, Zaher Omareen, and Nawara Mahfoud, English (London: Saqi, 2014), accessed June 1, 2020, https://bit.ly/3eB-8M2a.
423 *Goodreads*, "Syria Speaks: Culture and Art for Freedom," eds. Malu Halasa, Zaher Omareen, and Nawara Mahfoud, Arabic (Beirut: Dar El Saqi, 2014), accessed June 1, 2020, https://bit.ly/2BesSki.
424 *Google Scholar*, "Syria Speaks: Art and Culture from the Frontline," eds. Malu Halasa, Zaher Omareen, and Nawara Mahfoud (London: Saqi, 2014), accessed June 1, 2020, https://bit.ly/3dnN41u.

> The story started at the end of 2011 when I arrived in the UK. A group of Syrians, including myself, secured funding from Prince Claus for our initiative, called "Culture Resists."[425] The materials of the book were used as an exhibition of posters and paintings and its result was a booklet that explained about culture in Syria. We distributed thousands of copies free of charge in the UK and Europe. This booklet and exhibition were the genesis of the current book on the market that drew the attention of the Danish Center of Culture and Development. Prince Claus was primarily responsible for funding the book. The main aim of the book is to raise awareness of the revolutionary culture and art of Syria and it was an archival act to document the revolution.[426]

> The importance of the book lies in its being the first attempt to document the revolutionary language and symbolic products of the Syrian revolution. Despite the fact that all three editors—Zaher Omareen, Malu Halasa, and Nawara Mahfoud—had good intentions of voicing the revolutionary culture as a tool of resistance, the Arabic and English editions pursued different policies to express this. This can be seen in: (1) paratextual elements; (2) visual elements; and (3) omissions, facts and numbers.

Paratextual Elements

Syria Speaks focuses on Syrian cultural resistance. To begin with the title, in Arabic this is *Sūryā Tataḥaddath: al-Thaqāfa wa-l-Fann min ajl al-Ḥurriyya* [Syria Speaks: Culture and Art for Freedom]. The English version keeps the same main title but changes the subtitle to *Art and Culture from the Frontline*. This is an important contrast, since according to the Oxford Dictionary, "frontline" means "the military line or part of an army that is closest to the enemy."[427] With this change, the English subtitle immediately loads the Arabic one with a conflictual element that is not present in the Arabic book. We might assume that, because of the change to the English title, which does not include the word "freedom" as in the Arabic version, the book itself contains violent and/or military dimensions that are not present in the Arabic.

An examination of both covers (Figure 18) reveals significant differences. The Arabic book displays a young girl playing on a swing with a rope made from barbed wire, creating a paradoxical mixture of innocence and brutality. The image could be interpreted in multiple ways, including to represent life under Baath rule as a child, or the bad conditions children suffered after 2011. The small gray circle might reflect the conditions that Syrian children are experiencing. The photo of the girl is also in black and white; the girl might be seen as staring unhappily at the

---

425 Prince Claus, "We Are the Prince Claus Fund," official website, accessed June 1, 2020, https://princeclausfund.org/we-are (no longer available). The Prince Claus Fund for Culture and Development is a Dutch fund established in 1996 and named after Prince Claus of the Netherlands. It has annual awards for cultural projects.

426 Zaher Omareen (co-editor of *Syria Speaks*), interview with Eylaf Bader Eddin, March 8, 2019.

427 *Cambridge English Dictionary online*, "Frontline," accessed July 18, 2023, https://dictionary.cambridge.org/dictionary/english/frontline.

viewer and asking him or her to act. However, going back to the original post on the official Facebook page of the initiative of "The Syrian People Knows its Path", the caption of the photo says: "When the south of Damascus smiles, The Sieged south of Damascus."[428] The post highlights the regime's siege of the south of Damascus and cutting off food provision for civilians. The English cover replaces the young girl on the Arabic cover with a youth carrying a slingshot in front of a blurry background, which might be understood as the wreck of a destroyed street and a primitive instrument of defense, despite the book's goal of showing the cultural and revolutionary art situated within the resistance against the regime.[429] This goes hand-in-hand with the connotations of violence implied by the word "frontline" in the subtitle. The book states that the child shown on the cover is "based on Zaytoun, the little [Palestinian] refugee, originally designed by Muḥammad Ṭayib and inspired by the shelling and hardships inside [the] Yarmuk Palestinian refugee camp in southern Damascus."[430] Connecting the Syrian context to the Palestinian struggle against Israel might be seen as an indirect signal to the English reader, who has previously been "left with images of violence and stereotypical images of Palestinian terrorists and Israeli victims."[431] Combining the title with the cover art provides a better understanding of the paratextual elements of the book in Arabic and English, by considering the title as a caption for the cover photo. In his work *Image, Music, Text*, Roland Barthes emphasizes the salient role that captions can have in controlling interpretations of an image, arguing that "the text [caption] helps to identify purely and simply the elements of the scene and the scene itself. . . [it is] a description which is often incomplete."[432] It is not complete because the caption (in this case, the title) does not show the entire message intended by the image of the book, but comparing the images with the captions in Arabic and English helps us understand it more fully. In this way, the use of the word "frontline" and a different cover image convey more information by combining both the textual and visual elements of the cover. The English title contrasts with the Arabic one and suggests the book has more violent characteristics; it practices a type of violence by telling

428 Al-Shaʿbb al-Sūrī ʿĀrif Tariqu [The Syrian People Know Their Way], "When the South of Damascus Smiles, the Sieged South of Damascus," in The Syrian People Knows its Path, January 10, 2014, accessed June 20, 2023, https://shorturl.at/xNY26.

429 Halasa, Omareen, and Mahfoud, eds., *Syria Speaks*, 81. The caption of the cover in the book says that the child on the cover is supposed to represent a boy.

430 Halasa, Omareen, and Mahfoud, eds., *Syria Speaks*, 81.

431 Susan Dente Ross, "Unequal Combatants on an Uneven Media Battlefield: Palestine and Israel," in *Images that Injure: Pictorial Stereotypes in the Media*, eds. Paul Martin Lester and Susan Dente Ross (Westport: Praeger, 2003), 62.

432 Roland Barthes, "Rhetoric of the Image," in *Image, Music, Text*, essays sel. and trans. Stephen Heath (London: Fontana Press, 1977), 39.

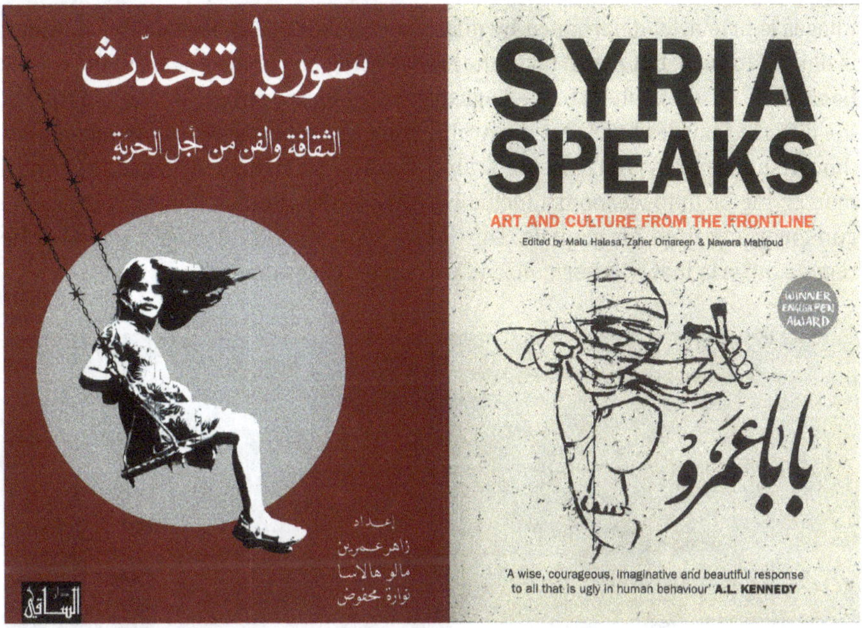

Figure 18: Arabic and English covers of Syria Speaks.

the viewer what to understand from it. Captions and titles encourage the viewer to read the cover in a specific way, if only because it "helps me [as a reader] to choose the correct level of perception, permits me to focus not simply my gaze but also my understanding."[433] In this example, power and authority are practiced not only to find an alternative interpretation, but to force the reader to take a particular perspective on perceiving and understanding the cover.

Visual Elements

The English and Arabic versions of *Syria Speaks* are presented in almost the same order but with slight changes to the visual content. This shows a new type of translation that works without original materials or replacing materials included from the original text. For example, the chapter entitled "Ongoing" contains not only a different number of images but also includes different ones.[434] The act of selecting

433 Barthes, "Rhetoric of the Image," 156.

434 Compare Sulafa Hijazi, "Ongoing," in *Syria Speaks: Art and Culture From the Frontline*, 10–16, eds. Malu Halasa, Zaher Omareen, and Nawara Mahfoud (London: Saqi, 2014); and Sulāfa Ḥijāzī, "Mustamirr," in *Sūryā Tataḥaddath: al-Thaqāfa wa-l-Fann min ajl al-Ḥurriyya*, 24–32, eds. Zāhir'Umārīn, Mālū Hālāsā, and Nawwāra Maḥfūḍ (Beirut: Sāqī, 2014).

images is primarily determined by nudity and politics. This does not mean that it is limited to such, but it also takes the risk of using one of the photos in the Arabic book and a different photo in the English version. Moreover, a close examination of the different images finds that the Arabic book excludes images containing nudity, though they are referred to in the Arabic text. The first image in Figure 19 shows a man giving birth to weaponry, including guns, a grenade, and knives, with a large gun emerging from his genitals. This image represents the violence sown by the regime in Syria that was normalized for younger generations.

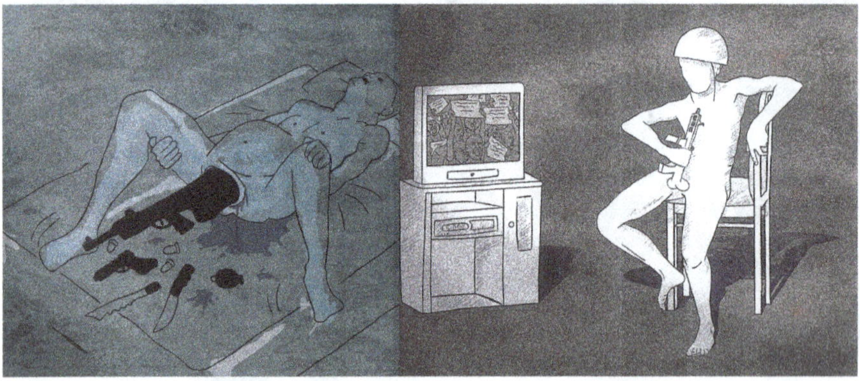

Figure 19: Images not included in the Arabic version of the "Ongoing" chapter.[435]

The second image in Figure 19 shows a man masturbating with a gun, which sheds light on the patriarchal characterization of Syrian society and the pleasure of violence. The Arabic text describes both photos without showing them. The non-inclusion of the images is not justified in the text, but Omareen explains that "it was the decision of the publisher. We were even forced to delete other images that hinted at some Gulf countries. We fought to keep them, but we reached a dead-end that might have caused us to stop publishing the Arabic book. The publisher believed that keeping those images in the Arabic book would be a reason to stop the growth of the market in some Gulf countries."[436] This response explains a great deal about the economic factors interfering with the content of the book, justified only by the possibility that a few Gulf countries might keep the book out of their market. The dual versions of the book can be seen in the examples below, which present a different version in English to the English market than in Arabic to the Arabic market.

435 Hijazi, "Ongoing."
436 Omareen, interview.

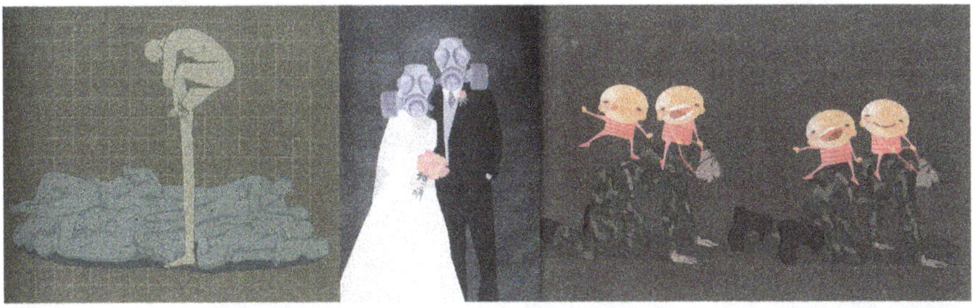

Figure 20: Images not included in the English version of the "Ongoing" chapter.[437]

Having examined the images not included in the Arabic version, I will now review the images that replaced them. In Figure 20, on the left, a man is trying to get away from the massacre beneath his feet, which reflects feelings of "fear, courage, hope, pain, guilt, weakness, and alienation."[438] While both texts mention the image, it is only reproduced in the Arabic version. The image in the middle of Figure 20, of a bride and groom wearing gas masks, shows "the paradox of life and death, considering that marriage as a symbol represents 'our need to stay alive,'"[439] defying death itself. The image on the right in Figure 20 shows children riding soldiers, which Hijazi describes as "their [the children's] way of avenging their stolen childhood."[440] The first image is explained and present only in Arabic, which is understandable because it is displayed in the Arabic version, while the third—children on the backs of soldiers—is only mentioned in the English text without displaying the image. In considering the reasons why these images are not displayed in the English version, this is likely because the readership is not familiar with such themes about Syria. For instance, the image on the left refers to the massacres perpetrated in Syria, the second represents repeated chemical attacks met with international silence, and the final one highlights how children are abused by being exposed to such violence. In addition to the graphic content that was prevented from being shown to the English-speaking audience, there is a level of simplicity in the images in the English version. This simplicity does not require deep analysis to present a thick and complex depiction of the Syrian context.

Another example of differing visual content occurs in the chapter entitled "Revolution 2011" by Khalil Younes. The image on the left of Figure 21 is displayed in the Arabic version, while the image on the right is shown in the English version. The

437 ijāzī, "Mustamir."
438 Ḥijāzī, "Mustamir," 15.
439 Ḥijāzī, "Mustamir," 28.
440 Hijazi, "Ongoing," 11.

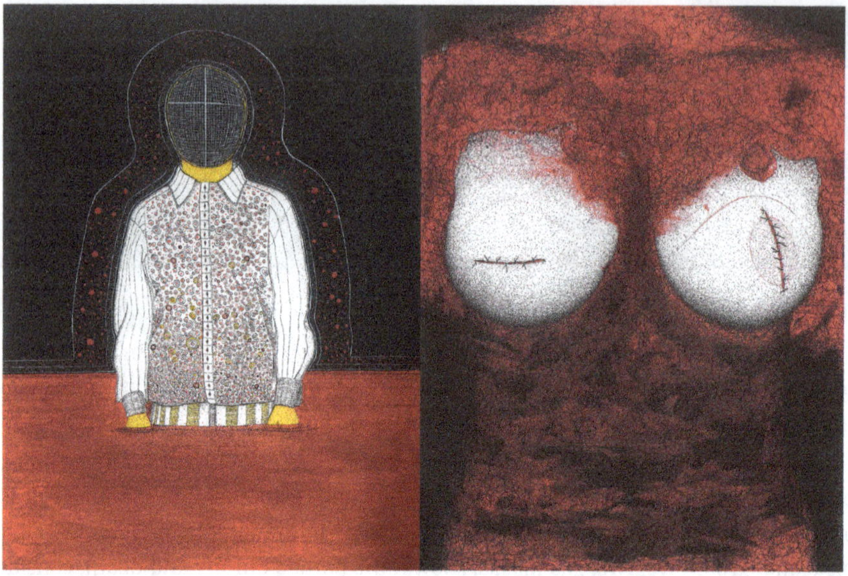

Figure 21: Different images included in the Arabic and English versions of the "Revolution 2011" chapter.[441]

image on the left shows a man standing in a waist-deep pool of blood. It is an untitled image, released in a collection called "Ink on Paper." The picture on the right is from the same collection and shows breasts with stitched incisions, surrounded by blood. The image clearly represents the violent massacre that happened in Hama in the 1980s, because it is entitled "Hama." The caption itself, however, is not sufficient to explain the brutality either of the painting or the regime to the English-speaking audience, providing only a very small window onto the atrocities of the regime.

A final example of the different visual content in the Arabic and English versions occurs in the chapter by Charlotte Bank, "Fann al-Iqnāʿ" [The Art of Persuasion].[442] The Arabic text displays seventeen posters while the English version includes eighteen, some of which are actually different. If we consider that this

441 On the left: Khalīl Yūnis, "Thawra 2011" [Revolution 2011], in *Sūryā Tataḥaddath: al-Thaqāfa wa-l-Fann min ajl al-Ḥurriyya*, eds., Zāhir ʿUmārīn, Mālū Hālāsā, and Nawwāra Maḥfūḍ (Beirut: Sāqī, 2014), 55. On the right: Khalil Younes, "Revolution 2011," in *Syria Speaks: Art and Culture from the Frontline*, eds. Malu Halasa, Zaher Omareen, and Nawara Mahfoud (London: Saqi, 2014), 41.

442 Shārlūt Bānk, "Fann al-Iqnāʿ," in *Sūryā Tataḥaddath: al-Thaqāfa wa-l-Fann min ajl al-Ḥurriyya*, eds. Zāhirʿ Umārīn, Mālū Hālāsā, and Nawwāra Maḥfūḍ, 77–96 (Beirut: Sāqī, 2014); Charlotte Bank, "The Art of Persuasion," in *Syria Speaks: Art and Culture from the Frontline*, eds. Malu Halasa, Zaher Omareen, and Nawara Mahfoud (London: Saqi, 2014), 66–83.

chapter was translated from Arabic, we can say that only fourteen posters are transferred from Arabic into English.

In contrast to the chapters previously discussed, it is unclear why different posters were selected for each version when all of them express pacifist ideas and themes of civil resistance, particularly in the English version. On the left of Figure 22, the poster calls for civil disobedience, while the one next to it encourages disobedience by showing a key, which represents workplaces such as shops. The other two posters in Figure 22 focus on Bashar al-Assad as a criminal, asking him to step down. All four posters were designed during preparations for the "dignity strike" in November 2011, which was prior to the militarization of the revolution.

In the Arabic version, the different posters are more sophisticated in terms of their level of symbolism and expressing the exceptionality of the Syrian context and revolutionary event. The poster at the top of Figure 23 criticizes the opposition voices that asked for internationalization of what was happening in Syria in January 2012, asking activists and protestors to keep demonstrating even if it seemed ineffective at first.[443] The poster on the bottom left of Figure 23 sheds light on the intense protests seen in Syrian universities at the end of October 2011, particularly Qalamun University on October 31,[444] containing a subverted motto of Damascus University. The original one reads "Say, oh my lord, advance me in knowledge" (see Figure 24),[445] and the center of the logo shows that Damascus University follows this Quranic verse literally: according to its website, "The purple color symbolizes the Damascene berry, which is a kind of fruit that is only found in Damascus. Hence, it was chosen to be the distinctive color of the university to express its uniqueness. . . As for the lamp, it symbolizes knowledge in different cultures and radiates the light of science and knowledge."[446] The poster replaces the word "knowledge" in the Quranic verse with "revolution," an act of subversion that shows not only daring in making the change to the Quranic verse, but also challenges the knowledge that a Baathist institution might offer to Syria. It thus sets itself in opposition to all types of knowledge produced under the Baath regime. Instead of a lamp, there is a Molotov cocktail, representing the revolutionary act at

443 *Al-Sha'bb al-Sūrī 'Ārif Tariqu* [The Syrian People Know Their Way], "*bostar I'ādit al-Tadwīl*" [The Poster of Re-internationalization], Facebook, uploaded January 19, 2012, accessed July 18, 2023, https://bit.ly/2yr2CSY.

444 "*Aḥrār wa-Ḥarā'ir Jāmi'at al-Qalamūn*" [The Free Men and Women of Qalamun University], *Anonymoussyrian* (YouTube Channel), uploaded November 1, 2011, accessed July 18, 2023, https://www.youtube.com/watch?v=XopDu-Q7Jqc.

445 Damascus University, "University Logo," official website of Damascus University, accessed July 18, 2023, http://www.damascusuniversity.edu.sy/index.php?lang=2&set=3&id=706.

446 Damascus University, "University Logo."

Figure 22: Posters added to the English version of "The Art of Persuasion" chapter.[447]

the university, and the poster text reads: "Every day we go [to universities] to learn freedom."

The final image, on the bottom right of Figure 23, shows a child with hair formed by the leaves of a tree, his face by the sky and a blurry building. The image was released after the Daraya massacre, where many children were killed, on August 25, 2012.[448] It represents the rural parts of the city and its famous grapevines. The connection here between the massacred children and the brutality of the regime is seen through the regime's destruction of anything alive, whether children or trees.

Comparing the poster content in the English and Arabic versions, the former includes simple and clear posters that do not require a lot of effort to understand. On the whole, the posters in the English version focus on the revolutionary acts in general, without delving deeper into Syrian exceptionality. In the Arabic version, on the other hand, the posters display more depth and a more complex view of the Syrian revolutionary event. This is not a critique of the English or Arabic versions, but rather an exploration of the different themes in the posters chosen for the two versions. It is important to recognize, however, that this difference prevents the English-speaking readership from accessing a greater variety of themes related to the Syrian revolution, which is (or should be) the primary task for a book that seeks to give Syrians a voice. In addition to this, the inclusion or exclusion of certain images is an act of interpretation, regardless of what is said in the text. This level of translation has been seen especially after 9/11, with books and, more frequently, the

447 Bank, "The Art of Persuasion," 66–83.

448 Mona Mahmood and Luke Harding, "Syria's Worst Massacre: Daraya Death Toll Reaches 400," *The Guardian*, August 28, 2012, https://www.theguardian.com/world/2012/aug/28/syria-worst-massacre-daraya-death-toll-400.

Figure 23: Posters added to the Arabic version of "Alshaab alsori aref tareko" ["The Art of Persuasion"] chapter.[449]

media, creating "non-translation, mistranslation and/or disputed translation."[450] This means that no matter what the images show in the book, the text underneath them will not change or explain it, as seen in the above examples. The aim of the

449 Bānk, "Fann al-Iqnāʿ," 77–96.

450 Emily Apter, *The Translation Zone: A New Comparative Literature* (Princeton: Princeton University Press, 2006), 14.

Figure 24: Official logo of Damascus University.[451]

English book to provide visual materials seems pointless if compared to the Arabic book, since it does not give information—or enough information—about the materials it presents. The English book shows pictures, paintings, and posters and their captions but neither explains why other images included in the Arabic version are not present, nor paraphrases the content of the images used. In this sense, it does not succeed in offering a systematic, strategic, and logical way of presenting revolutionary visual art, to the extent that the change of the photos and poster constitutes a rupture or causes a lack of text cohesion. They are shown merely for the sake of it. The different versions even offer differing narratives of Syria in the text. Bank, for example, explains the motives for and themes covered by the poster initiative as being "...children and young resistance fighters,"[452] but the archive of the initiative does not show resistance fighters and specializes in civil and non-violent resistance. The official Facebook page of the initiative contradicts Bank's description, stating: "... Everybody is invited for a dialogue for deciding the destiny of the country. Everybody is welcome to renounce violence. One thing not accepted here [in this initiative] is sectarian discourses, and the ones that divide our society.

---

451 Damascus University, "University Logo."

452 Bank, "The Art of Persuasion," 75; *Al-Shaʾbbb al-Sūrī ʿĀrif Tariqu* [The Syrian People Know Their Way], "About," official Facebook page, accessed June 1, 2020, https://www.facebook.com/pg/Syrian.Intifada/about/?ref=page_internal (no longer available).

Everyone is equal under the law."[453] According to the declaration of the initiative, they renounce violence, which shows that it did not want to create art for fighters, contrary to what the chapter claims. Once again, the interpretation of the initiative is different from how the initiative would like to present itself, which adds a further layer glimpsing at the present violence that the author is witnessing in Syria at the time of writing the piece, rather than the time of the posters were created and released. This added dimension can be further understood once we look further at other examples of translation in the book where similar themes are dealt with in relation to violence, and militarization.

Contradictory Text

In addition to translation by images, the two versions in Arabic and in translation include a number of contradictory passages that depend essentially on adding more or deleting some words of the Arabic text. The book's introduction is full of examples where the Arabic text has different layers and meanings compared to the English one, with the two versions potentially going in different directions at the level of signification.

Table 15: A comparison of the translation of two texts.

Creativity is not only a way of surviving the violence, but of challenging it. Of course, none of them support the regime anymore; but they have lost their ability to back the revolution because it has become complicated. Those who participated in it have changed, as have its political perspectives. Since the beginning of the uprising in 2011, everything has been radically altered on the ground – except for its artistic identity. Many Syrians had thought their "Arab Spring" would be different from those in Egypt and Tunisia, and they began constructing a Syrian revolutionary identity through posters, performances, songs, theater and videos.[454]	الابداع ليس الطريقة الوحيدة للاستمرار في الحياة في ظلّ العنف فحسب، إنما هو طريقة فعّالة لتحدي العنف ذاته. بعد ثلاث سنوات طوال، أصابت خيبة الأمل كثيرين، لم يتراجع عن موقفه أو يبدّله أو يبدّل قناعاته، فالمسألة أخلاقية قبل أن تكون سياسية، ومبدئية قبل أن تكون براغماتية. لقد دخلت الثورة كثيراً من الأطوار الميدانية والسياسية، فتغيرت وغيّرت وجهات النظر الكثيرة التي ترتبط بها. وتبدّل كلّ شيء على الأرض بصورة هائلة منذ عام 2011. الجميع كان يعتقد منذ البداية، بمن فيهم السوريين أنفسهم، أن الربيع العربي سيكون مختلفاً في دمشق عنه في مصر وتونس واليمن، لذا سعى أنصار الحراك الشعبي منذ البداية إلى تشكيل هوية ثورية سورية خاصة مستخدمين البوسترات السياسية والعروض الفنية والاغاني ومسرح الدمى والمقاطع المصورة[455].

453 "About", in Al-sha'b al-Sūrī ʿĀrif Ṭariqu [Syrian People Know their Path], Official Facebook page, accessed, June 1, 2020. https://www.facebook.com/pg/Syrian.Intifada/about/?ref=page_internal

454 Halasa, Omareen, and Mahfoud, eds., *Syria Speaks*, vii-viii.

455 ʿUmārīn, Hālāsā, and Maḥfūḍ, eds., *Sūryā Tataḥaddath*, 7.

I have included one sentence before and after the text above to show that it is the same text that is transferred from Arabic into English and to show more of the context of this passage. Although each passage of text above partially mirrors the other, it could be read differently in both languages. For example:

لم يتراجع عن موقفه أو يبدّل قناعاته، فالمسألة أخلاقية قبل أن تكون سياسية.[456]

This sentence can be translated literally as "no one has changed his/her attitudes or beliefs regarding the revolution because it is a moral issue, first, before it can be considered a political one," but the English text reads "they have lost their ability to back the revolution." The translator's choice here changes the Arabic interpretation of the text to the extent of contradicting it: while the Arabic text says that no one has changed his/her attitude, the English one indicates that they have lost the will to back the revolution. The difference between the texts is not due to a loose interpretation, but changes the opinions and situation of people who supported the revolution, implying that no one is supporting it anymore. According to this interpretation, activists and artists have two choices: either they continue to support a revolution that has changed, or they support the regime. This is how the revolution developed from a peaceful democratic one to take another direction entirely, but the change in the course of the revolution does not mean that people stopped supporting it, as the English text claims.

In the introduction to the English version, we are told that instead of "all Syrians" [*al-jamīʿ*], the text applies, from the beginning, only to "many Syrians."[457] The Arabic text sees itself as a representation of all Syrians, while the English text asserts that some Syrians are not represented. The English text also refers to "Exploring a new language of inclusiveness,"[458] while the Arabic text reads "lugha shāmila jadīda li-maʿna al-waṭaniyya wa-l-intimāʾ." The importance of *waṭaniyya* and *Intimaʾ* [patriotism and belonging] is that these were among the most important concepts for which Syrians called during the protests in 2011. Deleting those two terms is an erasure of the demands most often made by pro-revolutionary demonstrators. Another important omission from the English text is the passage in the Arabic text declaring the reasons for, and objectives of, *Syria Speaks*. The Arabic introduction says:

إن كان هذا الكتاب يطمح لإيصال رسالة ما ، فهو يطمح لأن يكون مقدمة لسلسلة ابحاث وكتب تتناول هذا الطيف الواسع من الفنون والمنتجات الثقافية التي افرزتها الثورة السورية والتي يستحق كلّ منها دراسة متبصرة.[459]

---

456 ʿUmārīn, Hālāsā, and Maḥfūḍ, eds., *Sūryā Tataḥaddath*, 7.

457 Halasa, Omareen, and Mahfoud, eds., *Syria Speaks*, vii.

458 Halasa, Omareen, and Mahfoud, eds., *Syria Speaks*, iv.

459 ʿUmārīn, Hālāsā, and Maḥfūḍ, eds., *Sūryā Tataḥaddath*, 15.

The literal translation of this sentence is: "If this book aspires to convey a message, it is that of being an introduction to a series of studies and books that will deal with the broad spectrum of art and cultural products that were generated by the Syrian revolution. They deserve to be deeply studied." This important sentence cannot be found in the English book and exists only in the Arabic version.

The dimension of religion also shows itself in contradictory ways in the Arabic and English versions. The Arabic book refers to "*bayna ṣadīqay ṭufūla yantamīyān ilā ṭā'ifatayn mukhtalifatayn*" [between childhood friends of two different sects],[460] while the English text specifies the sects to which they belong, describing "friendships between Alawi and Sunni childhood buddies."[461] This provides a clearer understanding of how each discourse (Arabic and English) sees the cartoon in question here, which is dependent in the first place on the common narratives of each discourse. The Arabic version acknowledges that Syria has different sects and ethnic groups, but the English clearly specifies the exact sects meant by the text, which is a sectarian interpretation of the Arabic text into English. This example shows the influence of the culture and discourse of the translator: their emphasis reduces the conflict to one between sectarian rivals, which shows that a limited but general perspective does not mitigate the idea of sectarianism in Syria.

It is also important to note that each text provides different numbers associated with critical events and incidents. The chapter in the English version entitled "Hama 82" says that an estimated 10,000–25,000 people were killed,"[462] while the Arabic version changes this figure to "more than 30,000."[463] While the exact number is unknown, the difference between the Arabic version and the English is striking and reflects the sources of their discourse: the English text references official numbers and statistics from English-language sources, while the Arabic version refers to Arabic sources.

One final example of the addition of extra information or perspective to the English translation is the insertion of the phrase of "instead of a gun" in the translation:

For Syrians and non-Syrians alike, there are many reasons to wake up every morning and reach for the pen, the easel, the camcorder or the laptop – instead of a gun.[464]	هناك حُكماً الكثير من الأسباب بالنسبة للسوريين أو غيرهم للنهوض كلّ صباح والتقاط قلم أو ورقة رسم او آلة تصوير والنضال بها وعبرها وصولاً لى تحقيق الحرية والكرامة والعدالة.[465]

460 ʿUmārīn, Hālāsā, and Maḥfūḍ, eds., *Sūryā Tataḥaddath*, 12.
461 Halasa, Omareen, and Mahfoud, eds., *Syria Speaks*, xii.
462 Halasa, Omareen, and Mahfoud, eds., *Syria Speaks*, 1.
463 ʿUmārīn, Hālāsā, and Maḥfūḍ, eds., *Sūryā Tataḥaddath*, 17.
464 Halasa, Omareen, and Mahfoud, eds., *Syria Speaks* (English), xv.
465 Halasa, Omareen, and Mahfoud, eds., *Syria Speaks*, 15.

The above English version adds violence and leaves out what are arguably the most important revolutionary words: *al-ḥuryya, al-karāma, al-ʿadāla* [freedom, dignity, and justice]. These three terms are the pillars of the revolution, but are replaced with "instead of a gun," a clear act of interpreting the Arabic text to fit the English discourse.

Overall, it is clear that the book focuses more on domesticating the revolutionary language to an English-speaking readership, providing the reader with an easier understanding of the source material at the cost of a loss of nuance and layered meanings. In some cases, relating to literary texts such as passages from short stories, novels, interviews, and articles, the book is successful in offering a better translation than the elements examined above. However, with regard to the revolutionary language in particular, there is an oversimplification of the material presented that serves neither the interests of the source language culture nor the importing dominant culture. In *Syria Speaks*, this act of domestication goes further, creating a quasi-translation utilizing elements that do not exist in the Arabic text. These include censored sexual artistic hints, represented by deleting all of the work with nudity from the Arabic version, a decision made by the publisher rather than the book's editors. The issue of religion is demonstrated through an interpretation of an Arabic sentence that specifies two unnamed sects as Alawi and Sunni, and the issue of politics by the decision to use simple posters in the English version, rather than the more sophisticated ones in Arabic. Each element is used differently to satisfy different audiences. For instance, nudity is generally not well-received in the Arab world, and thus any images containing nudity were removed in the Arabic version but added in the English book. This act of transferring the nudity-containing content only to the English readership is dependent on believing not only that such types of art are well-received in English but not in Arabic, but also that simply by their presence, the book would not be made available in some Gulf countries, and the economic prospects of the book would therefore be irreparably damaged.[466] Simple

466 Mark Williams, "MENA Publishing in the 2020s: Obstacles and Opportunities in the Arab Book Markets," *The New Publishing Standard*, September 30, 2019, accessed July 18, 2023, https://bit.ly/2Ap4HiW; Mirzā al-Khuwaylidī, "al-Raqāba fī al-Kuwayt Tamnaʿ Akthar min 4 Ālāf Kitāb" [Censorship in Kuwait prevents more than 4 thousand books], *Al-Sharq al-Awsaṭ*, September 13, 2018, accessed July 18, 2023, https://bit.ly/3g688vi. Gulf countries are considered to be a growing market for publishing houses and are, unsurprisingly, the largest market for Arabic publications. In fact, six international book fairs held in Arabic markets routinely attract more than a million people, while only one international book fair in Europe has that many attendees. At the same time, however, government censorship has increased over the past few years, resulting in hundreds or thousands of books being prevented from being displayed at these events. A recent example of book censorship in the Gulf countries occurred at the 2018 Kuwait Book Fair, when thousands of books were prevented from entering the country.

and easily accessible images are presented only in the English version, as seen in the chapter by Charlotte Bank, while also inserts violent elements and a religious focus. The examples presented above—relating to paratextual elements, visual elements, omissions, slogans and terms—show that the English and Arabic books do not mirror each other to a specific extent. They translate themselves based on the norms of their own social contexts, discourses, and narratives. I will return to these two different versions of the book in the conclusion to this chapter in a way that facilitates readership-anticipated preferences.

The Story of a Place: The Story of a People

The final book I will examine that transfers the revolutionary language and revolutionary events in Syria is *The Story of a Place: The Story of a People.*[467] Published in 2017, it attempts to collect materials and document the revolution from 2011 to 2015, with entries ordered alphabetically by city. The book is the culmination of an online project called "Creative Memory of the Syrian Revolution" (CMSR),[468] and was created in Arabic, English, and French, and funded and published by Friederich-Ebert-Stiftung.[469] It consists of a work team led by Sana Yazaji (the project and website leader and founder of CMSR) in conjunction with senior researcher Nada Najjar, and English translations by Rana Mitri. According to the book cover, the same team worked on both the English and Arabic versions. The book is divided into chapters, one for each city and arranged alphabetically, beginning with an historical introduction and account of the key events and the first demonstrations. The book also provides one or more images for each city as examples of its best known visual products. In terms of the book's visibility, it has a page on Goodreads, but the Arabic version has only one review and the English one has no reviews at all.[470] Google Scholar does not include any entries for either book, or mention its existence in the data.

467 CMSR, *Qiṣṣat Makān*; CMSR, *Story of a Place*.
468 CMSR, official website, accessed July 18, 2023, https://creativememory.org/ar/archives/.
469 Friederich-Ebert-Stiftung is a German political association affiliated with the Social Democratic Party; see https://www.fes.de/.
470 Goodreads, CMSR, *Qiṣṣat Makān Qiṣṣat Insān: Bidāyāt al-Thawra al-Sūriyya 2011–2015*, sup. Sana Yazaji, (Beirut: Friederich Ebert Stiftung, 2017), accessed January 6, 2023, https://www.goodreads.com/book/show/49479772; *Goodreads, "The Story of a Place, The Story of a People: The Beginnings of the Syrian Revolution (2011–2015)*, sup. Sana Yazaji, trans. Rana Mitri (Beirut: Friedrich-Ebert-Stiftung, 2017), accessed January 6, 2023, https://www.goodreads.com/book/show/49478474-the-story-of-a-place-the-story-of-a-people.

Paratextual Elements

The translation is rendered word-for-word from Arabic into English and tries to utilize neutral language to describe the events in Syria, despite its bias toward the revolution—the latter being clearly evident in the decision to document and collect only what related to pro-revolutionary Syrians. The covers of the Arabic and English books are identical, with only slight differences in the titles that can however be interpreted in differently in each language. For example, the Arabic title is *Qiṣṣat Makān Qiṣṣat Insān* [A Story of a Place, A Story of a Human], while in English it is *The Story of a Place: The Story of a People*. This represents a slight change in meaning because of the different ways of translating *insān*. This interpretation represents the Syrian people, which is clearer for the English readership than the more multi-dimensional word *insān*. However, the word *insān* also has more contextual dimensions evoked by the revolution. For example, Muḥammad ʿAbd al-Wahhāb, who was one of the first refugees at the Turkish-Syrian border in 2011, famously described the situation in Syria with the words: "*Ana insān mānī ḥayawān wa hal ʿālam kilhā mitlī*" ["I am a human, not an animal. All of those people are like me"].[471] This was the first time after the 2011 protests that a Syrian had referred to himself and other Syrians as human beings, protesting the inhuman treatment of Syrians by the regime. The importance of this word is thus not its literal meaning but as a slogan that has many layers of meanings reflecting Syrians before 2011. Human, in this context, means a person who has the right to life and the minimum conditions required to live with dignity, and the use of the pronoun "I" illustrates that Syrians can be individuals and citizens instead of subjects of the regime. When I asked Sana Yazigi about this interpretation, she replied: "There is no direct relationship between this famous sentence and the title, but we cannot deny that this sentence was in our subconscious and used unintentionally in the title."[472] This is why one interpretation of "Qiṣṣat Insān" narrates the long struggle of Syrians in which they tried to become human in the eyes of the regime. The Arabic title thus gains more plurality on the level of interpretation while the English one is clear and direct.

The Arabic version of the book is more organized and methodologically introduced than the English one, especially in terms of its format and style. For example, the Arabic introduction is well-ordered, with separate titles and subtitles that are not included in the English version. This is a result of how the archiving method is described in the English and Arabic versions: the Arabic one lists the steps and phases

---

471 *Zamān al-Waṣil*, "*Ṣāḥib ʾIbārat Anā Insān mānī Ḥīwān*" [The Person Who Said I am Human Not Animal], June 12, 2014, accessed July 18, 2023, https://www.zamanalwsl.net/news/article/50719.
472 Sana Yazigi (CMSR team lead), interview with the author, July 4, 2019.

in subtitles, while the English version displays the information in paragraphs. The English book presents a shallow overview of the methods and does not limit the timescale of the study. The Arabic one delineates three phases of the act of archiving, while the English generalizes the act without specifying the phases, challenges, research methodologies, or tag selection criteria. Not having these listed as subtitles does not mean that they are not included in the English text, but it does mean that this research act of archiving is not assigned the same importance. To provide an example:

تمثلت أبرز التحديات:

1- اقتصار المراجع والمصادر على شبكة الانترنت، وأحياناً بشح المعلومات عن مكان ما
2- تعدد الروايات المكتوبة – لدرجة التعارض- عن المنطقة الواحدة وماشهدته من أحداث.
3- كون بعض الروايات عبارة عن تجارب شخصية لكتّابها
4- في روايات أخرى كانت الروايات المتواجدة منحازة تبعاً لوجهة نظر هذا الطرف أو ذاك في تفسيره لأحداث الثورة السورية وتحولاتها"[473].

The English translation reads:

> Among the challenges are: the fact that references and sources are restricted to the internet; lack of information about a certain place; the existence of contradictory accounts of a single event or region; the personal individual nature of some stories; the biases present in some stories.[474]

As we can see from this comparison, the challenges appear more serious and official when they are displayed as separate bullet points rather than as short phrases as part of a single sentence. The English translation merely summarizes the steps, while the Arabic one offers a detailed explanation of them.

Visual Translation

At the level of visual translation, the book offers different patterns of visual products about Syria from 2011 to 2015. To begin with the cover photo, there is no change between the versions; an identical cover is used for both. At the level of the visual contents of the book, the book provides several images for each city. These various selections do not represent the visual revolutionary products of the cities but are purposeless, with no reason provided for using a particular image. For example, a town such as Kafr Nabl was famous worldwide for its extraordinary banners, meaning that showing only one banner might not be enough. The same thing is true of other cities such as Aleppo or Damascus where students were very active, and there was a variety of visual products. In the book, however, a visual product is

---

473 CMSR, *Qiṣṣat Makān*, 11.
474 CMSR, *Story of a Place*, 11.

presented simply because it is from a particular city, with no importance assigned to what it might represent.

Following on from the randomness of image selection, I have divided the visual contents of the book into two groups: partially translated images, and untranslated images. This type of translation is rendered through images, rather than translation of words or phrases, which undoubtedly creates ambiguity for the reader in extracting any interpretation from the images shown, especially where they contain textual elements.

Partially Translated Visual Content

With regard to the visual examples from every city, the books are consistent in using the same photos in the Arabic and English versions. The only obstacle the English-speaking reader might encounter is in understanding the content of certain photos, since some of them are not fully translated or not translated at all. For example, as we see in Figure 25, which displays one photo from a collection of photos that are not fully translated, only the main content of the banner is translated, rather than the entire thing.

The English translation is full of examples of partially translated and untranslated photos. In the above image for example, the translation provided is: "Religious extremism and political despotism are two sides of the same coin."[475] After this sentence is an untranslated sentence that says *lā yakfī an naqlib al-ʿumla yajib an nastabdiluhā* [It is not enough to flip the coin, but we must replace it].[476] "Must" is underlined and written in bold to emphasize changing the situation of Daesh and the regime. In my opinion, one meaning of the image is not only to show the similarities between Daesh and the Assad regime, but also to show the desire of Syrians to replace both the despotism of the regime and religious extremism.[477] The way the photo is translated shows the rejection of Assad and Daesh but does

475 CMSR, *Story of a Place*, 343.
476 CMSR, *Story of a Place*, 342.
477 "Daesh," as used here, is an alternative term for ISIS. The reason for using Daesh instead of ISIS is that it highlights that it is not a state but a terrorist organization. Daesh is a neologism coined from the initial letters of al-dawla al-Islāmiyya fī al-ʿIrāq wa-l-Shām [The Islamic State in Iraq and the Levant]. Despite its apparent similarity to ISIS, in Arabic it has shifted from an acronym to an active word with derivations. For example, *Dāʿishī* is an adjective meaning "belonging to Daesh" and *Daʿshana* is a verb that means to perform a barbarian act or to become a member of Daesh. Due to this negative meaning of the acronym in Arabic, other acronyms have been derived with a similar rhythm like *ḥālish*—the acronym of the Lebanese Hizbullāh—and *Mālish*—the Iranian and Iraqi militias fighting with the regime.

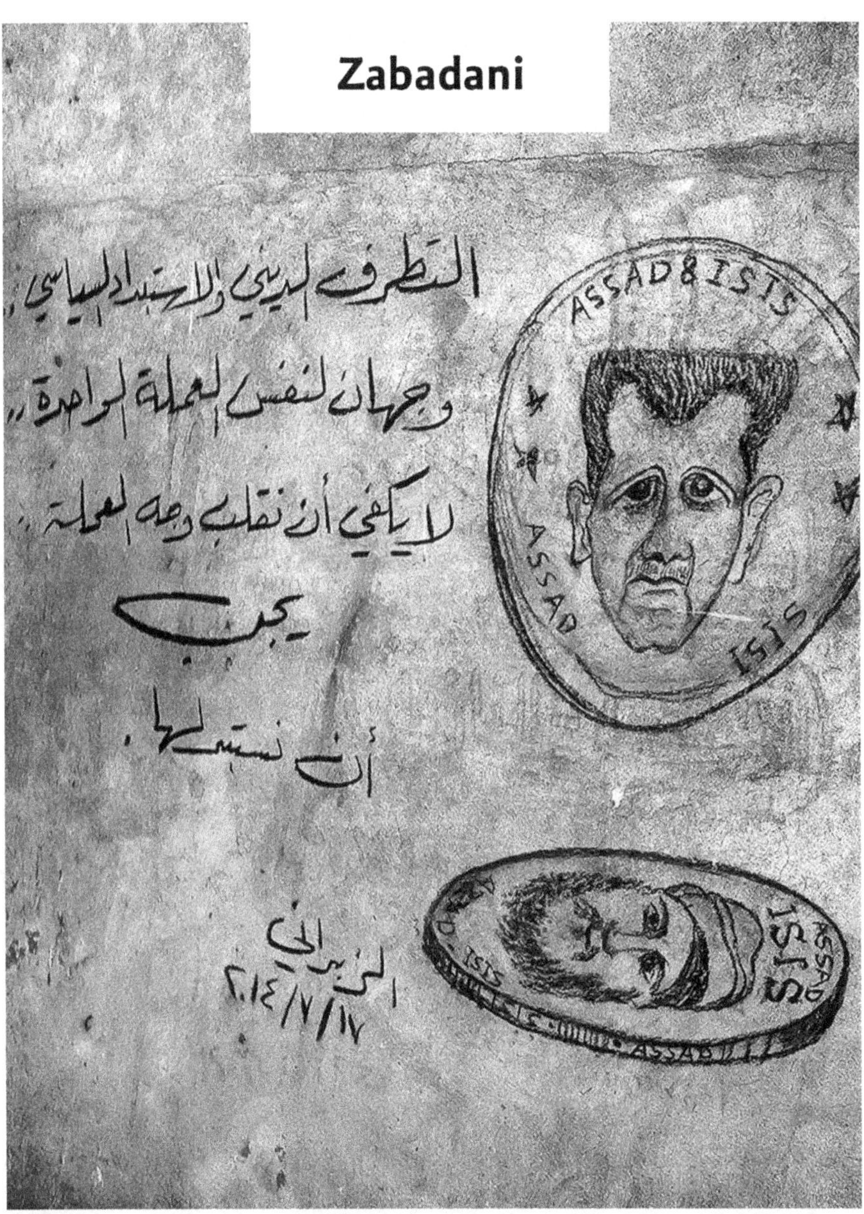

Zabadani
التطرف الديني والاستبداد السياسي
وجهان لنفس العملة الواحدة
لا يكفي أن نقلب وجه العملة
يجب
أن نستبدلها.
الزبداني
١٧/٧/٢٠١٤
ASSAD&ISIS
ASSAD
ISIS

Figure 25: One of the partially translated banners from Zabadani says: "Religious extremism and political despotism are two sides of the same coin. It is not enough to flip the coin, but we must replace it." There is a coin on the right, with one side showing Assad and the other Daesh.[478]

---

478 CMSR, *Story of a Place*, 342.

not show Syrians' desire to change both. In addition, the text makes the implicit claim that Daesh would simply reproduce the Assad regime, especially if read at the time it was published: 2014, when Daesh declared a caliphate in the state. This image represents a decisive repudiation of Daesh and the regime. At that time, the declaration of the Daesh caliphate provoked doubt among some activists about the direction of the revolution and its consequences. Some thought that the most important task was to eradicate the Assad regime and that everything thereafter would be easier to replace. Contrary to this suggestion, the image assertively rejects both options. In other words, the translation remains within a binary discourse and does not reveal the revolutionary change to which the early opposition aspired and sought to achieve. It shows that the regime and Daesh are similar, and that the main objective was to replace them, rather than merely denounce them.

Another example of partially translated photos and banners that creates ambiguity solely for non-Arabic speakers is shown in relation to Binnish.[479] This photo was taken during a strike held by civil defense activists in Binnish. The challenge of this photo is that it shows many activists, each holding a banner. In a book like this, it might have been more meaningful if the photo contained one translated banner or translated all the banners. The banners are all legible and convey a message that the book does not transfer in the English caption of "Activists from Civil Defense, Binnish, 2014." The message of the banners was one of opposition to the funders of civil defense. I believe this is an important message because of activists' criticism of the so-called "revolutionary institutions" working in liberated areas like Idlib and Aleppo in 2014. In addition to this, it is surprising for anyone familiar with Binnish that only this banner is translated. Binnish was not only known for notable revolutionary banners, but was the only Syrian city famous for using moving, interactive banners. These banners contained tens of small banners, which would be flipped at a specific moment to create a new banner that resembled mosaic art.

The third image (Figure 26) is from Amuda in 2013, and was produced after the explosion that killed Muḥammad Saʿīd Ramaḍān al-Būṭī, Mufti of Syria (1929–2013), in Damascus. Here, Al-Būṭī—the family name of the Mufti—is combined with the Syrian dialect word *būṭ*, which originally meant boot or shoes and is derived from English. The context of this word is the military boot in Arabic. One reading of the banner creates a connection between Al-Būṭī and the military boot—representing the regime—and accuses it of killing the Mufti. Al-Būṭī was famous for denouncing demonstrations, and was totally opposed to protesting against the regime. This reference makes a strong accusation that it was the regime who killed him. The

479 CMSR, *Story of a Place*, 60.

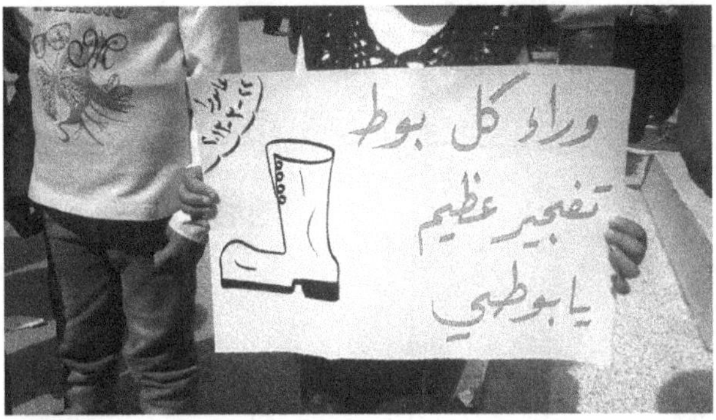

Figure 26: A banner in Amuda which reads: "Behind each military boot [būṭ] a great explosion, Al-Būṭī!".[480]

military boot, as a concept in 2011, became an icon of the regime, representing the protectors of Syrians in the eyes of regime supporters. It was of course seen differently by the revolutionary side, since whenever the army entered a city, many civilians were killed. For this reason, the military boot became a symbol of the killing machine of the regime. The English translation, "behind every military boot is a great explosion,"[481] overlooks the social and cultural dimensions of the banner itself. Moreover, the musicality of the banner text is missing, along with the connection between the family name of the Mufti and the military boot. It also fails to reflect the fact that the city of Amuda is populated by both Kurds and Arabs, making the banner exceptional because it was created somewhere with a majority Kurdish population. There is also a small hint, which is unclear on the banner but can be read next to the name of the city, that the Mufti himself was Kurdish. The translation of the banner precludes readers' understanding of the other dimensions involved, leaving them to interpret the banner alone, with no hints or clues.

Untranslated Visual Content

The second element of the book's visual content I will explore further comprises images that are presented in the English version without explanations of their content, only an indication of the artist/owner.

This collection of images confirms that the translations were produced literally word for word as in Arabic, but without translating their textual content. The caption for the first image in the English translation is "Jana Traboulsi, Houla," while the text in the photo says "This girl is called Houla. She has the same name of

480 CMSR, *Story of a Place*, 24.
481 CMSR, *Story of a Place*, 24.

Figure 27: A drawing by Jana Traboulsi: "This girl is called Houla. She has the same name of all children." It refers to the children massacred in Houla.[482]

all children." The image tries to shed light on the children massacred by the regime in 2012: thirty-nine children were killed in one of the most heinous massacres at that time and the first in which a large number of children were killed.[483] Including the image with only the name of its artist prevents the English audience from understanding the content. In addition to this, although the drawing is about Houla the artist is Lebanese. It is not of course the case that only the people of Houla have the right to write or create art about this massacre, but since one of the book's

482 CMSR, *Story of a Place*, 164

483 ʻAbd al-Raḥmān Khiḍir, "*Sabʻa Sanawāt ʻalā Majzarat al-Ḥūla. . . Najī Yarwī al-Maʼsāt*" [Seven Years Since the Massacre. . . A Survivor Narrates the Tragedy], *al-ʻArabī al-Jadīd*, May 25, 2019, accessed July 18, 2023, https://bit.ly/2Kmf4GM.

objectives is to provide an overview of the revolutionary acts and products for which each city was famous, displaying a drawing created elsewhere takes away this valuable opportunity from the city, preventing it from showcasing its symbolic goods. This city did in fact witness a very active wave of demonstrations, which is why the massacre was perpetuated as an act of vengeance by the regime. Without translating the textual content of the photo, I do not believe the English reader would be able to understand it. This example and those that follow are translations of images by images, frequently produced when translators have a scarcity of information about the topic and lack contacts in the field. And yet, according to the introduction, the authors of this book spent years researching, investigating, and conducting interviews with many people on the ground in these cities.

The second example is the caricature chosen for Palmyra, which again credits the artist but does not translate the textual content. The choice of this caricature is highly significant because it shows a different perspective from the one widespread in English-speaking media outlets. When the Assad regime destroyed Syria's cultural heritage there was no response or outcry from the international community, but when Daesh did the same, it was internationally denounced. The caricature tries to show the similarities between Assad and Daesh in destroying Syria both

Figure 28: Mwafaq Katt "Taddmur," a caricature showing Palmyra ruins through different historical periods: Roman, Assad, and DAESH.[484]

484 CMSR, *Story of a Place*, 248.

archeologically and with regard to its people. Usually, only the archeological side and relationship to cultural heritage are covered in English, but not the human suffering of the people there. The caricature shows three important phases in Palmyra's history. The first is the kingdom of Palmyra, represented by Zenobia, and bears the writing "Ruins of Zenobia." It can be understood as a sign of empowering women in Syria, since history demonstrates that a woman can be a queen, ruling a kingdom, without a king. It also shows how the city was prosperous, as the remaining archeological ruins demonstrate. The second phase, in the middle, is entitled "Ruins of Assad" and shows what the Assad regime was good at during its political reign: one of the archeological columns is destroyed—a symbol of neglect of Syrian heritage—and Palmyra has become symbolized by its prison. The last phase of the caricature is "Ruins of Daesh," and shows the destruction of the remaining ruins of

Figure 29: Stamps from the Syrian Revolution, August, 2012, Mūḥasan City: A special edition for the first year of the glorious Syrian revolution. [485]

485 CMSR, Story of a Place, 238.

the Zenobian kingdom and the existence of the prison of Palmyra, situating it similarly to the violent mechanisms of Assad. This comparison between the Assad and Daesh prisons illustrates their similarities with regard to brutality and violence. With this rich content and dependence on the captions inside the image in order to understand it, the English readership will not be able to glean the meaning of the photo. Its title, "Palmyra," may only reinforce a Western understanding of Daesh as barbaric and destroying Syrian heritage without showing its similarity to Assad.

The final example of an untranslated image comes from the city of Mūḥasan. Despite the fact that this stamp essentially depends on the text within it, none of it is translated in the English version, with the text instead only crediting the designer and explaining that it is a stamp from the revolution. As a stamp and a photo, most of the visual elements are unclear due to the lack of translation of the stamp itself, especially the text providing an explanation of the place and the stamp. Mūḥasan is a composite name, comprised of *mū* and *Ḥasan*: *Mū* is "the land from which the water of an overflowing river recedes,"[486] while Ḥasan is "[a] man who lived in the region."[487] This explanation is provided in the general introduction to the city in the English version, but without explaining that it is one of the textual elements of the stamp. The stamp shows a fighter plane, an image connected to two stories: the first of the resistance of Mūḥasan destroying French fighter jets in the colonial period,[488] and the second of the shooting down of the first regime aircraft by this city, which downed a total of eight air fighters.[489] As a result of these events, the stamp shows that Mūḥasan was called the "Bermuda Triangle" of the revolution and the cemetery of regime military airplanes. The comparison between these stories shows the significance of the airplane symbol on the stamp and treats the regime and colonialism alike. In addition to this, it is a symbol of the "Great Syrian Revolution," as the stamp shows, referring to the one that began in July 1925. The wave of demonstrations after March 2011 was called the Syrian revolution, but the Syrian war for independence is called the "Great Syrian Revolution." For this reason, the stamp is described as "a special issue for the Great Syrian Revolution" and is dated August 2012. All of these textual elements are missing from the translation, which thus prevents the English-speaking reader from understanding the content of the visual material presented.

---

486 CMSR, *Story of a Place*, 239.
487 CMSR, *Story of a Place*, 239.
488 *'Ayn al-madīna*, "Taqrīr khāṣṣ: Mūḥasan Taḥt a-Nār" [Special Report: Mūḥasan Under Fire], May 6, 2013, accessed July 18, 2023, https://bit.ly/2QtVxaT.
489 Firās 'Alāwī, "Mūḥasan: al-'Āṣima al-Ṣughrā" [Mūḥasan, the Smaller Capital], *al-Jumhuriyya*, November 4, 2016, accessed July 18, 2023, https://www.aljumhuriya.net/ar/35970.

Slogans

The final part of the book analyzed in this chapter is the way in which slogans are translated into English. Generally speaking, the act of selection is based on how far a slogan had spread to other places and whether it was repeated in most of the cities. Demonstrations in Syria typically began with slogans such as "huriyya" [Freedom] or "silmyya" [Peaceful], but other slogans were usually chanted rather than the general revolutionary ones. The book focuses only on those that were repeated in demonstrations and excludes others. Going to the source provided in the book—demonstrations themselves—the protests used not only the slogans included in the book but also three or four more that distinguished each demonstration and city from one another. For this reason, the selection in the book has not been made with the goal of showing a large number of different slogans but rather simplified to recognize that in most of the cities, as noted, demonstrators chanted "Freedom… freedom" and "Peaceful demonstration… Peaceful demonstration." In this section, I will offer a thick translation for the translated slogans, as I have done with the other revolutionary products in this chapter.

Table 16: Slogans and their translations.

Slogans and their translations		
Number	**Translation, page**	**Slogan, page**
1	Freedom, freedom, 172	حرية حرية،23
2	God, Syria, Freedom and nothing else, 172	الله سورية حرية وبس، 23
3	Where's the zeal? 172	وين النخوة، 23
4	Graffiti: Freedom forever despite you al-Assad, 67	جرافيتي: حرية للأبد غصب من عنك يا أسد، 38
5	He who kills his own people is a traitor, 317	يلي بيقتل شعبو خاين63
6	He who kills his own people is a traitor, 76	خاين يلي بيقتل شعبو130
7	Hey Bouthaina Shaaban, the Syrian people are not hungry, 317	يابثينة ياشعبان الشعب السوري مو جوعان، 63
8	We don't want forever… Assad, leave, 317	مابدنا للأبد أرحل عنا يا أسد، 63
9	Peaceful demonstration… peaceful demonstration, 140	سلمية سلمية، 94
10	One, one, one… Syrian people are one, 154	واحد واحد واحد. . . الشعب السوري واحد، 100
11	The people and the army, hand in hand, 166	الشعب والجيش ايد وحدة، 109
12	Death over humiliation, 96	الموت ولا المذلة،130
13	We choose death over humiliation, 199	الموت ولا المذلة، 254
14	The Syrian people will not be humiliated, 76	الشعب السوري مابينذل، 138

As most of these slogans are repeated in different parts of the book, here I have noted their first appearance and selected others. In this section I try to shed light on the translations of certain slogans and offer other possible translation(s) for them. It should be noted that the translation process for slogans in the book consisted of meticulously finding English alternatives for Arabic expressions and sometimes, as with most of the book, the translation takes a word-for-word approach. For this reason, in most cases, a literal translation fails to transfer dimensions other than the linguistic one or equivalence. Slogan 1, for example, reads "Freedom, freedom," which is a very accurate equivalent for the Arabic, but obscures the performative aspects of the slogan itself. One possibility might be to show the other dimensions of "freedom" for which demonstrators were calling. While it is relatively easy to access the meaning of a term like "freedom," an English-speaking audience might understand it differently and associate the term with other connotations, especially if it is compared to their most recent encounter with it. The freedom demanded by Syrians was not solely liberation from tradition (though it might be part of the term), but liberation and emancipation from the tyranny of the Assad regime. The term "freedom" was also used by both sides in the context of the revolution, with regime supporters referring to "freedom" in the contest of any act of vandalism or any crime, asking: "Is this the freedom you asked for? Is this it?" This other meaning and dimensions of freedom in the Syrian context for supporters of the regime is a contradictory one reflecting their loyalty.

Slogan 2 is totally divorced from its cultural and political history when translated as "God, Syria, Freedom and nothing else." In addition to the complexity of "freedom" explained above, this slogan underwent a long process of development during Assad's reign.[490] It was created specifically to destroy Assad's legend and replace his name with "freedom." Here, "nothing else" means no more of Assad's family in the new Syria. I would suggest a more nuanced translation for this slogan as: "Freedom. . .We have had enough of Assad; We want a free Syria" or "God and Freedom. . .Assad leave." The reference to God in the translation(s) here and in the table above refers to the one of all Syrian religions, which simultaneously carries the meaning of religious tolerance as well, and also subverts the regime's slogan of "God, Syria, Only Bashar." While the reference to God in the slogan seems therefore to give it a religious connotation this is not in fact the case, since its aim is more to subvert the slogan of the regime.

Slogan 3 shows the difficulty of translating a word like *nakhwa*, since the word "zeal" limits or eliminates its other cultural dimensions. First of all, this word was used to question the morals of people watching the demonstrations and to get

---

490 See previous chapter on the revolutionary language.

through to *shabiha* who were suppressing demonstrators. *Nakhwa*, in this slogan, was an invitation for passersby to join demonstrators, referred to by the Arabic term *faz'a*. The literal meaning of *nakhwa* is a sense of panic or great need to help someone, especially when they are being treated unfairly. Demonstrators chanted it as *wīn al-nakhwa* and followed it with *wīn*, which asks "where is your dignity" (i.e., how can you watch us demonstrating and not help us in what we are doing?). This slogan could also be simply translated as "Hey! Come join us" since one of its meanings is an invitation for people to join in, mixed with their feelings and emotions. For this reason, the translation of "zeal" does not fully cover all facets of the slogan.

In Slogan 4, the graffiti is also translated as "despite" for *ghasb min 'annak*, which is certainly an apt translation choice, but does not reflect the firm refusal of the Arabic word. When it is said as *ghasb min 'annak*, it means against one's will or done by force. While "despite" reflects one dimension of the graffiti, it does not show the way in which Syrians wanted to forcibly extract "freedom" from Assad's regime. Another element is that the translation personalizes the graffiti to Assad himself, in contrast to a deeper meaning that reflects all suppression perpetrated by authoritarian regimes. An alternative translation might be "we will gain freedom by force," or "whatever it costs us, we will have freedom." Here, the Arabic slogan says "forever," which is a reference to the cult of Assad, subverting the eternity of the Assad family in Syria and the idea that they would stay forever, as illustrated in pro-regime slogans like "Hafez al-Assad forever."[491]

Slogan 5 is explained above in relation to the translation of Samar Yazbek's book and is rendered differently in this book, but the translation may have missed one dimension of the connotation of putting "traitor" at the beginning or at the end of the phrase.[492] Slogan 7, meanwhile, is decontextualized. The impetus for it was a decision by Buthaina Shaaban, a consultant for Bashar al-Assad, to increase people's salaries in 2011: the initial reaction of the regime to the Daraa demonstration, regardless of the fact that people were not protesting for higher salaries but for the release of their children who had been arrested in March 2011. The book mentions this slogan only once and states that it was repeated in May 2011, almost two months after the Shaaban speech promising reforms on March 23, 2011. This does not mean that it was not chanted in the interim, but that it was a confirmation of the demonstrators' reaction to Shaaban as a spokesperson for the regime. It has been translated word-for-word, but the context may be lost. I would thus

491 See the previous chapter on Hafez al-Assad and "eternity/Forever" in slogans.

492 See the previous section on translating this slogan and the difference created by putting "traitor" at the beginning or at the end.

suggest a more rhythmic slogan that an English-speaking reader might enjoy and that also reflects certain dimensions of the original slogan in Arabic. One possible translation is "We are neither poor nor hungry. . . We demonstrate even if you are angry." This translation reflects more of the economic and political dimensions of the event itself and carries a rhythmic sense to it in English. The adjective "hungry" in the Arabic slogan not only reflects the economic dimension but has a potential cultural interpretation as well, since there is a Syrian proverb saying "He is rich but hungry." This connects hunger not with food or the luxuries that one might have but indicates a person who has no dignity or is always in need of more. This gives "hunger" the meaning of being greedy and suggests that, while the person described is not poor, they are acting that way. Demonstrators using this slogan tried to show that the regime's reforms made them seem both greedy and poor, in the economic sense of the word. For this reason, the connotations of this slogan function not only to subvert the regime's reforms but also to dismantle the mask of the reform: the increase in salaries that the people of Daraa did not need because they were not "hungry."

Slogan 8 is translated literally, but does not explain the concept of "Assad as the immortal leader."[493] It is not meant literally, for when demonstrators said, "We do not want forever," it meant the reign of the Assad family was over and the era of a free Syria was nigh. Slogan 9 provides a good explanation of its intent as "Peaceful" (*silmiyya* means peaceful), by referring in the slogan itself to the fact that the demonstrations are peaceful ones. It was one of the most repeated slogans throughout the waves of protest seen in Syria and confirms the peacefulness of protestors and the act of protesting, contrary to the images of armed protestors killing civilians presented in the regime media.

Slogan 10 is a good example of showing the context of the slogan by mentioning the reason and place it was developed: Homs. Homs is a multi-religious city, and the regime took advantage of this by warning protestors not to convert activism into sectarian violence. To show their unity in response, demonstrators chanted: "One, one, one; Syrian people are one." In addition to the word-for-word translation, this slogan can also be translated as "Syrians are united against the regime's plans." Here I refer to the regime because it was a response to the regime's intent to escalate the violence, which is what ultimately happened. A neatly domesticated translation of Slogan 11 is produced by replacing *īd waḥdi*, which means "one hand," with "hand-in-hand." This translation increases the elegance of the slogan and alludes to the meaning of the Arabic expression of "one

493 See Chapter 1 on the development of the Assad cult in slogans.

hand" as simultaneous unity and cooperation. Another possible translation might be "united together."

The last three slogans are connected by one word: *al-dhull*. Although "al-mawt wala l-madhalla" is a short and simple slogan, it is mentioned in two different contexts with two different translations. According to the book, the first was in Daraa and the second in Latakia. In both contexts, the slogan was written the same but pronounced differently; the people of Daraa pronounce the interdental *dh* (like "th" in "the"), while in other cities it was pronounced "z." The former city was the first to chant this slogan, followed by all of the other Syrian cities. This example, as many others, shows how slogans in dialect were able to travel from one geographic location to another. Both translations are acceptable and well-written in English, but the meaning of *dhull* cannot be explained in any English translation without reference to its cultural and historical meaning. First, this slogan evokes the famous words by Husayn, the son of Ali, that "hayhāt min adh-dhull," meaning that *dhull* should be far away because we do not accept it. The word is also connected to the death of Husayn, who faced death refusing *dhull*. Since any translation could be appropriated into our modern times and the context of authoritarian regimes, I would not suggest any translation other than the two offered in the book. The final slogan is also related to humiliation and the concept of *dhull* in Arabic. Adding a musical aspect to the slogan itself in English, I might suggest it be translated as "Not humiliated, not subjugated" as a double condemnation of the long-standing violent practices of the regime that served to humiliate Syrians. Another layer of this slogan that cannot be left out is the first slogan referring to *dhull* in the first demonstration by Syrians in Old Damascus in February 2011.[494] Moreover, as is the case for the other slogans, this slogan was not performed in a void: the reaction of the regime to it and others is one layer of its interpretation.[495]

By illustrating all of these different types of translation examples, I do not seek to detract from the efforts that have been made to render the revolutionary language in English. I have however shown how translating the revolutionary language in Syria is a very complicated act, and how translators do not have the final decision on what and how to translate, since other factors decide what the final product looks like. The various examples taken from the three books discussed above illustrate that they offer a language that is easy for the English-speaking readership to access via domestication that avoids conveying more cultural facets of the source, that would be more observed by foreignization. Foreignization is misplaced only to evoke negative aspects and dimensions, as seen in Weiss' trans-

494 For further explanation of this slogan in the revolutionary performance, see Chapter 3.
495 See Chapter 1 on the slogans of the regime.

lation of the "Free Virgin Women," or inserting "Allahu Akbar" out of context in order to add more foreign layers to the translation. The translation either by images without translated text, inserting elements of Islamization, appropriating themes and concepts for a different audience, and sectarianizing layers goes hand in hand with clear violence to the source text by the target text. Publishing houses, discourse and media, the translation market, the influence of the translator's ideology, and the narratives they bring with them are the reasons for the different interpretative translations for the revolutionary language. Thick contexts, rich layers, historical pasts, and cultural and political dimensions are mostly omitted from the three books. As we shall see below, the act of translation is done very differently by activist translators, local agents, and through individual initiatives from a non-English-speaking place to expose the English-speaking readership to the revolutionary language.

Translation as An Act of Exportation from a Peripheral Language

As seen above, translation from the center and on the margin is unable to present the thick contents and contexts of this language for numerous reasons, including translation difficulties, publishing house policies and marketing, and dominant and dominated discourses. As such, it would be impossible to mention a slogan like "Hafez, curse your soul" in an English translation performed from the center. Given these conditions of the translation market, were any initiatives engaged in translation as activism or self-translation into foreign languages in Syria?

The sociological approach to translation, as Johan Heilbron and Gisèle Sapiro see it, consists of three main elements. The first is the "structure of the field of the international cultural exchange."[496] The second aspect is related to "the dynamics of exchanges,"[497] which locate translation between two additional elements: politicization, and commercializing the act of translation in the market. This is characterized by the three translations discussed above; the act of translation is politicized in the first place and commercialized by appropriating the translated texts to the English-speaking readership. Usually, this act is characterized by joint funding of projects with a government body or represented by a state and an external organization. In the case of Syria, however, such funding is absent because since 2011, to the best of my knowledge, the regime government has not participated in

496 Heilbron and Sapiro, "Sociology of Translation," 93.
497 Heilbron and Sapiro, "Sociology of Translation," 97.

funding cultural products, which have been funded only externally. This aspect is represented by the translation of the three books discussed above: *A Woman in the Crossfire, Syria Speaks,* and *The Story of a Place.* Finally, a sociological approach to translation that describes the translation market scene and its products looks at "the agents of intermediation and the process of importing and receiving in the recipient country."[498] This section examines translation as an act of exportation from a peripheral language (Arabic), by looking at attempts made by agents other than the ones above who have produced translations into English—individually or collectively—to offer other readings of Syria. I describe it as a peripheral local act of translation, as the translators and agents who are responsible for transferring the Arabic text (revolutionary language) into English are given a narrow space that is not included in the market of the dominant language (English). Despite having produced their products in the dominant language, they cannot take up a space bigger than the one occupied by the peripheral language. This act of translation can be described as marginal, nearly invisible, and unheard, because these agents do not have the tools to access the main discourse. While there have been some Syrian initiatives, collective works, and individuals who have translated Syria in 2011, because they were not connected and did not have the network to access the English translation market, their work has remained in the same unseen sphere in which it was created. Examples include the Free Syrian Translators, the *Mundassa* website, and individual activists who created their own virtual space on social media and blogs.

Free Syrian Translators

A co-founder of the Free Syrian Translators explained the initiative's main objectives to me. Browsing its now-inactive website shows all of the items translated from the original languages into all of the languages available: Arabic, English, French, Dutch, German, Italian, Spanish, Kurdish, and Russian. The Free Syrian Translators worked not only with textual elements but all types of materials, including videos, interviews, articles, initiatives, posters (invitations or events) and news (a handful of items). Heilbron and Sapiro note that "The dominated languages export little and import a lot of foreign books, principally by translation,"[499] and this is seen clearly on the website, which includes 422 items imported from English and 150 items exported to English. If we take a closer look at the number of items, we

498 Heilbron and Sapiro, "Sociology of Translation," 95.
499 Heilbron and Sapiro, "Sociology of Translation," 96.

find that there were only 26 items, including 17 visual ones, translated from Arabic into English during the period November 6, 2011 to April 1, 2012.[500] During the same period, 215 items were translated from English into Arabic. The numbers on this website alone, specializing in translation, support the description of the act of importing and exporting as set out above.

The first translated item was a speech by Burhan Ghalioun on the occasion of Eid al-Adha.[501] I investigated how this speech was received in the English-language news and how it was translated or mentioned compared to the translation produced by the Free Syrian Translators. The results of a Google search with the use of a time-frame tool, which shows results at specific times for specific keywords (in this case the keywords were "Burhan Ghalioun" and "SNC Syrian National Council"), showed that he addressed Syrians in a speech but that most English-language news outlets highlighted different parts of the speech. For example, *The New Republic* magazine printed Ghalioun's speech in full, through the lens of the potential future Kurdish reaction to joining the SNC, in an article entitled "Why Syria's Kurds will determine the Fate of the Revolution."[502] Ghalioun's speech says the following about the Kurds, as translated by the Free Syrian Translators:

> Syria will never be the home of discrimination, injustice, and insularity. Rather it will become a home for a united Syrian people. With no more majority or minority but rather citizenship and equality. Treating all its people equally without any national, sectarian, doctrinal, or regional considerations. The new Syrian constitution will protect all minorities and reserve their rights, and the Kurds will be given their rights that have been deprived of and the discrimination they suffered from.[503]

As seen above, the speech mentions the Kurds in only one short sentence, as one component of Syria. In contrast to the Arabic speech, however, the entire English article discusses the situation of the Kurds in Syria from a sectarian perspective, analyzing the Syrian minorities and majority, which is not the objective of a political speech describing the future Syria envisaged by the Syrian National Council (SNC) after the fall of the regime.

---

500 This is the date of the first translated item from English to Arabic.
501 Burhan Ghalioun, "The Speech of the National Syrian Conference's President-subtitled," Eid al-Adha, trans. The Free Syrian Translators (YouTube channel), uploaded November 6, 2011, accessed July 18, 2023, https://bit.ly/2yrb2K4; Carnegie Middle East Center, "Burhan Ghalioun," accessed July 18, 2023, https://bit.ly/3cV33Us. Burhan Ghalioun is a Syrian academic affiliated to the Sorbonne in Paris. In 2012, he was the first chair of the Syrian National Council.
502 Michael Weiss, "Why Syria's Kurds will Determine the Fate of the Revolution," *The New Republic*, November 16, 2011, https://newrepublic.com/article/97493/syria-kurds-national-council.
503 Ghalioun, "Speech," Eid al -Adha, trans. The Free Syrian Translators.

On November 6, 2011, the London-based *Tamil Guardian* published an article that included only one sentence from the Ghalioun speech: "We will not negotiate on the blood of the victims and martyrs [. . .] we will not be deceived. The National Council will not allow the regime to bide for time."[504] The rest of the speech, with all of the guaranties Ghalioun offered for the future Syria, was disregarded. The speech was eight minutes long and showed the future of Syria desired by the Syrian council; its importance lay in showing the attitudes of the Syrian opposition to Syrians and the international community.

Syria Comment, one of the most well-known blogs in English about Syria, described the mixed reactions to the speech, posting excerpts of responses from other bloggers and the media.[505] The first was a very positive and relaxed response from *Maysaloon*, which stated that the speech was very reassuring to Syrians.[506] The second one criticized the speech, asking for guarantees that Alawi and Christians would be protected after the fall of the regime. Despite the fact that the speech addressed all minorities and exposed the huge grievances of the Kurds, reactions by other bloggers or media outlets noted by the *Syria Comment* did not feel that it was sufficient. These reactions again reflect the emphasis on sectarianism in the Western reception of anything produced from Syria and its conversion to include sectarian meanings.[507] In this specific post, the blog does not appear to take a position, but shows two different attitudes: one from a Syrian blogger and the other from the *Los Angeles Times*.

Another translated text is an interview with a leader of the al-Ashtar Battalion of the Free Syrian Army (FSA),[508] originally published in *Al-Hayat* on January 27, 2012.[509] The Arabic and English texts are very close regarding the unity of terms.

---

504 *Tamil Guardian*, "Syrian Death Toll Rises Despite Arab League Deal," November 6, 2011, https://www.tamilguardian.com/content/syrian-death-toll-rises-despite-arab-league-deal.

505 Joshua Landis, "Ghalioun Addresses Syrians for 'Eid Holiday in Bid to Establish Authority" in *Syria Comment* (blog), November 6, 2011, accessed July 18, 2023, https://www.joshualandis.com/blog/ghalioun-addresses-syrians-for-eid-holiday-in-bid-to-establish-authority/.

506 *Maysaloon* (blog), http://www.maysaloon.org/, accessed July 18, 2023; for full text of response to speech, see Maysaloon (blog), "Is a New Syria Possible," November 5, 2012, accessed July 18, 2023, https://www.maysaloon.org/search?q=ghalioun. *Maysaloon* is a Syrian blog that reviews Syrian news and criticized the Western interpretation of the Syrian revolution.

507 Landis, "Ghalioun Addresses Syrians;" Patrick J. McDonnell, "Opposition leader tells Syrians: 'Days of tyranny are numbered,'" *Los Angeles Times*, November 6, 2011, https://lat.ms/2Xhk1qC.

508 Razan Zaitouneh, "'And Thus We Turned to Military Struggle,' The Commander of Al-Ashtar Battalion in Al-Rastan," *Free Syrian Translators*, February 1, 2012, accessed July 18, 2023, https://bit.ly/3cUyaPS.

509 The article has been deleted from the Al-Hayat archive but is saved on the *SouriaHouria* website: Razan Zaitouneh, "*Qāʾid Katībat al-Ashtar fī al-Rastan: Hākadhā Taḥawwalnā ilā al-ʿAmal al-ʿAskarī*"

The word *thawra* is translated as "revolution," and it shows that the Free Syrian Army in Rastan is protecting demonstrators, not attacking the regime. The fighters are not "rebels," as described by most English-language websites, but referred to as "soldiers" and individuals who have defected from the regime. In addition to this, "soldiers of the battalion" is translated as "military personnel." This translation of the "Free Syrian Army" (as it is called in Arabic), is rarely found in English books, or the general English discourse. Members of the Free Syrian Army are instead generally translated as "rebels," a translation rejected by the revolutionary discourse in Arabic for many reasons. First of all, being a "rebel" means that you believe the other side (the regime) has legitimacy, was elected by the people and came to power legally, which is not the case with Syria: Hafez al-Assad came to power through a coup, and his son literally inherited Syria after him by changing the constitution, which otherwise would have prevented him from becoming president due to his age.[510] Second, the word "rebel" is associated with violent acts of resistance that were not seen as a general scene in the early months of the Syrian protests by the FSA.

Looking at the terms and vocabulary used by the Free Syrian Translators to translate articles from English into Arabic in the months of October, November, and December 2011, we can see that terms like "civil war" were used in the very early stages of the protests in October (the official start date for translating English articles into Arabic). It is important to say here that the English articles did not describe the demonstrations as a civil war, but this was seen as one potential future for Syria in the early stages. This reflects the act of article selection by the Free Syrian Translators and does not necessarily mirror all English-language journalism. However, as Reem Assil notes, "the act of selecting of articles was done by voting for the top-viewed articles in English on the revolution," indicating that the idea was prevalent in the English-speaking media.[511] Despite the fact that this initiative was working for the revolutionary side and saw their translations as activism, all of the English terms that are usually perceived as negative, such as "civil war," "rebels," "war," and "conflict" were translated according to the English context using a literal translation unchanged based on the revolutionary discourse or as they would have been in a translation from the center. As such,

---

[The commander of Al-Ashtar Battalion in Al-Rastan: "And thus we turned to military struggle"], *SouriaHouria*, January 1, 2012, accessed July 18, 2023, https://bit.ly/35Im48P.

510 *Oxford English Dictionary online*, "rebellion," accessed July 18, 2023, https://bit.ly/2XmTVTl. The *Oxford English Dictionary* defines rebellion as "an attempt by some of the people in a country to change their government, using violence." This is why I argue that the act of "rebellion" gives legitimacy to the regime and disregards the pacifist demonstrations of protestors.

511 Reem Assil, interview.

the terms were not translated as those used only in the revolutionary discourse. For example, what an English newspaper called "civil war," the Free Syrian Translators translated into Arabic as "ḥarb ahliyya." Contrary to what we saw in *Syria Speaks*, "niẓām qamʿī" was translated as "repressive state." The above examples show a different way of translating Syria in 2011 from English to Arabic and vice versa. All of the texts, videos, and visual items are translated as a mirror of the source language text, without inserting elements or changing terms into the revolutionary discourse.

Mundassa

Mundassa was the first platform that gave Syrians a place to exchange ideas and opinions.[512,513] The official launch of the English-language version of the website can be dated to the first post on July 8, 2011. The website is organized differently in English than in Arabic and has tabs for opinions, videos, and cartoons. The first post on the website explains the reason for the name *mundassa*, which is the female form of *mundass* [[male] infiltrator], used by Bashar al-Assad to describe demonstrators.[514] The opening post then goes on to ask an important question about perceiving Syria and its news, "How to understand what is happening in Syria away from [*BBC*, *CNN*, etc.]? My answer is you can go to Facebook and search for the English language news pages that talk about the 'Syria revolution.'"[515] Zead—likely a pseudonym—suggests looking for websites in English that talk about the "Syrian Revolution," implying that English-language websites that do not call what is happening in Syria a "revolution" do not need to be read. *Mundassa* is politicized, rather than neutral. Other articles are critical of English-language articles on Syria, such as an article published in the *Washington Post* in July 2011 which hinted at a coming Syrian civil war, quoting an unknown businessman who said "it is on the edge of civil war."[516] The English-language post on *Mundassa* explains how this

512 Unfortunately the website has been deleted and I am not able to cite specific links to the pages referred to; my only source is an archived version of the website itself, which, unfortunately, in its archived form prevents me from citing links. I will use the available information for the posts I am referencing without links. Please see screenshots of the post in Appendix 7.
513 For more information about *Mundassa*, see Chapter 2 on the revolutionary language.
514 Arabic nouns are gendered, with the base form being male. Thus, *mundass* would be (male) infiltrator.
515 Zead Raed, "Syria, What is missing," *Mundassa*, July 13, 2011 (personal research archive). Please see Appendix 7.
516 Liz Sly, "Sectarian Violence in Syria Raises Fears," *Washington Post*, July 18, 2011, July 18, 2023, https://wapo.st/3e7jJs1.

article feeds into the regime propaganda, portraying demonstrators as gangs who demand the right to kill other sects.[517]

Below I will provide examples of how Syria in 2011 was written about and translated into English on the *Mundassa* website. Some posts are translations of articles and others were written originally in English without translation in order to target the English-speaking audience. I will analyze examples of articles originally written in English, especially ones including revolutionary products like photos or slogans.

In an article entitled "What is happening in Syria," Arabi—a pseudonym—analyzes the Syrian protests. When it comes to demonstrations and slogans, the writer explains the full context of the slogans, giving a broader perspective of the oppressive practices and strategies of the regime. For example, Arabi notes that demonstrators said: "We are all one hand." The article makes the context clear in that, at the time, the regime was promoting sectarian clashes among protestors, who thus chanted this slogan to subvert the regime's story. Later, another slogan—"No to Sectarianism"—was added, and is accompanied by the same act of contextualization. Arabi says: "all Syrian social classes and ethnicities are involved. Assyrian Christians demonstrated despite Syrian security forces... in Qamishli 3500 Kurdish protested..."[518] He continues to mention all of the cities in which demonstrations took place in order to explain the meaning of this slogan for protestors from different cities and different ethnic and religious backgrounds who are participating in protests. This act of thick translation not only translates the slogans of demonstrators on the ground, but also provides context and a detailed explanation of who was protesting, illuminating the regime's agenda of promoting other narratives, which were then generally picked up by the English-language media in general while the revolutionary narrative was silenced.

Another example that embeds translation inside the text on the website can be seen in an article by Manar Haidar that discusses the upcoming battles in Aleppo and Damascus in 2012 and the potential total collapse of the regime.[519] This article seeks to prepare people to back the Free Syrian Army in the battle for Damascus and reminds people who protested against the regime of the values of the revolution. Haidar writes: "We have to prove that our revolution is for dignity and that the Syrian people are one. Otherwise, we would be taking our country to the

517 *Mundassa*, "Is it really 'Sectarian Violence' or Another Cheap trick?" July 19, 2011 (personal research archive), see Appendix 7.

518 Rabi'e Arabi, "What is Happening in Syria," *Mundassa*, August 14, 2011 (personal research archive), see Appendix 7.

519 Manar Haidar, "On the Fall of the Regime and Before Statement Number 1," *Mundassa*, July 31, 2012 (personal research archive), see Figure 36.

unknown. . . A civilized victory in popular revolutions is usually demonstrated in two points: tolerance and protecting government establishments."[520] With this short explanation, the banner shown below (Figure 30) is self-explanatory in context and literally states: "The difference between justice and vengeance is the same as the difference between revolution and the regime." This example and others translated and intentionally contextualized the revolutionary products by analyzing a topic and supporting it with a revolutionary example.

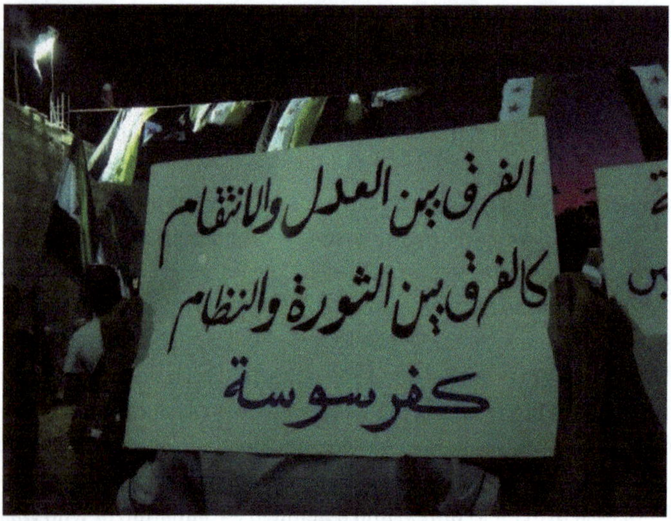

Figure 30: A banner in Kafr Sūsa, Damascus: "The difference between justice and vengeance is the same as the difference between the revolution and the regime".[521]

The function of the photograph and the text is to complement each other. The article does not provide a direct literal translation, but rather a thick translation of the revolutionary products it discusses.

Individual Initiatives

The above examples are the result of initiatives created by activists to translate and transfer the Syrian voice into English. A similar collaborative initiative is Sham News Network, which began as a normal coordination Facebook page to provide

520 Haidar, "On the Fall of the Regime."
521 Untitled photo of a banner from Kafr Sūsa, Damascus, *Mundassa* archive.

news about the revolution and developed into a professional news agency inside Syria.[522] It was officially founded on July 6, 2011.[523] Despite the fact that it became a source of information for many media outlets, its goal was to transfer information from Syria to the world. Its work was not mere translation but rather sought to display short, descriptive translations of Syrian events.

Figure 31, for example, shows an evening demonstration in Damascus, with a subtitle simply describing the general scene of the footage, without translating the slogans and banners carried by demonstrators. In this case, the translation does not transfer the context of the demonstration but only provides evidence that there was a demonstration on this date in Damascus. The goal of this initiative was not to translate but to produce footage for easy use by the Western media. It is likely that the foreign media did not deeply analyze the demands of demonstrators but rather simply mentioned that a revolutionary act took place, with an accurate date and place. The screenshot in Figure 31 is taken from one of the Sham News Network videos and shows that the subtitle provides a short description about the event but not the full context of the event itself. Informing an English-speaking audience reader that the protest took place during Ramadan does not contextualize this timing: since Ramadan is a difficult month for demanding activities, such as protesting, due to the long hours spent fasting, one interpretation is that this shows the persistence of people in wanting to topple the regime despite such challenges.

Figure 31: Sham News Network: A screenshot from a YouTube video of a demonstration.[524]

522 See the previous chapter on the functioning of the revolutionary coordination Facebook pages in covering the revolutionary events.

523 Sham News Network, "Who We Are."

524 Sham News Network, "SNN: Evening Demonstration Against Al-Assad's Regime, August 6, 2012," August 6, 2012, video, accessed July 18, 2023, https://www.youtube.com/watch?v=oxOvXWMUsaY.

Another example comes from the blog by Alisar Iram (pseudonym), who devoted four years to translating important works from Arabic into English. One of her posts is a translation of Yassin Haj Saleh's article "Farewell to Syria, For a While."[525] In this, Iram uses a literal translation for every term while translating. Sometimes she decides to foreignize certain terms, coining the adjective "Assadi" for Assad. She also insists on calling the organization known elsewhere as the Islamic State or ISIS, Daesh. The term Daesh not only gives the name of the organization an ugly sound, but further gives the impression in Arabic that it is a sinister monster or *ghūl*. This impression comes from the similar atrocities caused by Daesh and *ghūl* with no differentiation between the supposed state and the beast. However, each letter of this shortform refers in Arabic to the Islamic State in Iraq and Sham. Similarly, in 2014 *Hizbullah* was called *Ḥālish:* another acronym with negative connotations.[526] Moreover, the use of this term delegitimizes their claim to be a Muslim Caliphate or an Islamic legal committee. Another example from the same blogger is about the change in the direction of the revolution.[527] In a comparison of two texts, she adds additional words and sentences in parentheses to make herself clear to the English-speaking readership. For example, while explaining the reasons for the revolution, she says:

> The Revolution of the Syrian people was not kindled against the shrines of the House of the Prophet and his beloveds (revered by the Sunna and the Shiaa alike) or for the revival of the caliphate or in support of the Sunna against others in blatant sectarianism.[528]
>
> The Arabic text:
>
> "ثورة الشعب السوري لم تقم من أجل هدم الاضرحة والاساءة الى اهل البيت، احبَاء النبي محمد، ولم تقم من اجل احياء الخلافة او حتى من أجل نصرة اهل السنة على غيرهم."[529]

We see in this example that she adds the comment, "revered by the Sunna and the Shiaa alike" to provide additional information for the English-speaking reader who is informed that it is a religious conflict. In her Arabic text, she says "for Sunni and others," and it is clear that she means both Sunni and Shia alike. If it had been translated like the Arabic text, it might have been understood as "Sunni and other affili-

525 Alisar Iram, "*Fī Wadā' Sūriyya Mu'aqqatan,*" *Alisariram* (blog), October 12, 2013, accessed July 18, 2023, https://bit.ly/2QVD2Mx.

526 "*'Araftum 'Dā'ish' . . . fa-man 'Mā'ish' w-'Ḥālish'*" [You Already Know Daesh (ISIS), Who is Mā'ish and Ḥālish'?] in CNN Arabic, February 25, 2014, accessed July 18, 2023, https://arabic.cnn.com/middleeast/2014/02/25/syria-lebanon-isis.

527 Alisar Iram, "da'ūnā lā nuḥaqqer 'nfusunā walā thawratanā" [Let Us Not Debase Ourselves Or Our Revolution], *Alisariram* (Blog), June 6, 2013, accessed July 18, 2023, https://bit.ly/33lT6tL.

528 Iram, "Let Us Not Debase Ourselves."

529 Iram, "Let Us Not Debase Ourselves."

ations of Sunni," but the parenthetical phrase makes sure that the English-speaking reader would understand it as meant. Moreover, she adds "sectarianism" in the English text to negate it as a reason for the revolution, which, as previously discussed, is a common misconception for English-speaking readers.

Conclusion

When it is considered as an act of importation to the center from the periphery, the act of translation happens not in a vacuum but from and to a discourse that is regulated and situated to suit it, even if it takes a slightly different shape than the present discourse on the Syrian revolution. Although only three books including the revolutionary language have been translated into English, this is a great success from a translation point of view, compared to other political events in the Global South that have been written about through the translational lenses of the revolution. Despite the exceptional nature of these translations and the existence of support through the three funders, the differences in the source and target language source texts are mainly focused on main repetitive elements: Islamization, violence and militarization, simplicity, and sectarianism through falling into the trap of the English-language discourse regarding the Syrian revolution in 2011. Two books were published in translation before the original Arabic text. Malu Halasa explained to me that "this was the policy of the publishing house. They think that its market will be bigger if it is published in English first."[530] This suggests that such works are meant to be translated as a source language text and that their existence in Arabic does not interest an Arabic readership who already has a more detailed picture about the situation in Arabic, as shown by Goodreads and Google Scholar. Producing translations without highlighting the source language of the translated text operates an intentional loss of the source of the knowledge which obscures its roots and opens onto more ambiguous interpretation of the translated works. It confirms that the only important audience of the book is the English-speaking one. One necessity that Syrian authors are working through in their exile with their lack of competency in the languages of their host countries is producing their works directly through translation without having an Arabic text for their work, or releasing it months later, or never having their books in Arabic. This need is established by the demands of funders who require translated books rather than Arabic books to release in the book market through different artistic residencies in exile, due to the low demand of Arabic books in exile. While this appears to display the good

530 Halasa, interview.

intentions of bringing new authors into the new cultural field, it does so only in theory, since they are still labeled with the place they came from. This produces a new phenomenon of refugee authors, academics, and playwrights, etc.: a phenomenon that has the superficial aim of integrating knowledge from outside but in fact limits them and boxes them into their art and translation community ghettos.

This stands in contrast with the work of activist translators or translation from the periphery toward the center, which although it does not reach the center, follows a similar approach to thick translation. As Sapiro observes, we can see—for example with the figures from the Free Syrian Translators website—that the periphery imports more translations than it exports. This enables the dominant discourse to select texts for translation and, with the participation of some internal agents, to shape the translation in the dominant discourse as being produced from the center, for the center, about the margin. Translation in this sense recognizes the dominant English-language discourse and even uses the exact terms that this discourse produces about the revolution such as "civil war" and "conflict," even if it does not suit the discourse into which they are employing their translation. Deep contexts, thick histories, and multiple layers of this language are found in this type of translation, as seen in the way in which a simple slogan or a banner is developed over dozens of years, affected by political events, reshaped by political shifts, and then finally molded in its final phase before being translated. The act of translation from the center to the periphery translates only the final phase, without taking into consideration the pre-processes or pre-making of the slogan, and without unpacking its contents.

If seen as a violent act of translation, this violence, demonstrated in the above research, functions with "an imperialist appropriation of foreign cultures for domestic agendas," that domesticates the source language text and offers an easy interpretation for the target language audience.[531] Once the text is close to the readership, it automatically engages with the dominant discourse narratives and tries to offer a different story about the same event but one that is similar if it is scrutinized meticulously as done in the three books. One book alone does not represent this discourse, but it is confirmed by the repetitive statements of the three books in translation. This does not, however, confirm any bad intentions on the part of the translators, editors, or writers of the translation. Despite the attempts of the translators of the three books including the language of the Syrian revolution to give a voice to Syrian activists and those who supported the revolution, translations

531 Lawrence Venuti, "Translation as a Social Practice: or the Violence of Translation," in *Translation Horizons: Beyond the Boundaries of Translation Spectrum*, ed. Marilyn Gaddis Rose (Binghamton: State University of New York, 1996), 211.

produced by the center on the periphery are characterized by the above violence that works contrary to "locus translation." It is very naive to draw an accusation like this, but it is a more complicated matter that is embedded in the domination of narratives and discourses, publishing houses' policies and censorship, and/or unintentional acts of experimental practices of translation.

In contrast, activist translators perceive thicker contexts, contextualized translation, thick translations, and plural ones, when it comes to translating themselves or from English to Arabic. The aim of this latter is to keep as much of the foreignness of the text as possible in order to accumulate resistant layers of translated texts against "ethnocentrism, racism, cultural narcissism, and imperialism."[532] Such translational practices can be seen in the Anglo-American translation theory that erases the cultural characteristics of the source language text and adds domesticated concepts such as Islamization, violence, and others to the target language text for the English-speaking reader. This translation does not reflect the essence of the discourse of activists and shows it either superficially or in a modified way. For this reason, compared to the activist translations, the translations of the language of the Syrian revolution represented in this research and illustrated in the three books discussed above show a reduction that prevents an English-speaking readership from understanding the history of this language. It is unsurprising therefore to see that activist translation is more detailed than funded translations or the work of professional translators despite their attempts to take a stand for the revolution.

532 Venuti, "Translation as a Social Practice," 196.

Conclusion

باسم النظام يتم توليد الخوف, وبفعل الخوف يتم فهم النظام نفسه
ممدوح عدوان، من حيّونة الإنسان

"On the name of the regime, fear is generated and by fear itself, the regime can be understood"

Mamdouh Adwan, *Ḥaywanat al-Insan* [Human's Animalization]

As my research has shown, one could argue that the language of the Syrian revolution was a revolution itself on many levels. At the political level, it was used to revolt against and oppose the regime's language, and became the medium used by revolutionists. It was also a revolution of language, with new terms, expressions, and innovative uses observed in the first year of the revolution.[533] In addition, this language was revolutionary because it was able to emancipate itself in 2011, after decades of gestation and marginal maneuvers defying the language and discourse of the regime. This process of emancipation took several decades, but in the end, the language used in the Syrian revolution had separated from the dominant discourse.

During the decades of gestation, pro-revolution Syrians in the oppositional discourse created strategies to secretly challenge the discourse of the regime. One technique, acting "as if," was used by Syrians to make the regime believe it had their support.[534] Under the "rules" of such performances, oppositional Syrians in the regime spheres were allowed to engage in a degree of criticism. Such permitted (and limited) criticism can be seen in certain symbolic products created under Hafez al-Assad, which miriam cooke terms "commissioned criticism."[535] Importantly, this technique was no longer effective after 2011, and varied geographically across Syria. For the regime, the performance of "as if" was perceived as a sign of obedience and its domination over Syrians. While the regime's interest lay in the final product of the "as if" performance—obedience—there was a simultaneous oppositional discourse that did not engage in "as if" performances, and criticized the regime beyond the limits it permitted. An additional component of dominance was prison as a form of symbolic violence, with the regime using this to rule Syrians. However, I argue that these schemas of dominations were not the only elements that produced the compliance of Syrians. Gradually, between the 1980s

533 For more information on the use of the term "revolution" in this research, see the introduction.
534 Lisa Wedeen, *Ambiguities of Domination: Politics, Rhetoric, and Symbols in Contemporary Syria*, 2nd ed. (Chicago: University of Chicago Press, 2015), ix.
535 miriam cooke, *Dissident Syria: Making the Oppositional Arts Official* (Durham: Duke University Press, 2007), 72.

https://doi.org/10.1515/9783110767698-006

and 2011, more political prisoners were released than were arrested. This explains that prison, acting "as if," and the limited spaces of criticism were not the only techniques used by the regime.

I argue that the symbolic violence represented by discourse and language was a more effective instrument to coerce Syrians into compliance, through the consumption and reuse of a language in which they were immersed from childhood through to adulthood. This does not mean that the regime stopped terrorizing Syrians with the violence of killing and imprisoning, but its domination fed on the echo of atrocities that were committed only when needed, as in Hama, Qamishli, Sweida, and then in 2011. The time between each massacre secured a long period of discipline among Syrians, before another massacre had to be committed in order to maintain this balance with the help of the discursive products. Despite all of the different modes of violence that the regime practiced in the last half century, an oppositional discourse was developing slowly but very steadily within the dominant discourse. This was the case until March 2011, when this discourse was shaped more clearly by the language of the revolution as the regime lost control of most spaces, which were contested and reclaimed by activists. The language of the revolution then emerged.

Born from the discourse of the regime, as illustrated in many examples in the previous chapters, the revolutionary language and its discourse took advantage of this shared history and development in a strategic fashion. The performances of the revolutionary language that established itself in 2011 could often be seen to imitate the dominant discourse. In other examples, the regime's production of symbolic products tended to imitate the revolutionary ones. The same practices used by the regime to dominate Syrians were repurposed by the revolution to emancipate itself from the regime discourse and solidify into a new oppositional discourse. The mutual practices of the regime and oppositional discourses illustrate how the revolutionary language should be understood not only on the basis of its current context but also from its own development, a history that goes back to 1970.

Despite the revolutionary language's defiance of and emancipation from the regime language and dominant regime discourse, it was not as successful in translating itself for an English-speaking audience. From the earliest days of the protests, activists wondered why Syria was a neglected revolution. In contrast to Egypt and Tunisia, the Syrian revolution was translated without taking into consideration the thick past of the revolutionary language and its sophisticated, multilayered nature, which left it open to multiple, significantly different interpretations. As I show in the previous chapter, Arabic is one of the least translated languages into English, making up less than 1% of all works in the English translation market. Translations of books about Syria from Arabic into English make up only a small fraction of that 1%, which includes all the books from all Arab countries translated into English.

This illustrates the typical relationship between central and peripheral languages in knowledge exchange as approached sociologically in translation studies. Central languages export more than they import, which is the case for the Arabic-English translation flow. All of this led to translating the Syrian revolutionary language by combining translations with concepts and themes that the English-speaking readership is accustomed to perceiving in general discourses and narratives, such as sectarianism, Islamization, and violence. This made it an act of domestication rather than foreignization, by which translation itself is a hindrance that does not give the reader deep access to the thick contents of the language.

As this illustrates, this book does not solely focus on translation in the literal sense of what makes a good or bad translation. A translation is a matter of interpretation that becomes a representation of a specific narrative in a discourse. When different terms and labels are used in one language and discourse to differentiate between the discourse of the regime or the revolution, the translational act cannot be a neutral act once it is translated into English. What translation was not able to clearly show was the contest between narratives and discourse in Arabic itself, before the transfer into English. Describing the revolution in Arabic as a "crisis," or referring to "events," draws directly on the regime's discourse. In this sense, translating the Arabic equivalent of "revolution" as a "conflict" or "crisis" sticks to the regime discourse of perceiving the revolution as a temporal event caused by illegal demands from external powers to destabilize the country. How much is lost through the act of translation if it deactivates all of the struggling dimensions of this dynamic? In this context, translation is a political tool that must engage with activism, since in their introductions, all of the translated books discussed state that they want the stories of Syrians to be heard. If this is the case, how were they heard? Were activist agencies voiced? If yes, how? And what labels were used?

Despite the clear political agenda of the three books to support the Syrian struggle against the Assad regime, the examples discussed show that the translators were unable to emancipate themselves from the English dominant discourse. The translators, initiatives, and publishing houses were crystal clear about their intention of demonstrating what was not said about the Syrian revolution and its language as an act of advocacy, or in order to empower pacifist activists. However, either for the sake of simplifying the complexities of the Syrian situation for a general English-speaking readership, the indirect influence of English-language domination, narratives, and media discourse, the commercial imperative of publishing houses to sell more copies, the lack of linguistic knowledge of Syrians who were unable to have access to their translated texts, or the translators' beliefs that their innovative translations could make a difference in translation, translations of the language of the revolution into English did not reflect its complicated nature and did not unpack its multiple layers.

In some cases, English-language domination over not only translators, but also academics, generalizes terms and views. At academic events, I sometimes had difficulty, for example, making my audience understand what I meant by the Syrian "revolution": I would be asked if I meant to say the "civil war" or "conflict." When raising such queries, academics and translators do not question their own position, the geographic place from which they are looking at Syria, their linguistic skills, or the embodied discourse from which they speak. Greater awareness is needed to justify and explain the grounds that a translator is using in order to engage with such a work of translation, which starts by meticulously studying the needs for such a translation and drawing attention to possible slips that he or she might make unintentionally while translating a text which sees itself in its original language as being from the revolution. By selecting a combination of repeated terms, a translator not only shows their reliance on a specific discourse or narrative, but "activates quite a different narrative framework"[536] that shows a political reality and frames and reframes a political event such as the revolution in Syria.

When it comes to revolutionary times, translations demonstrate how translational choices affect not only the final product if compared to the source language text, but also how they fit into the English-language discourse, since it is impossible for the translator to remain neutral. Translation is rendered from one native discourse represented in the source language text into a new one in the target language text.

In contrast to translations from the center on the margin, activist translators, despite their limited ability to distribute their translations to the center, were more careful in offering English texts with their source language texts, placing particular importance on the cultural, social, and political context of the language of the revolution. This act of translation, rendered by activist translators, is very similar to thick translation: a translation that gives more contextual detail to the source language text and makes readers delve more into new dimensions that the English translation lacks while translating, transferring, and voicing the voiceless. Contrary to this, however, and despite the sincere attempts to produce English translations from the center on the margin, the way in which they were translated reproduced numerous similar dimensions that activists did not want to include when representing themselves, for example by showing a simple posters in *Syria Speaks* compared to a complex image in the Arabic book, or adding violent terms or elements to show violence that were not present in the Arabic text.

---

536 Mona Baker, "Reframing Conflict in Translation," in *Critical Readings in Translation Studies*, ed. Mona Baker (London: Routledge, 2010), 157.

A thick translation practice requires not only an understanding of the implications of translating a text from one language into another, but also an awareness that the translator must keep in mind at all times while translating a text. In times of revolution and war, a translator cannot be neutral and remain in the middle. The role of the translator is to empower the cause for which the translation is created. Here I am referring to an ethical responsibility that a translator has to write about those voices and for them to be voiced. However, this responsibility is shared equally with the authors, who in some cases do not know the language into which their work is translated: they agree to the translation in order to reach the largest possible audience without being able to engage critically with it in detail.

This ethical responsibility is shouldered equally with scholars who are researching Syria and revolutionary moments, for the importance of translation is manifested not only in its use as a tool by which to transfer a text from one language into another, but also in the writing and re-writing of events, uprisings, and revolutions opposed and eradicated by dictatorships in order to win not only over Assad's destroyed Syria, as now appears to be the case, but also on narratives that might dominate if this responsibility is not implemented in practice as instruments of research. As I see Syria now in 2023, the act of writing and re-writing is equal to translation in its resistance. In this sense, all research that is produced on the revolutionary language and protests in Syria represents a consequential archive, a huge part of which has been lost already through Assad's distortion of Syrian narratives.

Appendices

Appendix 1
Political Prisoners 1984 to 2011

A brief review of the annual Human Rights Watch reports illustrates the relationship between violence and the regime's control of Syrians. The most significant period of arbitrary detention before 2011 was during the 1980s, when the Muslim Brotherhood movement revolted against the regime. The annual report from Amnesty International in 1984 reported 3,500 persons arrested.[537] In 1990, Human Rights Watch reported more than 7,000 political prisoners, the largest number in the history of Syrian imprisonment at that time.[538] Two years later, Human Rights Watch said that 2,826 prisoners had been released and 52 new prisoners arrested.[539] In 1995, 260 were released and 100 arrested.[540] The number of prisoners released increased the following year, reaching 1,500 with few new arrests.[541] The year 1997 indicated an improvement in human rights, as few arrests were reported.[542] In the following three years, 1998 to 2000, 330 prisoners were released, and the only arrests reported were those of a few well-known activists.[543] Then Bashar al-Assad became president of Syria. During the first decade of Bashar's reign, about 1,200 political prisoners were released, and around 600 were arrested prior to 2011 and the revolution.[544] The below graph illustrates the number of the arrested and released in the time of frame of 1984 to 2011.

537 Amnesty International and John G. Healey, "Syria: The Amnesty Report," *The New York Review of Books*, January 19, 1984, accessed July 18, 2023, http://www.nybooks.com/articles/1984/01/19/syria-the-amnesty-report/.

538 Human Rights Watch, "Human Rights Watch World Report 1990 – Syria and Syrian-occupied Lebanon," January 1, 1991, accessed July 18, 2023, http://www.refworld.org/docid/467fca3ec.html.

539 Human Rights Watch, "Human Rights Watch World Report 1992 – Syria and Syrian-occupied Lebanon," January 1, 1992, accessed July 18, 2023, http://www.refworld.org/docid/467fca5c1e.html.

540 Human Rights Watch, "Human Rights Watch World Report 1995 – Syria and Syrian-occupied Lebanon," January 1, 1995, accessed July 18, 2023, http://www.refworld.org/docid/467fcab8c.html.

541 Human Rights Watch, "Human Rights Watch World Report 1997 – Syria," January 1, 1997, accessed July 18, 2023, http://www.refworld.org/docid/3ae6a8ac28.html.

542 Human Rights Watch, "Human Rights Watch World Report 1998 – Syria," January 1, 1998, accessed July 18, 2023, https://goo.gl/HNER9g.

543 Human Rights Watch, "Human Rights Watch World Report 1999 – Syria," January 1, 1999, accessed July 18, 2023, https://goo.gl/JY9s6E; Human Rights Watch, "Human Rights Watch World Report 2000 – Syria," December 1, 1999, accessed July 18, 2023, https://goo.gl/H2hDiA; Human Rights Watch, "Human Rights Watch World Report 2001 – Syria," December 1, 2000, accessed July 18, 2023, https://goo.gl/yHJ2t7.

544 Human Rights Watch, "Human Rights Watch World Report 2002 – Syria," January 17, 2002, accessed July 18, 2023, https://goo.gl/qR743f; Human Rights Watch, "Human Rights Watch World Report 2003 – Syria," January 14, 2003, accessed July 18, 2023, https://bit.ly/3aLEpVK; Human Rights

https://doi.org/10.1515/9783110767698-007

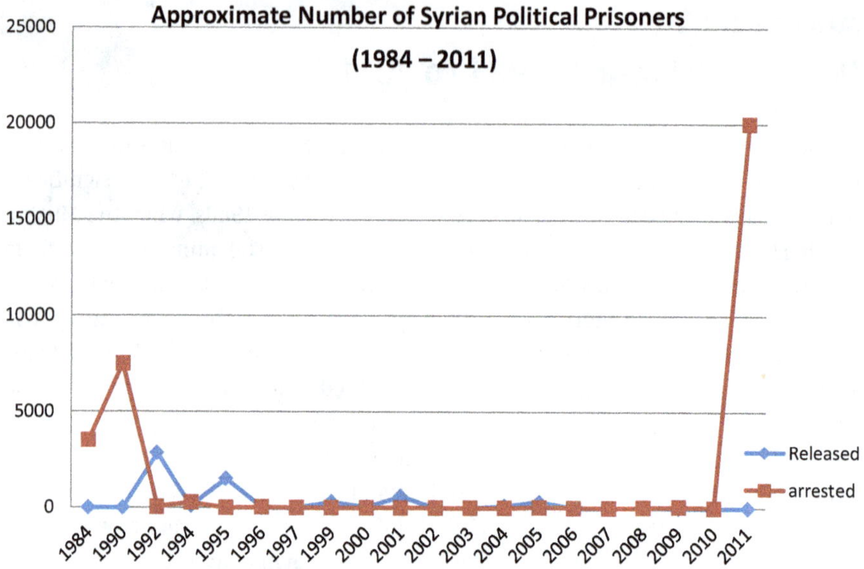

Figure 32: Approximate number of Syrian political prisoners during the Assad regimes (data adapted from Amnesty International, "Syria: The Amnesty Report," and Human Rights Watch "World Report," 1990–2011).

Watch, "Human Rights Watch World Report 2004 – Syria," January 1, 2005, accessed July 18, 2023, https://goo.gl/tGFyny; Human Rights Watch, "Human Rights Watch World Report 2005 – Syria," January 1, 2005, accessed July 18, 2023, https://goo.gl/TbepmZ; Human Rights Watch, "Human Rights Watch World Report 2006 – Syria," January 18, 2006, accessed July 18, 2023, https://goo.gl/UgTzve; Human Rights Watch, "Human Rights Watch World Report 2007 – Syria," January 11, 2007, accessed July 18, 2023, https://goo.gl/v2Cvut; Human Rights Watch, "World Report 2008 – Syria," January 31, 2008, accessed July 18, 2023, https://goo.gl/csgJyW; Human Rights Watch, "World Report 2009 – Syria," January 14, 2009, accessed July 18, 2023, https://goo.gl/V5Pc5r; Human Rights Watch, "World Report 2010 – Syria," January 20, 2010, accessed July 18, 2023, https://goo.gl/PBCoz1; Human Rights Watch, "World Report 2011 – Syria," January 24, 2011, accessed July 18, 2023, https://goo.gl/MtNH2A.

Appendix 2
Revolutionary Songs Over A Year

All of those songs can be listened and watched online on the website of Creative Memory of the Syrian Revolution: https://creativememory.org/ar/collections/eylaf-bader-eddin-songs (See Appendix 6 for the songs).

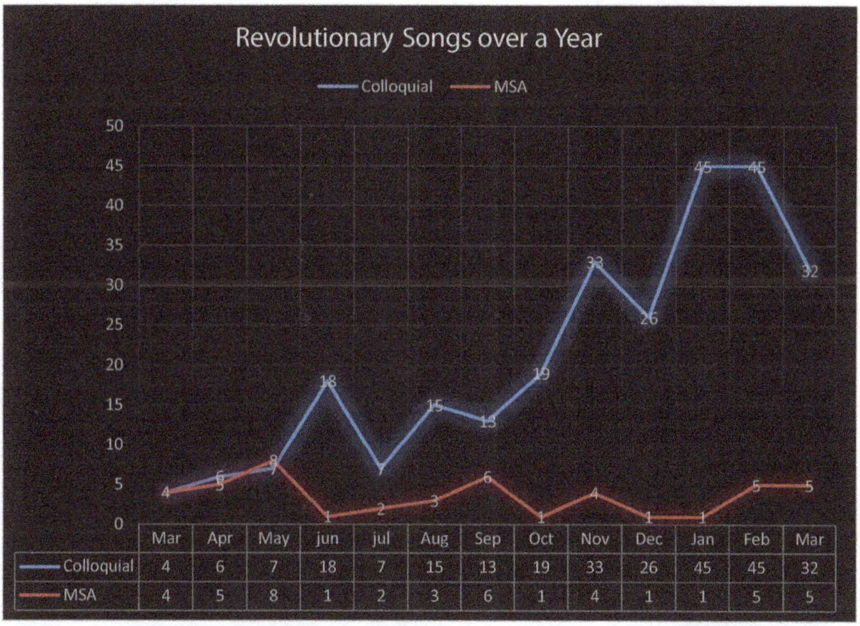

Figure 33: Graph showing the language used in revolutionary songs.

https://doi.org/10.1515/9783110767698-008

Appendix 3
Publishers of English Translations From Arabic

Abe Books
Akashic Books
Amana Publications
Amazon Crossing
Anchor Press
Antibook Club
Arabia Books Ltd
ARC Publications
Archipelago Books
Atlantic Books
AUC Press / Hoopoe
BAO Editions
Black Swan
Bloomsbury /US/UK/Qatar Foundation/ HBKU
Brepols Publ
Brigham Young Press
Brill
C Hurst & Co Publisher Ltd
Canongate Books
Center for Middle Eastern Studies, University of Texas and Austin
City Lights Publishers
Clockroot Books
Comma Press
Copper Canyon
Cosimo Classics
Curbstone
Darf Publishers
Diwan Press
Ediburgh University Press
Emblem Editions
English PEN
Everyman's Library
Farrar, Straus and Giroux
Fons Vatae
Forgotten Books
Fourth Estate
Free Press
G Duckworth
Ganta Books
Garnet Publishing
Georgetown University Press
Graywolf Press

https://doi.org/10.1515/9783110767698-009

Harper Collin
Harper Perennial
Harvill Secker
HausPublishing
Hutchinson
IB Tauris
Interlink
International Islamic House
Islamic Book trust
Islamic Foundation
Islamic Text Society
Jonathan Cape
Leiden University
Louisville
Lynee Rienner Publishers
Maclehose Press
Marlboro Press
Massachusetts Review
A Midsummer Night's Press
Miracle Publishing
Moment Digibooks Limited
Neem Tree Press Limited
New Directions
Northwestern University Press
Nufal
NYU Press (New York University Press)
NYRB Classics
Olive Branch Press
One World Publication
Oneonta
Pantheon
Penguin Books
Periscope Books
Picador Paper
Pluto Press
Princeton
Random House Trade Paperbacks
Reading University Press
Rider
Routledge
Sandsone Press
Saqi Books
Seagull Books
Seven Stories Press
Sheep Meadow Press
Sternberg Press

Sunflower Books
Syracuse University Press
Telegram Books
The Feminist Press
The International Institute of Islamic Thought
Tupelo Press
Turtleback Books
University of California Press
University of Chicago Press
University of Texas Press
Verso
Vintage
Wavelande Press
Yale University Press
Zaidan Foundation
Zed Books

Appendix 4
Music Group 1 (Regime → Revolution)

"سوريا يا حبيبتي"

"Syria, My Beloved"

Cover version by al-Mundassīn	Original version used by the regime
أعيدي لي كرامتي أعيدي لي حريتي Give me back my dignity. Give me back my freedom.	أعدتِ لي كرامتي أعدتِ لي حريتي You gave back my dignity. You gave back my freedom.
بالحب والوفاء وصرخة الإباء أنير درب ثورتي With love, commitment and the scream of disobedience, you enlighten the path of my revolution.	بالحرب والكفاح وشعلة الجراح تنير درب ثورتي With war, struggle, and the flame of wound you enlighten the path of my revolution.
شعبنا يسير لنصره الكبير، مبشراً بعودتي ورافعاً كرامتي مجدداً هويتي Our People are on the way to their great victory, heralded by me coming back raising up my dignity and confirming my identity.	بعثنا يسير لمجده الكبير، مبشراً بعودتي ورافعاً كرامتي مجدداً هويتي Our Baath is on the way to the great glory, heralded by me coming back raising up my dignity and confirming my identity.
الآن أمضي إلى غدي حريتي ملك يدي Now, I am reaching my tomorrow and my freedom is in my hand.	الآن إني عربي، يحق لي اسم أبي ومن أبي Now, I am an Arab. I have the right to hold the name of my father.
رصاص بندقية لن يقتل الحرية في أمتي الأبية The bullets of a rifle will not kill freedom in my proud Umma.	رصاص بندقية مصنع حرية للأمة الأبية The bullets of a rifle is the freedom factory of the proud Umma.
يا شعبنا يا فخر كلّ ثائر يا قلعة الأحرار والحرائر Oh, our People! You are the pride of every revolutionary and the castle of the freemen and freewomen.	سورية يا درب كلّ ثائر يا قلعة الأحرار والحرائر Oh, Syria! You are the path of the revolutionary and the castle of the freemen and freewomen.
برغم كل القمع والمجازر صمودكم حرّك الضمائر Despite all of the suppression and massacres [by the regime], your resilience has touched everyone.	صمودك العظيم في البشائر تزف للآمال والضمائر Your great resilience [against Israel] is a good omen for our hopes and expectations.
أعاد لي كرامتي أعاد لي حريتي [The above-mentioned suppression and massacres of the regime] gave me back my dignity and gave me back my freedom.	أعيدي لي كرامتي أعيدي لي حريتي Give me back my dignity. Give me back my Freedom.

"Syria, My Beloved" by Ghurabāʾ Grūb	سوريا يا حبيبتي (جروب غرباء)
O Syria! My revolution won't be allayed; it will keep flaring up.	سوريا حبيبتي لن تستكين ثورتي حتى تعود ثورتي
We shall keep moving forward: our slogan is Peace. O, God, bless my revolution.	وسنمضي للأمام، شعارنا السلام، يا رب بارك ثورتي

https://doi.org/10.1515/9783110767698-010

(continued)

"Syria, My Beloved" by Ghurabā' Grūb	**سوريا يا حبيبتي (جروب غرباء)**
O, Damascus! Rage in freedom and pride! O, Aleppo! Rage towards freedom, in the footsteps of Homs which has hoisted the banner of eternal revolution, calling on the free people to safeguard the Home.	هبّي دمشق حرة أبية، هبي يا حلب إلى الحرية، أمضي فهذه حمص الأبية، حملت لواء الثورة الابية، تنادي يا أحرار هبوا لنحمي الدار
Bringing glad news of my triumph and hoisting my banner. Syria, my beloved.	مبشرا بنصرتي ورافعاً رايتي، سورية يا حبيبتي
No matter how buildings are subject to fierce artillery shelling, I shall keep steadfast in defense. Disguising masks have dropped, you 'false resistor, for the sake of being in service to the Jews.	مهما قصفت الدور بالمدافع، قسماً سأبقى صامداً أدافع، سقط القناع أيها الممانع، لخدمة اليهود تبقى طائع
O, Coward! You who sold Golan Heights! I swear by Allah, I won't abandon my revolution. Syria, my beloved.	يا أيها الجبان يا بائع الجولان، الله أكبر انني عن ثورتي لن انثني سورية يا حبيبتي
O you, Syria's thug! I voluntarily offer my blood to ransom free women no matter what atrocities you perpetrate. Almighty God is powerful over everything. We pledge to martyrs that we shall not swerve from our path. Pound the aggressor's prisons and shatter them by hand. Syria, my beloved.	يا طاغي الشام بأرض جائر، دمي يهون في فدى الحرائر مهما ارتكبت بشعبي المجازر، الله ربي قادر وطاهر نعاهد الشهيد، واللهي لن نحيد، دكّوا سجون المعتدي وحطموه باليدي، سورية يا حبيبتي

Music Group 2 (Revolution → Regime)

"Yā ḥīf" [Alas/Shame] (Original), by Samih Choukair	**ياحيف (الثورة) سميح شقير**
Alas/What a shame	يا حيف اخ ويا حيف
Bullets rained on the disarmed people, alas.	زخ رصاص على الناس العزّل يا حيف
Oh, how do you [the regime] arrest flower-aged children?	وأطفال بعمر الورد تعتقلن كيف
While you are the son of my country, you are killing my children. Your back is to your enemy and you attack me with your sword.	وانت ابن بلادي تقتل بولادي. وظهرك للعادي وعليي هاجم بالسيف
This is what happened... Alas... Oh, my mother... Alas... This is Deraa... Alas.	وهذا اللي صاير يا حيف. بدرعا ويا يما ويا حيف
Oh, mother, when the youth heard that freedom was getting close, they went out to chant for it.	سمعت هالشباب يما الحرية عالباب يما. طلعو يهتفولا

(continued)

"Yā ḥīf" [Alas/Shame] (Original), by Samih Choukair	ياحيف (الثورة) سميح شقير
Oh, mother, they saw rifles. . . They said they were our brothers. They would not fire on us.	شافو البواريد يما. قالو اخوتنا هن ومش رح يضربونا
Oh, mother, they shot us with live bullets. . . We died at the hands of our brothers in the name of the security of the country.	ضربونا يما بالرصاص الحي. متنا. بايد إخوتنا باسم امن الوطن
Who are we? Ask history. Let history read our page.	واحنا مين إحنا واسألوا التاريخ. يقرا صفحتنا
One word of freedom, oh mother, trembled the jailors' authority.	مش تاري السجان يما كلمة حرية وحدا هزتلو اركانو
When the crowd chanted, he became like a stung person, throwing fire on us.	.ومن هتفت لجموع يما اصبح كالملسوع يما. يصلينا بنيرانو
It is we who said, "the one who kills his people is a traitor. . . no matter who he is."	واحنا اللي قلنا اللي بيقتل شعبو. خاين. يكون من كاين
The people are like fate, when it asks, it should be answered. The people are like fate. . . the hope is obvious.	والشعب مثل القدر . من ينتخي ماين. والشعب مثل القدر. والامل باين

"Alas/Shame!" Cover version by William Hasan (1)	ياحيف وليم حسن
You betray your people and friends! Shame on you! How dare you kill your home's soldiers?	تغدر اهلك وأصحابك خاين ياحيف وجنود بلادك تقتلهن كيف
Syria is your home, and your grandchildren's dreams will come true with the blade of a sword.	سوريا بلادك وأحلام أحفادك بالمحبة منبنيها وبحد السيف
The youth heard that freedom was at the threshold, an opportunity they made use of to pull down the enemy's plans and to spoil their plots.	سمعت هالشباب يما (أمي) الحرية عالباب يما (أمي) فرصة ستغلوها خطتها لتخريب يما (أمي) لمصلحة الأعداء يما (أمي) خطة نفذوها
Under the pretext of freedom, they devastated the country.	باسم الحرية يما مرافق كلا يما يما دمروها
Some dollar-worshipping people beside them have sold their principles.	وحدهم (بجانبهم) أوغاد من أجل الدولار يما (أمي) أخلاقهن باعوها
Syria is your home, and your grandchildren's dreams will come true with the blade of sword.	سوريا بلادك أحلام أحفادك بالمحبة منبنيها وبحد السيف
On Syria's land, the dearest of all lands, are terrorists with arms in their hands that killed our siblings, claiming it peaceful.	على أرض الشام يما (أمي) وبأغلى الاوطان يما (أمي) عناصر أرهابية، بايدون سلاح يما (أمي) قتلوا أخوتنا يما سموها سلمية
We, the Home army, have a clean sheet of deeds. We were martyred for our siblings. Those who sell their country are traitors whose fathers are so too, whoever they are!	نحن جيش الوطن هل بيضا صفحتنا استشهدنا لأخوتنا ويلي يبيع الوطن خاين ابن خاين ويكون مين كاين

(continued)

"Alas/Shame!" Cover version by William Hasan (1)	**ياحيف وليم حسن**
Pro-Assad demonstrations erupted in Syrian cities. Free people chant we ransom you, (Syria, the emblem of freedom) with our souls.	طلعت مسيرات يما للأسد تأييد يما بالمدن السورية هتفوا هل الأحرار يما بالروح منفديك والله يا رمز الحرية
You, roses of Syria, vividly carry the glimmer of hope.	وأنت يا ورد الشام فيك الأمل بادي
We are horsemen, threatening the aggressors.	واحنا معك فرسان يا ويلو لمعادي
Syria is your home and grandchildren's dreams. We shall build it all.	سوريا بلادك وأحلام أحفادك ، ايد بأيد منبنيها مدينة وريف

"Alas/ Shame" Cover version by Sām Ḥisīkū (2)	**سام حسيكو ياحيف**
Shame on those who burn a flag.	ياحيف عال يحرق علم يا حيف
Shame on those who wound a country.	يا حيف عال يجرح وطن يا حيف
Shame on those who betray our army, the country's backbone.	يا حيف عال يغدر بجيشنا وهوي عماد الوطن يا حيف
Shame! the army has been betrayed.	يا حيف والجيش انغد يا حيف
Shame! It is an irony of fate.	يا حيف نسخر هالقدر يا حيف
Shame on the army of betrayal.	يا حيف من جيش الغدر ياحيف
Shame on those who drink from a well and throw stones into it.	ياحيف يشرب مي ويرمي فيها حجر
Shame on a bad-quality art.	ياحيف يا فنان على فن انكسر

"Alas/Shame" pro-regime cover version by Sumar al-Hamwi (3)	**ياحيف (النظام) سومر الحموي**
Shame	يا حيف اخ ويا حيف
How are [these] leaders ruling in the name of Arabs?	حكام باسم العرب عم تحكم كيف؟
Syria is the enemy and Israel is the guest! How?	سورية العدو صارت واسرائيل ضيف كيف؟
The Arabic Qatar became Hebrew, accompanied by Saudi Arabia, which is using its tongue as sword.	قطر العربية، صار عبرية، ومعها السعودية بلسانها تهجم بالسيف
Shame. . . This is a conspiracy against the Syrian People. . . Shame.	هذا التآمر ياحيف. عالشعب السوري ياحيف
Peaceful and reforms, oh mother, we chanted with them.	سلمية واصلاح يما . وهتفوا للحرية يما، نادينا معاهم
But we saw Salafists and weapons. The sedition was getting close. So we challenged them.	سلفية وسلاح شفنا، والفتنة عالباب يمه، قمنا نتحداهم

(continued)

"Alas/Shame" pro-regime cover version by Sumar al-Hamwi (3)	**ياحيف (النظام) سومر الحموي**
They betrayed us. Blood turned water. It is a sedition with our brothers' incitement for the sake of Israel.	غدرونا يمه والدم صاير مي، فتنة، بتحريض أخوتنا، من أجل إسرائيل
They changed our Qibla[545]. . . Read the history. . . the *'Irāb*[546] betrayed us.	غيرنا قبلتنا، وأقرؤوا التاريخ، الأعراب غدرتنا
Oh mother, one word from the USA united all of the traitors.	مش تاري الخوان يمه بكلمة من أمريكا التموا كل أركانو
Qaraḍāwī, 'Ar'ūr incited and stung his brothers like a hornet.	قرضاوي وعرعور أصبح، أصبح كالدبور يلسع ويحرض أخوانو
The traitor is the one who kills his own family? Do we die for his own sake?	والخاين قال يلي بيقتل أهلو ومالو؟ نموت كرمالو
The People are against sedition. . . it has unity and is aware. The people know that it has its own army.	الشعب ضد الفتن.. الوحدة رسمالو والشعب يملك الوعي والجيش رسمالو

"Alas/Shame" (pro-regime), Nawwar Haidar (4)	**ياحيف (النظام) 2 نوار حيدر**
Shame	يا حيف اخ ويا حيف
They used to say peaceful and they killed, alas.	كانو يقولو سلمية وقتلوا ياحيف
The army of my country is killed, how?	.جيش بلادي يتقتل كيف
From out of my country you incite them by saying freedom, but you attack me with a sword.	من برا بلادي حرضت ولادي، تطلق حرية وعلي هاجم بالسيف
Oh, mother, this is the situation now, alas, you give a hand to your enemy.	هذا اللي صاير ياحيف بلادي يا يمه ياحيف وايدك للعادي ياحيف
Under their demands they hid their intention, mother, "Sedition," they chanted for it.	تحت مطالبهم يمه خبو نيتهم يمه، فتنة يهتفولا

545 Qibla is the direction toward which Muslims pray five times a day.
546 *The Quran*, Sura (9: 97), trans. Quran Project at King Saud University, accessed July 18, 2023, http://quran.ksu.edu.sa/translations/english/202.html?a=1332. 'Irāb and Arab have the same root in Arabic. The difference is that "'Irāb" was used to describe Arabs who attacked the prophet and tried to kill him. The Quran says "The Bedouins are stronger in disbelief and hypocrisy and more likely not to know the limits of what [laws] Allah has revealed to His Messenger. And Allah is Knowing and Wise" (9: 97).

(continued)

"Alas/Shame" (pro-regime), Nawwar Haidar (4)	**ياحيف (النظام) 2 نوار حيدر**
In the name of Jihad, They killed our brothers who protected us with their chests.	باسم الجهاد يمه، قتلو اخوتهم يلي بصدورهم يحمونا
It is you who said whoever kills his people is traitor? No matter who is he?	وأنت بتحكي يلي بيقتل أهلو خاين، يكون مين كاين
They destroyed Al-Aqsa Mosque. Where were you? Sleeping? They profaned the Quran. Wage Jihad! Still sleeping? They distorted your religion. Wage Jihad! Still Sleeping?	دمروا الأقصى، وين كنت نايم، دنسوا القرآن، جاهد كنت نايم، شوهوا دينك، جاهد كنت نايم
The traitor, oh mother, doubts his brother. Who doubts, he betrays.	مش تاري الخوّان يمه يشك بخيو يخون يمه، من شكو بخيو خانوا
The traitor asked God for liberation? Who were your soldiers?	من العالي بطلب يمه يحررها، مين جنودك كانوا
It is you who said whoever kills his people is a traitor? No matter who is he?	وأنت بتحكي يلي بيقتل أهلو خاين، يكون مين كاين
They destroyed Al-Aqsa Mosque. Where were you? Sleeping? They profaned the Quran. Wage Jihad! Still sleeping? They distorted your religion. Wage Jihad! Still Sleeping?	دمروا الأقصى، وين كنت نايم، دنسوا القرآن، جاهد كنت نايم، شوهوا دينك، جاهد كنت نايم

Music Group 2 (National Anthems)

The Guardians of the Homeland (The Official Syrian National Anthem)	**النشيد الوطني السوري الرسمي**
The Guardians of the Homeland! To you is our salutation! Nobles spurn humiliation The den of Arabism is sacredly unapproachable The Throne of suns is too high to assault.	حماة الديار عليكم سلام أبت أن تذل النفوس الكرام عرين العروبة بيتٌ حرام وعرش الشموس حمى لا يضام
Damascus gardens are constellations so much like high skies It is a sun-lit land Matchless land and sky it is!	ربوع الشآم بروجُ العلا تحاكي السماء بعالي السنا فأرضٌ زهت بالشموس الوضا سماءٌ لعمرك أو كالسما
Hopes are pinned; hearts beat fast over a country-unifying flag It certainly has its people's black color of their eyes And so it has martyrs-blood ink	رفيف الأماني وخفق الفؤاد على علم ضم شمل البلاد أما فيه من كلّ عينٍ سواد ومن دم كلّ شهيد مداد؟

(continued)

The Guardians of the Homeland (The Official Syrian National Anthem)	النشيد الوطني السوري الرسمي
Fierce pride and glorious past Souls are offered lavishly Al-Walīd and Al-Rashīd belong to us We have every reason to rule and build the country	نفوسٌ أباةٌ وماضٍ مجيد وروح الأضاحي رقيبٌ عتيد فمنا الوليد ومنا الرشيد فلم لا نسود ولم لا نشيد

National anthem used in a demonstration (first modified version)[547]	النشيد الوطني السوري المعدّل في المظاهرة
The regime's guardian are tyrants and ignoble. You killed proud souls.	حماة النظام طغاة لئام قتلتم، أهنتم نفوساً كرام
You bombarded houses, spilled blood, masked suns and spread darkness.	قصفتم بيوتًا أرقتم دماء حجبتم شموساً، نشرتم ظلام
The Levant land suffers and prays God lifts the Affliction.	ديار الشآم تعاني الأذى تناجي العظيم لرفع البلا
The tyrant army is ignoble and hateful. It antagonizes the people and the sky.	فجيش الطغاة لئيم حقود يعادي العباد يعادي السما
The pulse of the land and its pray are against the tyrant who wreaks havoc.	ونبض البلاد وكل الدعاء على ظالم عاث فيها الفساد
The tyrant sold the people's land, befriended the invaders, displaces the people and spilled blood.	أما باع أرضاً ووالى الغزاة وهجر شعباً، أراق الدماء
Proud men and proud people confronting the enemies strictly.	رجال أباة وشعب عريق بوجه الأعادي شديد عتيد
Some are wounded or martyred. Certainly the expulsion of tyrants is close.	فمنا الجريح ومنا الشهيد فدحر العتاة قريب أكيد

National anthem (second modified version)	النشيد الوطني الثاني
The Guardians of the Homeland, peace be upon you. People want to topple the regime. Shedding the people's blood is forbidden. Come and defend a subjugated people.	حماة الديار عليكم سلام- الشعب يريد إسقاط النظام دم الشرفاء عليكم حرام- فهبّوا لنصرة شعب يُضام

547 "Ughniyat ḥumāt al-Diyār ʻalaykum Damār. . . Qataltum Ahantum Nufusan Kerām" [The Song of the Guardian of the Homeland, Destruction be Upon You. You Killed, Humiliated Proud People], uploaded September 7, 2011, accessed July 18, 2023, https://www.youtube.com/watch?v=iUBlxXb-N9d8.

(continued)

National anthem (second modified version)	**النشيد الوطني الثاني**
The Levant land suffers from sorrow and grieves to God for the hardened injustice. The wound of the Freewomen is unforgettable and the purity of Levant has been profaned.	ربوع الشآم تعاني الأسى- وتشكو إلى الله ظلماً قسا فجرح الحرائر لا يُنتسى- وطهر الشآم قد دُنسا
With determination, certainty, honest hearts, and a challenging peacefulness against tyrants, we will eradicate the unjust from everywhere and make patience our provision.	بعزم اليقين وصدق الفؤاد- وسلمية تتحدى العتاد سنقتلع الظلم من كل واد- ونجعل من صبرنا خير زاد
A people loyal to a glorious past made strong through [their] ordeals. The people will keep the pledge, Ibn al-Walīd, and will restore glory to the Umma.	فشعب وفيّ لماضٍ مجيد- وفي النائبات قويٌّ عتيد سيبقى على العهد يا بن الوليد – وللأمة المجد سوف يعيد

Appendix 5
Slogan Development (1970–2011)

Hafez Assad	Bashar Assad	Revolution	Development	Pro-regime response
إلى الأبد يا حافظ الأسد Assad Forever		حرية للأبد غص من عنك يا أسد Freedom forever, over your will Assad.		أبو حافظ (ثلاث صفقات) Abu Hafez (three claps)
قائدنا إلى الأبد الأمين حافظ الأسد Our Leader forever; the honest [trusted, faithful] Hafez al-Assad		يلعن روحك يا حافظ Hafez, Curse your Soul	يا حافظ سماع \طلاع\ وشوف، \صرنا\ نسبك عالمكشوف Hafez, come back to life and see, we curse you explicitly	يرحم روحك يا حافظ Mercy on your soul Hafez
لم يتبق لنا إلا أنت No one left for us but you	لم يتبق لنا إلا أنت No one left for us but you	يلعن روحك أنيسة Anisa, Curse your Soul		
حلك يالله حلك يطلع حافظ محلك Oh, God! It is time to give your place to Hafez.	حلك يالله حلك يطلع بشار محلك Oh, God! It is time to give your place to Bashar.	يلعن روحك يا حافظ على هالجش يلي خلفتو Hafez, Curse your Soul for this burro you gave birth to	جنو جنو البعثية، لما طلبنا الحرية ، يلعن روحك يا حافظ يا ابن الحرامية" Poor Baathists, They went crazy when we asked for freedom. Hafez, Curse your Soul, you are a thief like your father	

https://doi.org/10.1515/9783110767698-011

Appendix 6

Slogan Development (1970–2011)

Time Period	[illegible]	Revolution	Development	[illegible]
[illegible]	[illegible]	[illegible]	[illegible]	[illegible]

Appendix 6
A Musical archive: March 2011 to March 2012

For listening to songs, visit the following link: https://creativememory.org/ar/collections/eylafbader-eddin-songs. A few songs from the list were not included by Creative Memory of the Syrian Revolution as they violate the website's terms and policies.

#	Date	Song	Singer	Language	Purpose/ use	Downloaded
1.	22-03-2011	ياسوري كسّر القيود وانتفض	unknown	Modern Standard Arabic	Revolution	yes
2.	24-03-2011	درعانا تنادي	وصفي المعصراني	Modern Standard Arabic	Revolution	yes
3.	25-03-2011	الله أكبر يا سورية(قدس) هللي	unkown	Modern Standard Arabic	Revolution \ Palestine\ Regime	yes
4.	25-03-2011	سوري أنا	وصفي معصراني	Colloquial	Revolution	yes
5.	28-03-2011	اه على وطني	Unknown	Modern Standard Arabic	Regime\ Revolution	yes
6.	28-03-2011	ياحيف	Samih Shukair	Colloquial	Revolution	yes
7.	28-03-2011	بيان رقم واحد	Unknown(Later MCROCO)	Colloquial	Revolution	yes
8.	31-03-2011	شو منحبك يا بلادي	Yahya Hawa	Colloquial	Revolution	yes
9.	08-04-2011	الله سورية حرية وبس	unknown	Colloquial	Revolution	yes
10.	15-04-2011	اني اخترتك ياوطني	مارسيل خليفة	Modern Standard Arabic	Regime\ Revolution	yes
11.	16-04-2011	وطني أنا	Malek Jandali	Modern Standard Arabic	Revolution	yes
12.	17-04-2011	قامت ثورتنا	وصفي المعصراني	Colloquial	Revolution	yes
13.	18-04-2011	في سبيل المجد	unknown	Modern Standard Arabic	Regime\ Revolution	yes
14.	19-04-2011	أناديكم	مارسيل خليفة	Modern Standard Arabic	Regime\ Revolution	yes
15.	20-04-2011	بعشق اسمك سوريا	Samer Kabro	Colloquial	Regime\ Revolution	yes
16.	24-04-2011	حمص يا دار السلام	وصفي المعصراني	Modern Standard Arabic	Revolution	yes

https://doi.org/10.1515/9783110767698-012

(continued)

#	Date	Song	Singer	Language	Purpose/ use	Downloaded
17.	25-04-2011	ياصنّاع الفتن	Unknown	Colloquial	Regime\ Revolution	yes
18.	30-04-2011	الشعب السوري مابينهان	------------	Colloquial	Revolution	yes
19.	30-04-2011	على درعا شدينا	حسام عساف	Colloquial	Revolution	yes
20.	02-05-2011	عالجنة رايحين	-------------	Colloquial	Revolution	yes
21.	03-05-2011	ارفعي الرأس	يحي حوى	Modern Standard Arabic	Revolution	yes
22.	04-05-2011	تغربنا ولكن مانسينا	Yahya Hawa	Modern Standard Arabic	Revolution	yes
23.	05-05-2011	قالو عنا مندسين	Syrian Mundasin Band	Colloquial	Revolution	yes
24.	05-05-2011	يلي بيقتل شعبو خاين	Fajer Suria Band	Colloquial	Revolution	yes
25.	08-05-2011	سورية الله يحميها	unknown	Colloquial	Revolution	yes
26.	08-05-2011	البيان رقم 2	MCROCO	Colloquial	Revolution	yes
27.	16-05-2011	من يوقف حقداً أسود	Arrawabi Band	Modern Standard Arabic	Revolution \ Palestine\ Regime	yes
28.	19-05-2011	افعل ماشئت	AbdelQader Quza'	Modern Standard Arabic	Revolution	yes
29.	19-05-2011	إذا الشعب يوماً أراد	لطيفة التونسية	Modern Standard Arabic	Revolution \ Palestine\ Regime\ Arab World	yes
30.	23-05-2011	يا سوريا لاتنحني	أحمد السعدي	Modern Standard Arabic	Revolution	yes
31.	27-05-2011	درعا صهيل الحرية	المندسين السوريين	Modern Standard Arabic	Revolution	yes
32.	27-05-2011	حماة الديار	أحمد الشريقي	Colloquial/ Modern Standard Arabic	Revolution	yes
33.	30-05-2011	طالع روحو على كفو	حسام طحان فرقة الهدى الدولية	Colloquial	Revolution	yes
34.	03-06-2011	صاحت حوران	براء الصلخدي	Colloquial	Revolution	yes

(continued)

#	Date	Song	Singer	Language	Purpose/ use	Downloaded
35.	03-06-2011	الشعب السوري مابينداس	فرقة نسور سوريا	Colloquial	Revolution	yes
36.	04-06-2011	يمل	الدب السوري	Colloquial	Revolution	yes
37.	04-06-2011	شام على العقيقة اجتمعنا	Unknown	Colloquial	Revolution	yes
38.	07-06-2011	هي شامنا	فضل شاكر	Modern Standard Arabic	Revolution	yes
39.	08-06-2011	ما منحبك	sama band	Colloquial	Revolution	yes
40.	11-06-2011	صرخة حوران	Unknown	Colloquial	Revolution	yes
41.	16-06-2011	باسم الوطن راب	الرابر سوري حر	Colloquial	Revolution	yes
42.	17-06-2011	عتم الليل	عمر علوان	Colloquial	Revolution	yes
43.	18-06-2011	الشعب يريد	Unknown	Colloquial	Revolution	yes
44.	20-06-2011	خاين يلي بيقتل شعبو	Yahya Hawa	Colloquial	Revolution	yes
45.	22-06-2011	شام العزة	فرقة المندسيين السوريين	Colloquial	Revolution	yes
46.	23-06-2011	اهل الثورة	احمد الكردي	Colloquial	Revolution	yes
47.	24-06-2011	بدنا نعبي الزنزانات	Strong Heros of Moscow	Colloquial	Revolution	yes
48.	26-06-2011	صرخة سوري راب	النسر السوري	Colloquial	Revolution	yes
49.	27-06-2011	حط الكف على الكف	عبيدة السوطري	Colloquial	Revolution	yes
50.	28-06-2011	ياسمين الشام	يحي حوى	Colloquial	Revolution	yes
51.	28-06-2011	معكم	Unkown	Colloquial	Revolution	yes
52.	30-06-2011	ثوب الحرية	لقمان حوري	Colloquial	Revolution	yes
53.	09-07-2011	ريميكس اغنية يلا أرحل يا بشار	غير معروف	Colloquial	Revolution	yes
54.	10-07-2011	طالع عالموت	Yahya Hawa	Colloquial	Revolution	yes
55.	10-07-2011	شدّ الهمة وشدّ الحيل	أحرار الشام	Colloquial	Revolution	yes
56.	12-07-2011	يلا أرحل	Sozi free	Colloquial	Revolution	yes
57.	12-07-2011	قولوا الله يا رجال	عمر علوان	Colloquial	Revolution	yes
58.	14-07-2011	راب الثورة السورية	أنا مندس AntiVirus	Colloquial	Revolution	yes
59.	19-07-2011	يا سورية لاتسجلينا غياب	خاطر ضوا	Colloquial	Revolution	yes

(continued)

#	Date	Song	Singer	Language	Purpose/ use	Downloaded
60.	24-07-2011	دافعوا عن الحرية	Rem Banna	Modern Standard Arabic	Revolution \ Palestine\ Regime	yes
61.	30-07-2011	ارحل يا بشار	أحمد لشريقي	Modern Standard Arabic	Revolution	yes
62.	07-08-2011	سوريا متل النار	محمد صبح	Colloquial	Revolution	
63.	10-08-2011	انا الشهيد	شادي ابازيد	Colloquial	Revolution	
64.	10-08-2011	فتحت عيوني	Uknown	Colloquial	Revolution	yes
65.	17-08-2011	الله معنا	المعتصم بالله العسلي	Colloquial	Revolution	yes
66.	18-08-2011	قاظان بيتكون	unknown	Colloquial	Revolution	yes
67.	19-08-2011	ارحل	فرقة أحرار الشام	Colloquial	Revolution	yes
68.	20-08-2011	(بدي حرية (راب	-----------	Colloquial	Revolution	yes
69.	21-08-2011	اشتقت لأخواني	حسام طحان	Modern Standard Arabic	Revolution	yes
70.	22-08-2011	لاتقتلني بسلاحي	فرقة الهدى الدولية	Colloquial	Revolution	yes
71.	23-08-2011	سورية أم الشهيد	شباب تنسيقية القدم	Colloquial	Revolution	yes
72.	25-08-2011	غني غني للسجان	تجمع نبض	Colloquial	Revolution	
73.	26-08-2011	سوريا بدها	-------------	Colloquial	Revolution	
74.	27-08-2011	لارجوع إلى الوراء	-------------	Modern Standard Arabic	Revolution	yes
75.	28-08-2011	روحي لك فداء	حسام الطحان	Modern Standard Arabic	Revolution	yes
76.	28-08-2011	اغنية تخاذل أهل حلب	-----------	Colloquial	Revolution	
77.	29-08-2011	نشيد سورية الحرة	فرقة الهدى الدولية	Colloquial	Revolution	yes
78.	29-08-2011	مين بيحبك \ منحبك	شهد برمدا	Colloquial	Revolution \ Regime	yes
79.	31-08-2011	شو بيقربك حمزة	خاطر ضوا	Colloquial	Revolution	yes
80.	01-09-2011	الموت ولا المذلة	Gurabaa Group	Colloquial	Revolution	yes
81.	03-09-2011	اغنية حرية وبس لا تقولوا عني مندس	مجموعة معكم	Colloquial	Revolution	yes
82.	05-09-2011	حموية والحامي الله	أحرار الشام	Colloquial	Revolution	yes
83.	05-09-2011	(اجاك الدور (راب	-------------	Colloquial	Revolution	yes

(continued)

#	Date	Song	Singer	Language	Purpose/ use	Downloaded
84.	06-09-2011	سورية بدها حرية	Mohamad Kahlawi	Colloquial	Revolution	yes
85.	06-09-2011	الله معك يا شهيد	سوزي	Colloquial	Revolution	yes
86.	07-09-2011	أرض جدادك	----------	Colloquial	Revolution	yes
87.	08-09-2011	سورية لأجلك نصلي	وصفي المعصراني	Modern Standard Arabic	Revolution	yes
88.	10-09-2011	هدية عيد ميلاد بشار الأسد	MC Revo	Colloquial	Revolution	yes
89.	10-09-2011	الحرية	حسام طحان	Colloquial	Revolution	yes
90.	13-09-2011	ياسوري الله أكبر	عبدالرحمن السعدي	Colloquial	Revolution	yes
91.	14-09-2011	مأساة ثورة (RAP)	MC. Revi	Colloquial	Revolution	yes
92.	14-09-2011	الكاذبون على المدى	نايف الشرهان	Modern Standard Arabic	Revolution	yes
93.	15-09-2011	زمن الحرية	أحمد زميلي خيري حاتم سائد العجيمي محمد صبح	Modern Standard Arabic	Revolution	yes
94.	19-09-2011	خنساء جسر الشغور	عبيدة السوطري	Modern Standard Arabic	Revolution	yes
95.	22-09-2011	ماتت قلوب الجيش	وصفي المعصراني	Modern Standard Arabic	Revolution	yes
96.	23-09-2011	بلادي سورية	سوزي	Colloquial	Revolution	yes
97.	23-09-2011	نحن الشعب (السوري (راب	فرقة جبلاوية	Colloquial	Revolution	yes
98.	26-09-2011	فيدوني	المعتصم بالله العسلي	Modern Standard Arabic	Revolution	yes
99.	01-10-2011	كيفك انتا	سوزي	Colloquial	Revolution	yes
100.	03-10-2011	عزك باقي	أحمد الشريقي	Colloquial	Revolution	yes
101.	04-10-2011	البلد بلدك	وصفي المعصراني	Modern Standard Arabic	Revolution	yes
102.	07-10-2011	قلعة حرية	احمد الكردي	Colloquial	Revolution	yes
103.	08-10-2011	هيه يالله	-------------	Colloquial	Revolution	yes
104.	10-10-2011	شو ماقلك	حسين جبري	Colloquial	Revolution	yes
105.	10-10-2011	طلو البويلة	حسين جبري	Colloquial	Revolution	
106.	13-10-2011	آخر أيام البعثية	سوزي	Colloquial	Revolution	yes

(continued)

#	Date	Song	Singer	Language	Purpose/ use	Downloaded
107.	16-10-2011	نشيد جيش الاحرار	Unknown	Colloquial	Revolution	yes
108.	20-10-2011	أنا طالع اتظاهر	وصفي المعصراني	Colloquial	Revolution	yes
109.	20-10-2011	دكتور عيون راب	MC Roco	Colloquial	Revolution	yes
110.	24-10-2011	سكابا يادموع	وصفي المعصراني	Colloquial	Revolution	yes
111.	24-10-2011	كرمالك يا سوريا	عبيدة السوطري	Colloquial	Revolution	yes
112.	25-10-2011	بوطني	وصفي المعصراني	Colloquial	Revolution	yes
113.	27-10-2011	الام الحمصية	سوزي	Colloquial	Revolution	yes
114.	28-10-2011	عالهودلاة	وصفي المعصراني	Colloquial	Revolution	yes
115.	28-10-2011	شو جابك ع سوريا	سوري مندس	Colloquial	Revolution	yes
116.	29-10-2011	سألوني ليش بدك حرية	زهرة الحرية	Colloquial	Revolution	yes
117.	30-10-2011	أنا اسمي نظام	الدب السوري	Colloquial	Revolution	yes
118.	30-10-2011	صرخة بلادي	صلاح حرب	Colloquial	Revolution	yes
119.	03-11-2011	انت الحرية	أحرار الشام	Colloquial	Revolution	yes
120.	06-11-2011	طريق النصر	سوزي	Colloquial	Revolution	yes
121.	08-11-2011	نشيد همس الشهيد	----------	Modern Standard Arabic	Revolution	yes
122.	09-11-2011	ثوار ياريح الجنة هبي	ثوار الزبداني	Colloquial	Revolution	yes
123.	09-11-2011	نشيد أيام العيد	سوزي	Colloquial	Revolution	yes
124.	12-11-2011	بياع الجولان	ابو عزاب	Colloquial	Revolution	
125.	13-11-2011	دلعونة الحرية	وصفي المعصراني	Colloquial	Revolution	yes
126.	13-11-2011	يابي يابي يابي يا بشار	مجموعة من الاحرار	Colloquial	Revolution	yes
127.	13-11-2011	حمص العدية	صقر حمص	Colloquial	Revolution	
128.	13-11-2011	كلنا حمزة الخطيب	يوسف السوطري	Modern Standard Arabic	Revolution	yes
129.	15-11-2011	ياريت طلعنا	Unknown	Colloquial	Revolution	yes
130.	15-11-2011	ارحل عنا بلا حوار	حسين جبري ابو زهير	Colloquial	Revolution	
131.	16-11-2011	قلو آمان يا دكتور	زهرة الحرية	Colloquial	Revolution	yes
132.	17-11-2011	حرام عليك	عبدالباسط ساروت	Colloquial	Revolution	yes
133.	17-11-2011	أنا حمصي	صقر حمص	Colloquial	Revolution	yes
134.	18-11-2011	شدو الهمة	وصفي المعصراني	Colloquial	Revolution	yes

(continued)

#	Date	Song	Singer	Language	Purpose/ use	Downloaded
135.	20-11-2011	سلملي عليه	صقر حمص	Colloquial	Revolution	yes
136.	20-11-2011	بالآخر رح يرحل يا بلدنا	سوزي	Colloquial	Revolution	yes
137.	20-11-2011	الشعب والجيش	يمان سفلو	Colloquial	Revolution	yes
138.	22-11-2011	السفاح أبو نص لسان	فرقة المندسين السوريين	Colloquial	Revolution	yes
139.	22-11-2011	اغنية الثورة (السورية (راب	Unknown	Colloquial	Revolution	yes
140.	22-11-2011	حرام عليك من دون موسيقى	عبدالباسط ساروت	Colloquial	Revolution	yes
141.	23-11-2011	أغنية الشهيد غياث مطر	وصفي المعصراني	Colloquial	Revolution	yes
142.	23-11-2011	دوران الأيام راب	Mc Roco	Colloquial	Revolution	yes
143.	24-11-2011	جامعات بلا شبيحة	روزنامة الحرية	Colloquial	Revolution	yes
144.	24-11-2011	عشر شبيحة	سلاحف النينجا	Colloquial	Revolution	yes
145.	26-11-2011	سننزع قلب كلب الشام نزعاً	ابراهيم الماجد	Modern Standard Arabic	Revolution	yes
146.	26-11-2011	خمس شبيحة حد العين	صبايا حمص	Colloquial	Revolution	yes
147.	26-11-2011	يامو يامو	حسين جبري ابو زهير	Colloquial	Revolution	
148.	26-11-2011	مدينة الاحرار ((راب	الفاروق	Colloquial	Revolution	yes
149.	26-11-2011	الحرية	جعفر حوى	Modern Standard Arabic	Revolution	yes
150.	27-11-2011	مازوت	نص تفاحة	Colloquial	Revolution	yes
151.	27-11-2011	الوطن جنة	خاطر ضوا	Colloquial	Revolution	yes
152.	28-11-2011	بشار بآخر ايامو	صقر حمص	Colloquial	Revolution	yes
153.	29-11-2011	ارحل يانظام	الدب السوري	Colloquial	Revolution	yes
154.	29-11-2011	لاكتب عبواب الفرع	Unknown	Colloquial	Revolution	yes
155.	29-11-2011	طير الورور	سوزي	Colloquial	Revolution	yes
156.	02-12-2011	سورية الحرة	أحرار سوريا	Colloquial	Revolution	yes
157.	03-12-2011	ارحل ارحل يا بشار	صقر حمص	Colloquial	Revolution	
158.	04-12-2011	طالع عالحرية	فرقة شباب التغيير	Colloquial	Revolution	yes
159.	04-12-2011	الجيش الحر	وصفي المعصراني	Colloquial	Revolution	yes

(continued)

#	Date	Song	Singer	Language	Purpose/ use	Downloaded
160.	04-12-2011	خضرا يا بلادي	صقر حمص	Colloquial	Revolution	
161.	05-12-2011	الحالة تعبانة يا بثينة	صقر حمص	Colloquial	Revolution	yes
162.	06-12-2011	يا حرية	الدب السوري الدبة السورية	Colloquial	Revolution	yes
163.	06-12-2011	ايقاع الثورة والنصر	محمد كحلاوي	Colloquial	Revolution	yes
164.	06-12-2011	استميتوا	ابراهيم الماجد	Modern Standard Arabic	Revolution	yes
165.	11-12-2011	شارك في الإضراب	الدب السوري	Colloquial	Revolution	yes
166.	12-12-2011	بوثينة يا ختيارة	زهرة الحرية	Colloquial	Revolution	yes
167.	12-12-2011	يا سوريا يا بلادي	خاطر ضوا	Colloquial	Revolution	yes
168.	13-12-2011	زفة الشهيد	كرم عبدالكريم وآخرين	Colloquial	Revolution	yes
169.	15-12-2011	جنة جنة	عبدالباسط ساروت	Colloquial	Revolution	yes
170.	16-12-2011	هس	نص تفاحة	Colloquial	Revolution	yes
171.	17-12-2011	لا تبالي يا بلادي	سوزي	Colloquial	Revolution	yes
172.	20-12-2011	حيالله رجالك ياشام	جعفر و عبدالسلام حوى	Colloquial	Revolution	yes
173.	21-12-2011	نازل على حمص	ميير باند	Colloquial	Revolution	yes
174.	24-12-2011	بيشو	نص تفاحة	Colloquial	Revolution	yes
175.	26-12-2011	يسقط حزب البعث	أحرار الشام	Colloquial	Revolution	yes
176.	26-12-2011	ياجحيش	أحرار قامشلو	Colloquial	Revolution	yes
177.	26-12-2011	أغنية ميلاد الحرية	سوزي	Colloquial	Revolution	yes
178.	29-12-2011	أبو رقبة طويلة	Unknown	Colloquial	Revolution	yes
179.	29-12-2011	علمني حبك يا وطني	صقر حمص	Colloquial	Revolution	yes
180.	30-12-2011	اه لو الكرسي بيحكي	أصالة نصري	Colloquial	Revolution	yes
181.	31-12-2011	ياجزار	أحرار الشام	Colloquial	Revolution	yes
182.	31-12-2011	دندن معنا يا سوري	صبايا حمص	Colloquial	Revolution	yes
183.	01-01-2012	انت الحرة	أحرار الشام	Colloquial	Revolution	yes
184.	01-01-2012	معليش درعا	أحرار الشام	Colloquial	Revolution	yes
185.	02-01-2012	قربنا يالحرية	سميح شقير	Colloquial	Revolution	yes
186.	04-01-2012	حكم بشار ما نريده	أحرار الشام	Colloquial	Revolution	yes
187.	04-01-2012	عالرابية	حسين جبري أبو زهير	Colloquial	Revolution	yes

(continued)

#	Date	Song	Singer	Language	Purpose/ use	Downloaded
188.	04-01-2012	آخر أيامو	وائل	Colloquial	Revolution	
189.	04-01-2012	نتف شعرك يا بشار	صقر حمص	Colloquial	Revolution	yes
190.	04-01-2012	سورية سورية	أحرار الشام	Colloquial	Revolution	yes
191.	07-01-2012	معارض سوري	MC Roco	Colloquial	Revolution	yes
192.	08-01-2012	Arab Spring	Feat.Egtya7 , Hafsi ,SiadGjam ,Al farook , Mc Queener	Colloquial	Revolution	yes
193.	08-01-2012	مظاهرة	Mc Roco	Colloquial	Revolution	yes
194.	09-01-2012	حلم الشهادة	عبدالباسط الساروت ووصفي المعصراني	Colloquial	Revolution	yes
195.	10-01-2012	هيدا حكيك	حرائر الكسوة	Colloquial	Revolution	yes
196.	11-01-2012	علي صوتك علي	فرقة سما	Colloquial	Revolution	yes
197.	11-01-2012	حابب طمنكون	فرقة سما	Colloquial	Revolution	
198.	11-01-2012	هالزعران	بهيرة	Colloquial	Revolution	yes
199.	12-01-2012	أنا حمزة	فرقة سما	Colloquial	Revolution	
200.	12-01-2012	سوريا يا حبيبتي	فرقة غرباء	Modern Standard Arabic	Revolution	yes
201.	12-01-2012	ما منحبك	فرقة سما	Colloquial	Revolution	yes
202.	12-01-2012	/حمار-حالي-حاسس/	فرقة سما	Colloquial	Revolution	yes
203.	12-01-2012	لخبط هالجيش	زهرة الحرية	Colloquial	Revolution	yes
204.	12-01-2012	طلعت ريحتك	فرقة سما	Colloquial	Revolution	yes
205.	12-01-2012	/النوم-محلا-يا	فرقة سما	Colloquial	Revolution	yes
206.	13-01-2012	انا جندي من الجيش الحر	أبو زيد صقر الوادي	Colloquial	Revolution	yes
207.	13-01-2012	دقوا عالطناجر	صقر حمص	Colloquial	Revolution	
208.	13-01-2012	ست الحبايب سوريا	صقر حمص	Colloquial	Revolution	
209.	13-01-2012	يا طيره طيري يا حمامه	صقر حمص	Colloquial	Revolution	
210.	13-01-2012	لو يسكت	جعفر حوى	Colloquial	Revolution	yes
211.	13-01-2012	زينو الساحة	وصفي المعصراني	Colloquial	Revolution	yes
212.	14-01-2012	تنبيه من معارض	Mc Roco Ebn Siba3	Colloquial	Revolution	yes
213.	15-01-2012	كانوا الثوار كان	سوزي	Colloquial	Revolution	yes

(continued)

#	Date	Song	Singer	Language	Purpose/ use	Downloaded
214.	17-01-2012	نشيد حماة الديار	المندسين السوريين	Colloquial	Revolution	yes
215.	19-01-2012	شهداء الزبداني الأبية	unknown	Colloquial	Revolution	yes
216.	19-01-2012	سوريا بدها حرية بكل لغات العالم	unknown	Colloquial	Revolution	yes
217.	19-01-2012	وين الشرف والناموس	المندسين السوريين	Colloquial	Revolution	yes
218.	19-01-2012	الله اكبر عليك يا أسد راب	unknown	Colloquial	Revolution	yes
219.	19-01-2012	علمني حبك	محمد الحلبي	Colloquial	Revolution	yes
220.	20-01-2012	منقلو ارحل بيقول زيتونا	لطفي الحر	Colloquial	Revolution	yes
221.	22-01-2012	يا جيش الحر	صقر الوادي احمد الشريقي تغريد الرنتيسي	Colloquial	Revolution	yes
222.	22-01-2012	منكبك	صبايا حمص	Colloquial	Revolution	yes
223.	24-01-2012	اغنية المامبو للدابة	حرائر الكسوة	Colloquial	Revolution	yes
224.	25-01-2012	يا بو الدبابة الخضرا	صبايا حمص	Colloquial	Revolution	yes
225.	26-01-2012	طلعت يا محلى نورها	بنت حمص	Colloquial	Revolution	yes
226.	29-01-2012	ياطير خدني	عبدالباسط الساروت	Colloquial	Revolution	yes
227.	30-01-2012	خطة قدمكم	unknown	Colloquial	Revolution	yes
228.	30-01-2012	هالظالم الفرعون	بنت حمص	Colloquial	Revolution	yes
229.	01-02-2012	لالعن بيك يا بشار	الحمصية	Colloquial	Revolution	yes
230.	02-02-2012	صرخة مدينة حماة	سوزي	Colloquial	Revolution	yes
231.	03-02-2012	رصاصة الحرية	Mc Roco Ebn Siba3	Colloquial	Revolution	yes
232.	03-02-2012	قومي من تحت الردم – الثورة السورية	ماجدة الرومي (الأغنية الأصلية للبنان) ولكن هنا موضوعة للثورة	Modern Standard Arabic	Revolution	yes
233.	04-02-2012	أطفالي	Mc Roco Ebn Siba3	Colloquial	Revolution	yes
234.	04-02-2012	زجل ثوري	حرائر دمشق	Colloquial	Revolution	yes
235.	05-02-2012	الثورة دي	بنت حمص	Colloquial	Revolution	yes

(continued)

#	Date	Song	Singer	Language	Purpose/ use	Downloaded
236.	05-02-2012	تأخرنا عليكي حماة	ووصفي المعصراني	Colloquial	Revolution	yes
237.	07-02-2012	كلي حرية	وسام غمراوي	Colloquial	Revolution	yes
238.	08-02-2012	لن نستكين	أحمد الشريقي	Modern Standard Arabic	Revolution	yes
239.	08-02-2012	لو سالتك أنت حمصي	بنت حمص	Colloquial	Revolution	yes
240.	09-02-2012	داعس عالقرداحة	Mc Roco	Colloquial	Revolution	yes
241.	11-02-2012	يالله نهتف سوا	أحرار الشام	Colloquial	Revolution	yes
242.	11-02-2012	كرمالك يا شام	أحرار الشام	Colloquial	Revolution	yes
243.	12-02-2012	يا سوريا لا تحزني	أحرار الشام	Colloquial	Revolution	yes
244.	12-02-2012	يا رجال الشام قومي	قاشوش حرستا	Modern Standard Arabic	Revolution	yes
245.	12-02-2012	يا ثورتنا	قاشوش حرستا	Colloquial	Revolution	yes
246.	12-02-2012	حمص الصامدة	ووصفي المعصراني	Colloquial	Revolution	yes
247.	13-02-2012	إلى الشام	علي المغربي	Modern Standard Arabic	Revolution	yes
248.	14-02-2012	كرمالك حمزة الخطيب	أحرار الشام	Colloquial	Revolution	yes
249.	14-02-2012	حي الشهيد يا سوريا	ابراهيم السعيد أحمد الهاجري	Colloquial	Revolution	yes
250.	14-02-2012	قرب يومك	محمد دلعوب	Colloquial	Revolution	yes
251.	15-02-2012	حمص يا أم المرجلة	أحرار الشام	Colloquial	Revolution	yes
252.	15-02-2012	محل صغير	نص تفاحة	Colloquial	Revolution	yes
253.	15-02-2012	الجيش الحر دمّر رجال الأسد	خيري بلال	Colloquial	Revolution	yes
254.	16-02-2012	يالجيش السوري قلي شو صار	أحرار الشام	Colloquial	Revolution	yes
255.	16-02-2012	يلعن روحك يا حافظ بالانجليزية	ليلي آلن\ اسامة ادريس أبو برهان	Colloquial	Revolution	yes
256.	17-02-2012	يلعن روحك يا حافظ بكل الألحان	-----------	Colloquial	Revolution	yes
257.	18-02-2012	لأجلك يا مدينة الإباء يا حمص	سوزي	Colloquial	Revolution	yes

(continued)

#	Date	Song	Singer	Language	Purpose/ use	Downloaded
258.	19-02-2012	اجراموا متل أبوه	أحرار الشام	Colloquial	Revolution	yes
259.	19-02-2012	يا أم الشهيد	فرقة غربة	Colloquial	Revolution	yes
260.	19-02-2012	يا دمشق	خالد الحقان محمد الحسيان	Modern Standard Arabic	Revolution	yes
261.	20-02-2012	حيو ثورتنا	أحرار الشام	Colloquial	Revolution	yes
262.	20-02-2012	صياد ولاد الحرام	MC Roco	Colloquial	Revolution	yes
263.	21-02-2012	بين العصر والمغرب	حسين جبري	Colloquial	Revolution	yes
264.	21-02-2012	صيحة حرية	أحرار دوما	Colloquial	Revolution	yes
265.	22-02-2012	شو صار بوطني	MC Roco – Ebn Siba3	Colloquial	Revolution	yes
266.	22-02-2012	اعصار ثورجي	Mc Roco || || Ebn Sba3 || || Al-FarooQ	Colloquial	Revolution	yes
267.	22-02-2012	حمص تبكي	عمر العمير	Colloquial	Revolution	yes
268.	22-02-2012	زبداني السيف البتار	ثوار الزبداني	Colloquial	Revolution	yes
269.	22-02-2012	جيش أحرار	بنت حمص	Colloquial	Revolution	yes
270.	22-02-2012	ساقط ساقط يابشار	--------------	Colloquial	Revolution	yes
271.	23-02-2012	منصورين بعون الله (رد على أنا سوري (يا نيااللي	أحرار الشام	Colloquial	Revolution	yes
272.	23-02-2012	احرار سوريا	klash - big h 2012 كلاش	Colloquial	Revolution	yes
273.	24-02-2012	من بابا عمر إلى الخالدية	وصفي المعصراني	Colloquial	Revolution	yes
274.	24-02-2012	من أجلك سوريا	عمر وسارة وأمجد أبو الحاج	Colloquial	Revolution	yes
275.	28-02-2012	والله محتاجك يا خي	ووصفي المعصراني	Colloquial	Revolution	yes
276.	28-02-2012	لأسقط نظامك يا أبن الانجاس	فرقة أحرار الشام	Colloquial	Revolution	yes
277.	29-02-2012	مجدك يا سورية	أحرار الشام	Colloquial	Revolution	yes
278.	29-02-2012	انشودة يكفي قتل	مهند سلطان	Colloquial	Revolution	yes
279.	01-03-2012	إلنا الله يا سورية	أحرار الشام	Colloquial	Revolution	yes
280.	01-03-2012	دوما أم الشهداء	------------	Colloquial	Revolution	yes
281.	02-03-2012	يسقط النظام الدموي	أحرار الشام	Colloquial	Revolution	yes

(continued)

#	Date	Song	Singer	Language	Purpose/ use	Downloaded
282.	02-03-2012	وا سوريا	محمد الجبالي	Modern Standard Arabic	Revolution	yes
283.	03-03-2012	بركان حمص	مجاهد حوري زاده	Colloquial	Revolution	yes
284.	04-03-2012	هبت الشام	محمد اسراء	Colloquial	Revolution	yes
285.	06-03-2012	بدك ترحل يا بشار	فرقة غربة	Colloquial	Revolution	yes
286.	06-03-2012	بتعرف شو اللي صار، بليلة من آذار	------------	Colloquial	Revolution	yes
287.	07-03-2012	نيالك يا وطن	MC Revo	Colloquial	Revolution	yes
288.	10-03-2012	سوريا يا حبيبتي	المندسيين السوريين	Modern Standard Arabic	Revolution	yes
289.	10-03-2012	بعدك على بابي	سوزي	Colloquial	Revolution	yes
290.	10-03-2012	أنا أنا انسان أنا	سوزي	Colloquial	Revolution	yes
291.	10-03-2012	بكرى أحلى	Mc Roco	Colloquial	Revolution	yes
292.	11-03-2012	ثار الوطن العربي	مصطفى العزاوي	Colloquial	Revolution	yes
293.	11-03-2012	بابا عمرو	Mc Roco || || Ebn Sba3 || || Al-Mo3taKaL	Colloquial	Revolution	yes
294.	12-03-2012	نحنا معانا الله	أحرار الشام	Colloquial	Revolution	yes
295.	12-03-2012	ستي يا ستي	بنت حمص	Colloquial	Revolution	yes
296.	12-03-2012	يا سيف الأمة يا درعا	أبو زيد صقر الوادي (ألحان وكلمات)	Colloquial	Revolution	yes
297.	13-03-2012	نداء الحرية	حذيفة السوطري	Colloquial	Revolution	yes
298.	14-03-2012	اه منك يا حلب	أحرار الشام	Colloquial	Revolution	yes
299.	14-03-2012	نشيد عذري اليكم	بلال الكبيسي	Modern Standard Arabic	Revolution	yes
300.	14-03-2012	فاقد هويتي	Qabar & عبد الباسط الساروت	Colloquial	Revolution	yes
301.	15-03-2012	يا وطني الأجمل	سوزي	Colloquial	Revolution	yes
302.	15-03-2012	حيو الثورة	أحرار الشام	Colloquial	Revolution	yes
303.	15-03-2012	البطبوط	بنت حمص	Colloquial	Revolution	yes
304.	15-03-2012	سوريا يا بلادي	ابراهيم الدردساوي	Colloquial	Revolution	yes
305.	17-03-2012	حمص عاصمة الثورة	أنس السقا	Colloquial	Revolution	yes

(continued)

#	Date	Song	Singer	Language	Purpose/ use	Downloaded
306.	18-03-2012	دلعونا الثورة السورية	------------	Colloquial	Revolution	yes
307.	19-03-2012	سكابا يا دموع العين	أحمد الشريقي	Colloquial	Revolution	yes
308.	20-03-2012	يا أم الشهيد يا حنينة	أحرار الشام	Colloquial	Revolution	yes
309.	21-03-2012	انشق	عبدالسلام حوى	Colloquial	Revolution	yes
310.	22-03-2012	نحن ثرنا يا شآم	أحرار الشام	Modern Standard Arabic	Revolution	yes
311.	22-03-2012	صبراً يا نفسي	نايف الشرهان	Modern Standard Arabic	Revolution	yes
312.	22-03-2012	ياطير خدني	وصفي المعصراني	Colloquial	Revolution	yes
313.	23-03-2012	يا أم الشهيد يا حنينة	أحرار الشام	Colloquial	Revolution	yes
314.	23-03-2012	قادمون يا قرداحة	Mc Roco	Colloquial	Revolution	yes
315.	23-03-2012	من درعا طلعت حرية	--------------	Colloquial	Revolution	yes

Appendix 7
Mundassa Screenshots

Figure 34: The tabs of the website.

Figure 35: Zead Raed, "Syria, What is Missing?" Mundassa, July 13, 2011, personal research archive.

https://doi.org/10.1515/9783110767698-013

Is it really "Sectarian Violence" or another cheap trick?

Written by BF on 19 July, 2011 · 04:34 AM · No Comment Yet
Rating: +1 (from 1 vote)

While I was sipping my morning coffee I stumbled upon an article in the Washington Post titled "Sectarian Violence Rises Fears"

The article talks about the recent events in Homs, the syrian city in the middle of the country, and tries to describe the situation there. I have to admit the writer of the article (Liz Sly) has done a good job trying to gather information from different groups in the city. After all, we all know the media black out the syrian regime created all over the country making it almost impossible for any journalist to get a descent reliable story about what exactly is happening there. Most of the sources that feed the media these days are "anonymous people", youtube videos, facebook pages, and the regime media which makes the picture gloomy and unclear for all of us.

Homs is a great example of the syrian Mosaic. It represents how different sects in Syria have been living together in harmony for ages. Syrians are just like any other population in this world, among them you can find the the good, the bad, and the ugly. Therefore, no one can ever deny the existence of extremists within all groups and sects, and the fear of sectarianism and a civil war has been always in the back of the Syrians' heads, which perfectly explains why the uprising in Syria didn't start earlier and waited for so long under the military boot of this regime, and until finally standing up firm.

On the other hand, the Syrian regime knows exactly this fact about the country; that's why they kept waving the sectarian card in our faces since the begining of the uprising, and that's why I ask all the people to see the full picture before start judging the situation

Since the begining of the uprising, the syrian regime launched a campaign calling people of syria to stand firm against secterianism with the "Great leader Al-Assad". As we all know, things started in the city of Daraa where sectarianism doesn't exist whatsoever in the first place, so we all were laughing at such cheap campagin and no one actually from outside syria paid much attention to it.

As the uprising continued, the regime began using another tactic in their propaganda against the protesters. They start calling all of them "extreme Muslims", "members of Alqaida", "terrorists" and all other names that the Qadafi regime used before! This time also, no one from outside Syria took much notice of that. However, some people in syria started to express their fears from the extremists being able to reach power if Assad's regime fell. Understandably, most of these people were either secular or belonged to minority groups. I have to admit these fears are legitimate and even myself, being a non religous secular person, I had the same kind of fears in the back of my head and I went through different discussions with people from a different backgrounds at that stage to come up with the next point of view:

Syria needs to change. The change is imminent and I know for sure that this regime is not welling to give its power up easialy. Assad is willing to kill half of Syrian people to stay in power, and actually has already killed thousands in the streets and imprison thousands more. If we don't change now things will get even worse and in the near future we'll have nothing but povirty, illiteracy and extremism among Syrian people

The regime propaganda is actually targeting the syrian "intellects" aiming to keep them away from the protesters and have them to accept all the killing and atrocity in the streets. The regime obviously was playing the "fear" card and wanted us to stop supporting the protest and go back to the same old system with different slogan "reforms", even though its be-known that this regime is unable to "reform" itself.

Leaving the protesting streets and stripping them of their intellects and wise people who belong to all Syrian sects and groups is what the regime wants. If this happens, the street will be composed of clueless revolutionists with legitimate cause which will open the door for extremists to fill this leadership gap, and eventually transform this uprising into what the regime wants, a sectarian conflict. People in the streets may lack on-field leaders and brains to organize the uprising. That's why people of all groups and sects are required to join the revolution so we can make sure that in the future when the regime falls, no vacuum situation is going to be established on ground.

At the end of the day, many Syrian intellects from all sects actually went to the streets and supported the protests, such as Suhair Atassi, Mai Skaff, Fares Helo, Muntaha Al-Atrache, and we are very proud of them. Many of these people where threatened by Assad's security forces and some were actually imprisoned. Al assad regime wanted to show the world that the protesters as extremists, but that card was proven wrong again.

Nowadays, I think the regime is playing its final dirty card, the real sectarian conflict which the regime shall create on ground, and in which real people are going to be victims of, and most importantly will scare more people and push them away from the uprising. This exactly is what the "Christian businessman" who the writer of the article interviewed expressed of fears, and why he pulled himself out of supporting the protests. Real action is what's desperately needed for the regime right now so they can go ahead with the atrocities in Homs, Hama, Deir Ezour, Damascus and all other cities where the regime faces strong and solid opposition.

I believe that the syrian people will again prove the regime wrong, and will eventually call the bluff and turn the table upside down and win the "Game of Freedom"!

Figure 36: BF (pseud.), "Is it Really 'Sectarian Violence' or Another Cheap Trick?" Mundassa, July 19, 2011, personal research archive.

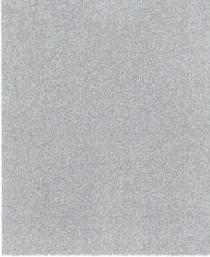

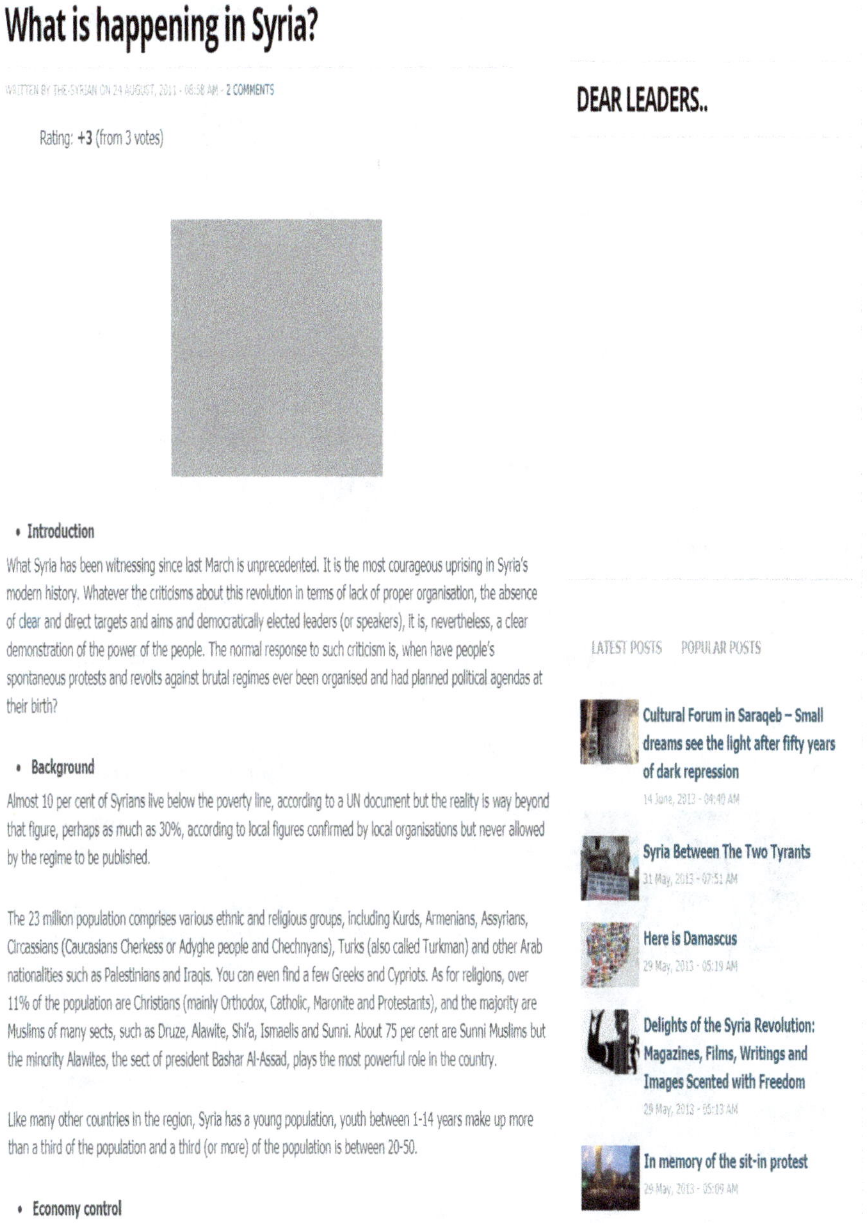

What is happening in Syria?

WRITTEN BY THE-SYRIAN ON 24 AUGUST, 2011 - 08:58 AM - 2 COMMENTS

Rating: **+3** (from 3 votes)

- **Introduction**

What Syria has been witnessing since last March is unprecedented. It is the most courageous uprising in Syria's modern history. Whatever the criticisms about this revolution in terms of lack of proper organisation, the absence of clear and direct targets and aims and democratically elected leaders (or speakers), it is, nevertheless, a clear demonstration of the power of the people. The normal response to such criticism is, when have people's spontaneous protests and revolts against brutal regimes ever been organised and had planned political agendas at their birth?

- **Background**

Almost 10 per cent of Syrians live below the poverty line, according to a UN document but the reality is way beyond that figure, perhaps as much as 30%, according to local figures confirmed by local organisations but never allowed by the regime to be published.

The 23 million population comprises various ethnic and religious groups, including Kurds, Armenians, Assyrians, Circassians (Caucasians Cherkess or Adyghe people and Chechnyans), Turks (also called Turkman) and other Arab nationalities such as Palestinians and Iraqis. You can even find a few Greeks and Cypriots. As for religions, over 11% of the population are Christians (mainly Orthodox, Catholic, Maronite and Protestants), and the majority are Muslims of many sects, such as Druze, Alawite, Shi'a, Ismaelis and Sunni. About 75 per cent are Sunni Muslims but the minority Alawites, the sect of president Bashar Al-Assad, plays the most powerful role in the country.

Like many other countries in the region, Syria has a young population, youth between 1-14 years make up more than a third of the population and a third (or more) of the population is between 20-50.

- **Economy control**

DEAR LEADERS..

LATEST POSTS POPULAR POSTS

Cultural Forum in Saraqeb – Small dreams see the light after fifty years of dark repression
14 June, 2013 - 04:40 AM

Syria Between The Two Tyrants
31 May, 2013 - 07:51 AM

Here is Damascus
29 May, 2013 - 05:19 AM

Delights of the Syria Revolution: Magazines, Films, Writings and Images Scented with Freedom
29 May, 2013 - 05:13 AM

In memory of the sit-in protest
29 May, 2013 - 05:09 AM

Figure 37: Rabi'e Arabi, "What is Happening in Syria?" Mundassa, August 14, 2011, personal research archive.

Growth in sectors such as financial services, construction, telecommunications, tourism, and non-oil industries and trade, are, relatively speaking, "diversifying" the economy, but most are under the control of the ruling family and the corrupt officials who were appointed by the regime. In Syria, people make jokes about corruption and they believe that had there been a global championship in corruption, Syrian officials would always have claimed gold medals!

- **Foreign Policy**

Internationally, Syria has good relations with Iran, including defence ties, and the two signed a memorandum of defence understanding in June 2006. Iran also considers Syria as its other end of the Shi'a "crescent" along with Hezbollah in Lebanon because the Assad family belongs to a branch of the Shi'a (Alawite).

In the past year, there has been substantial high level bilateral contact between Turkey and Syria – with several meetings and visits between Assad and Recep Tayyip Erdogan, the Turkish prime minister, and a range of agreements have been signed.

The ties with Russia are mainly built on strategic interests (on the Russian side) because Syria is almost the only Naval base and gateway for Russia in the Mediterranean. Apart from that, in Syria most weapons are imported from Russia, North Korea and a few other countries such as Iran.

- **Internal Policy: the corrupt family and regime**

Assad's father's internal policy from 1971-2000 was based on sectarianism, i.e., the rule of one sect. This sect was combined with those who are loyal from others religions, ethnicities and sects and members of the ruling Ba'ath party. All his supporters benefited financially from their positions and have helped in making corruption in all areas of life and on all levels the major characteristic of the country's infrastructure. In addition, they maintained a deliberate bureaucracy that can still be found in most of the government establishments. By creating such policies, Hafez Assad and later his son perpetuated their power and control and turned the country into a big farm where most people are looked on as nothing but cattle in the Assad's "ranch". However, the main policy to preserve power was through imposing fear and horror among the people.

Cultural Forum in Saraqeb
29 May, 2013 - 05:02 AM

"The Syrian Resistance" Declaring the Cleansing of Banyas
8 May, 2013 - 08:54 AM

The Obstacles that Face the Syrian Rebels to Capture Damascus
4 May, 2013 - 01:57 AM

Just letting you Know What Assad Supporters Have Been Saying
4 May, 2013 - 01:32 AM

Messages Across the Ocean. Syria and Boston.
21 April, 2013 - 04:33 AM

LATEST COMMENTS

- **Michel Kilo to Pope Benedict XVI: Extend your hand in the name of God, the Most Gracious, the Most Merciful (1)**
  - http://www.icivics.org/node/491176/: I'm truly enjoying the design and layout of... 0 0
- **He Who Justifies the Murder of a People is Killing the Human within Himself (2)**
  - Web Site: Hello to all, how is everything, I think every one is getting more from this web... 0 0
- **A Memory of A Year in the Purgatory: A Testimony about Prison and Journalistic Work (20)**
  - gutschein-platz.net: Thanks for sharing your info. I really appreciate your efforts and I... 0 0
  - midsummer: Remarkable issues here. I'm very satisfied to look your post. Thank you... 0 0
- **Declaration of Dignity-LCCsy (1)**

Figure 37 (continued)

You can perhaps imagine what happened to a person who once, mistakenly, wrapped a falafel sandwich in a piece of newspaper that had a picture of Hafez Assad. This story happened in the early 90s, and until now, the family of that falafel seller don't know anything about him. He vanished after he was kidnapped by the security, like thousands of other Syrians who have also disappeared for good.

Some might say that Syria is supposed to be a secular state and to some extent, a "socialist" country, with an anti-imperialist and anti-Zionist stance. This is true if you listen to the controlled Syrian media, which still uses patronising methods of old fashioned propaganda. However, this is not true if you ask the Syrians who live inside the country. They know the truth of those lies and "mottos" that the regime has continued to exploit and abuse and use over decades to keep emergency law in place and to keep the gates of quick-rich schemes open to his people and regime protectors, especially those who belong to the same sect as the Assad family. No one in Syria can question any person in power or in the security forces and especially the army, the majority of whom belong to the Alawite sect. They are above the law!

These revolutionary "mottos and slogans" have proved a good way to delude some people in believing that the bad economic situation is a result of Syrian "moral and patriotic" standpoints against Israel and the U.S.

Hafez Assad, during 30 years of being in total brutal command and power, managed to brainwash some regime supporters but all other Syrians knew that he was merely a guard dog of the northern Israeli borders (including the Syrian occupied Golan Heights), preventing any direct military action against Israel. Syrians know that if anybody is caught even trying to motivate people to claim the Golan Heights back, s/he will be arrested and may be executed under emergency law, using the excuse that this person was attempting to incite people against the regime and the "national security" of the country. Syrians know that since 1982, there has not been a single bullet shot against Israel despite all the violations committed by the Zionists against Syria. Instead, the regime deployed tanks, warships, helicopters and artilleries to crackdown and kill peaceful demonstrators. The Israelis even flew their F16 jets at low altitude above Bashar Assad's castle near Lataqia city and on another occasion they destroyed a scientific chemical lab in the east. They challenged Assad to retaliate but he didn't dare to even deliver a speech to his people about those incidents. Syrian people know that, as with all Arab leaders, their armies are built only to defend the leaders' seats and power, not the country or the dignity of the nation.

Bashar Assad took over the country after his father's death (in 2000), when his supporters passed the presidency to him in less than 5 minutes at the People's Assembly in Damascus. He inherited what his father, Hafez Assad, had constructed, that is, a well-designed policy of intimidation, which involved bullying and planting constant fear inside people's hearts. People still live in such panic until today although the wall of silence has cracked and the barrier of fear has started breaking down regardless of the lies about mythical slow "reforms" that are uttered by Bashar Assad in national media.

your info. I really appreciate your efforts and I... 0 0

- midsummer: Remarkable issues here. I'm very satisfied to look your post. Thank you... 0 0

- **Declaration of Dignity-LCCsy (1)**
  - embroidery services: It's really a cool and useful piece of information. I am... 0 0
- **What is happening in Syria? (2)**
  - nadia ak: i have a family member gone from here to syria at the moment fighting for... 0 0
- **The Forgotten Ones in Syrian Jails for Decades (2)**
  - Sarah abdallah: My name is Sarah abdallah I'm from Sydney Australia and I just... 0 0
- **Asma Al-Assad, Oh' Where are you? (11)**
  - Al-Mondass: how about you suck my healthy dick? 0 1
  - http://customessaysite.com/: Everyone should build their own social site at home. It... 0 1
- **What if Bashar Assad wins (5)**
  - Bill: From Biblical standpoint, Ezekial 38 & 39 shows a very cohesive Russia, Iran,... 0 0

Figure 37 (continued)

One of the important things worth mentioning is that Hafez Assad, and later his son appointed most relatives and other Alawite people to positions of power including the Army, Security, Media, Transportation, Economic organisations, Telecommunications, Public Health organisations and hospitals, Education establishments and almost every vital and sensitive position in the country. The Assad family only trusted their sect in the top positions (mainly security and army positions). The lesser positions went to his supporters and other sycophants. The final decision in every fundamental matter (internal/external policies and decisions) is always up to the president. The regime also managed to associate the concept of love and loyalty to the country with the love and loyalty to the president. So, if you are against him, you will be classified as a traitor, foreign agent, Zionist etc and consequently, against the country.

It is enough to take a short trip around any city in Syria to see the number of statues and pictures of the Assad family in every street, place, shop, official building and more, and if any place doesn't have a picture, the owner or the person who runs that place will be in serious trouble. People in Syria say that the cost of the statues, pictures, concerts, and events glorifying and praising this "God-Like" figure and his family since the 70s, would have saved at least 15% of the population from poverty. As for the national TV, it is almost dedicated to praise and "sanctify" him. Hundreds of songs were composed to eulogize him and his father.

Syrian demonstrators have lashed out at rampant corruption in the Syrian government. Among the many officials accused of being the worst "thieves" – Rami Makhlouf, the first cousin of Bashar al-Assad and the family business manager and the most powerful economic figure and "businessman" in Syria, reportedly controls as much as 60 per cent of the country's economy through an intricate web of holding companies. He is considered to be the "Don" of the Syrian "Cosa Nostra" and all other criminal security syndicates in the country have to have his permission for every extortion that takes place. In fact, the man can change any law within five minutes and re-tailor a new one to suit his own interests.

- **The tyrant versus people**

All that you see or hear now from Syria is people demonstrating and calling for their freedom and for the regime to step down after all these years of agonising oppression. The regime's response has been shoot to kill the protestors in public with no consideration for any kind of moral attitude or respect for human rights and people's freedom of speech or freedom of choice. The street demands democracy, a word that has never existed in the dictionary of Assad's kingdom library. The immediate response of the regime was inhuman arrests, ruthless kidnappings, vicious intimidation, and above all, bloodthirsty killings. Any person or group who dares to say no to the dictatorship or

Figure 37 (continued)

participates in any of the peaceful demonstrations will face a savage response. This means that people who get arrested will suffer all kinds of unimaginable torture. For example, skinning men alive is a common and known method. In addition, the security forces are known for being experts in sexual abuse, insults and beating the "detainees" not to mention the use of psychological distress methods. The security forces spare no one, children, women, old men and even animals. They can do whatever they want to any living being in the country without being questioned or even criticised and have been given full authority to exercise any action of terror. In addition, they all are armed with hatred for other sects and people with free political principles and thoughts.

That's exactly what Bashar Assad's father did in the past. So, for Syrians, it is 40 years of cruelty and ill treatment for those supposed to be "his own people".

- **The Uprising**

The spark for the intifada started in Dar'a (a city in the south), last March, when security forces arrested 24 children (aged between 7-13) because they chanted "Down with the regime" after they finished a school day. It was in fact an ongoing TV chant or a "catch phrase" that almost everybody in all Arab countries had heard for days and months while watching TV. The children were mainly imitating the people of Egypt who used to chant it in Altahrir Square in Cairo.

The families of the children along with other recognised heads of local families in the city went to the governor to complain and ask for the release of the children, especially after they heard that the children had been beaten up and some tortured by the security forces. The governor's reply was offensive and he called the head of the security of Dar'a, Atef Najib (the president's cousin). Atef came and told the people in a very insulting way that they would not see their children and it would be better to think of "getting new children" and if they couldn't, he could help by sending his men to "impregnate" the women of Dar'a city (he used the abusive 'F' word). He said it in a very challenging and revolting way.

That was enough to spread major protests in the city, which witnessed appalling suppression by Assad's security and army for over 23 days. Many people were killed, injured, kidnapped and arrested and the army besieged the city and cut the electricity, communications and water for over 20 days, and banned food (including dry milk for children) from entering the city. Since that time, the protests spread across the country until today.

Figure 37 (continued)

- **The Revolution**

In the early stages, protesters in all cities and villages were calling for reforms, freedom, release of thousands of political prisoners and dissidents and an end to corruption. All they were calling for was a simple decent life. As demonstrations were met immediately with live bullets, the rallies changed tone, calling for the fall of the regime.

The opposition movement has gained momentum in spite of a persistent and brutal military crackdown. Syrian security forces, henchmen, thugs and pro-government armed men have continued their repressive measures despite the so called "abolishment" of the 48-year-old emergency law, which was issued by Assad a few weeks after the Dar'a massacres and his pledge to implement reforms. None of these reforms have been implemented.

The revolution has spread to the whole country. People chant "no to sectarianism", "we all are one hand", "freedom for Syria" and many other slogans. All Syrian social classes and ethnicities are involved. In the east, Assyrian Christians demonstrated despite Syrian security forces raided their headquarters (the Assyrian Democratic Association) in Qamishli and arrested 12 of its main members. In Qamishli over 35000 Kurdish protesters took to the streets calling for an end to the regime, a call that was echoed across the north. This spread to almost every village and town in the east and south east to include Amouda, Dair Al-Zour, Mayadin, Albo Kamal, Ain al-Arab, Raqah and many other cities and towns. In the North, Edleb and all its surrounding towns and villages up to Jeser Alshoughoor, which suffered another savage crackdown by the Assad army and security. More than 14000 people fled Jeser Alshoughoor city to Turkey as refugees. Many parts of Aleppo city also joined in big rallies which also spread to Efrin and other northern areas. In the west, Lataqia, Jableh, Tartoos and Banias observed massive demonstrations but it was Banias that suffered the most from the dreadful response by the Assad security. However, the main cities that witnessed the biggest demonstrations were Homs, Hamah, Alqusair, and Tal Kalakh. People still remember the Hama massacre occurred in February 1982, when the Syrian army, under the orders of Hafez al-Assad, conducted a scorched earth policy against the town of Hama in order to quell a revolt by the people. The result was more than 25000 killed in less than two weeks.

In Damascus, demonstrations were almost in (almost) all suburbs and surrounding villages and towns as well as inside the city. Hundreds of people were killed by the Assad security and thousands were arrested and tortured. Most demonstrations took place on Fridays and some other days but for the last 3 weeks, during Ramadhan month, there have been demonstrations on a nightly basis in most of the above mentioned cities and towns.

The only people who never participated in any rally or demonstration were the Alawite who, instead, got armed by the security and Maher Assad, the president's infamous brother.

Figure 37 (continued)

- **Assad's killing machine continues**

No one can confirm the correct figures of casualties because the Syrian regime has banned all foreign media and restricted access to all troubled areas since the uprising began, making it impossible to get independent assessments. Defying restrictions and risking detainment, many ordinary Syrian citizens and political activists have taken it upon themselves to report to human rights organisations (Arabs and International) and recorded and posted videos of the protests online. The Syrian regime has banned all human rights and medical organisations from entering or working in the country. However, the revolution has established what is known as "Local Co-ordination Committees" who communicate with International media and other organisation via some technological means such as satellite mobile phones (e.g. Thuraya network) and Internet. Young Syrian revolutionaries keep developing basic but genius ways to bypass the controlled Internet proxy connection.

Human rights organisations believe that the number of people killed so far, since the start of the revolution, might have reached 2600 civilians, more than 6700 injured, at least 4200 missing and over 50000 people arrested, among them around 17000 who were released after few months or weeks. All were abused and tortured and so far, 71 have died under harrowing methods of torture. The figures don't include the army officers and soldiers who refused to shoot the unarmed peaceful demonstrators and were executed or shot either on the spot by the security snipers and henchmen or in prisons.

- **The "Shabbiha" (Henchmen) phenomenon**

The Assad regime uses groups of thugs called "shabihah", which mainly means henchmen. Most of them are Alawites and were members of a secret religious society called "Al-Mortadha" which was founded by Hafez Assad's brother Jamil. Their main jobs were to smuggle drugs, weapons, cigarettes, extract protection money from local small businesses, and control the ports in Tartoos and Lataqia so every local businessman has to pay before off-loading shipments. They also smuggled almost any material that was banned from sale in the country. The ban of those materials was deliberate in order to open the gates of wealth for the regime people and those thugs and especially for the Alawites in the security systems since the law doesn't apply to them. This is to gain their loyalty and link their living standards with the existence of the regime.

Figure 37 (continued)

These thugs are very loyal to Assad and not controlled by any official organisation but all take their orders from the heads of security branches and of course, from the main butcher of Syria Maher Al-Assad, the brother of Bashar, who controls the most powerful military regiments in the country. Two of these notorious regiments are the Republican Guard and the Fourth Regiment whose soldiers and officers are mainly Alawites.

- **The recent situation**

International organisations have described the human rights status in Syria as being one of the worst in the world, with security forces having a long history of harassing and imprisoning rights activists and critics of the government.

The Syrian government continues to patronise its people and the international community by blaming an "armed resurrection" by Salafi militant groups who come from 'outside' the country and 'unknown' local terrorist groups, 'Zionist' agents etc for the deaths of civilians and some security and army people. The way they outline these false claims is very comic and could be a subject of TV satire or comedy. The regime's media has fabricated some evidence of an insurrection by placing ammunition on dead civilians in troubled locations, but unfortunately for them, a soldier filmed the set up and the video was leaked and posted on youtube and Aljazeera TV. Any 12 year old child could tell how primitive and unintelligent the Syrian media and security is from the way they set up their moronic sketches.

However, real raw video footage shot by demonstrators reveals a completely different story and hundreds of harrowing videos have been posted on youtube so far. All are accessible by any person who has an internet connection.

It would take hundreds of pages to go through what the Syrian people and demonstrators have been through so far and what kind of atrocities the regime has inflicted upon them. Hence, I will refrain from listing other barbaric operations as the news agencies around the world have broadcast some of the horrors and killing that hasn't stopped until now.

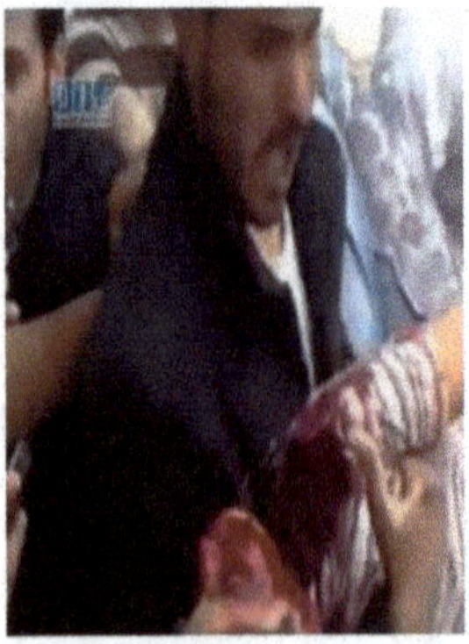

Figure 37 (continued)

- **Conclusion**

The question that imposes itself is "where is Syria heading to"? There are many scenarios, any of which could be the right answer, but if we are looking for a feasible solution to this situation, it is in the hands of Bashar Al-Assad. Unfortunately, it is obvious that he will favour power and control, wealth and theft to giving his people a peaceful life. This is evident because the only solution that he has come up with is violence and oppression and deploying as many army and security forces in every city and town in the country as he can. It seems clear that his main aim is to drag the country to civil war.

Another scenario would be a coup d'état staged by some Alawite generals to end the violence but keep the sect in power-an option that not all Syrians would be happy with.

With regard to the possibility for international military intervention, it was made clear by all Syrian demonstrators and opposition that they are against it. However, nobody is sure how long they can hold to this because their thresh-hold has started declining and their plan to keep the revolution peaceful will sooner or later end because of the dramatic increase in killing and arrests carried out by the regime. So, an armed resistance and international military intervention are still an option for the revolution development in future.

The regime knows that international military intervention will drag the whole region into an all out war. This will happen if Iran gets involved along with Hezbollah in south Lebanon. Israel will take advantage of this pandemonium to achieve it's "Biblical Armageddon" by destroying Iranian nuclear capabilities but they are likely to ask the Americans to do this dirty job. Israel will attack Hezbollah in the south of Lebanon in order to create a new situation on ground for a potential peace process according to Israeli rules and conditions. If an attack against Iran takes place by the NATO or US, Russia will not stand by watching and will get involved somehow, so we might witness a new apocalyptic mayhem in the Middle East. Of course, we should consider a Palestinian uprising or an Israeli invasion on Gaza as well.

As for China, her foreign policy depends on what US and Europe could bargain with, that is, what are the rewards that China can gain if it abstains at the Security Council? We have already seen Chinese delegations visiting the newly born country of South Sudan, so is this a devil's deal?

Lets all pray that reason and wisdom will prevail in the Middle East and people of Syria will win their heroic Intifada peacefully, facing the tanks of Assad's regime with their bare chests.

Rabi'e Arabi

Syria, 14/8/2011

Figure 37 (continued)

On the fall of the regime and before statement number 1

WRITTEN BY FREESYRIA ON 31 JULY, 2012 - 09:28 AM - **NO COMMENT YET**

Rating: **0** (from 0 votes)

On the 18th of July and after targeting the top security figures designated to deal with the crisis in Syria, news and analyses were flooding the media.

Major media networks hosted numerous political and military analysts and the smallest strategic details became common knowledge to everybody. Social media pages, even those supporting the regime, were also filled with details, witnesses and news of defections that reached the establisher of what is called "The Syrian Electronic Army". The defection of such a personality bears a lot of meanings and indications, especially that it is him and people like him who contributed to the great divide and tension among people in the country.

The big names that were reported killed that day are the same reported killed in a previous poising operation, which means that the revolution succeeded in targeting them. Even if the operation on the 18th of July wasn't a revolutionary act and was actually planned by the regime to get rid of these people for reasons pertaining to it, it is still a revolutionary success and the result of a revolutionary act.

One of the reports according to the Free Army in Damascus claims that the poising operation was indeed successful; however, the regime chose to delay announcing these people dead for 2 reasons. Firstly, it didn't want to portray any weakness at the time. Secondly, it chose that day to announce the killing of a number of its most highly-ranking officials to use it as a justification and excuse for an unprecedented assault on Damascus on such a decisive day.

It has been said that it would be quite decisive and a point of no return for the Free Army to arrive in Damascus. It would also decisive for the revolution whose success may actually depend on that. with the recent news on Aleppo, similar analyses would certainly apply to the second most major city in the country.

However, whoever thinks that Damascus or Aleppo battles would be resolved and ended militarily is quite mistaken and deluded. Both sides, the revolutionaries on one hand and the regular army and security forces on the other, had engaged in what can be called preliminary battles in other cities. Their biggest concern would be these two cities as the Free Army wants to reach them while the regular army is fighting to block it from entering them. Once Damascus or Aleppo are in the battle and both armies are in, people's morale and support would be the decisive factor.

DEAR LEADERS..

LATEST POSTS POPULAR POSTS

0 (19) **Amnesty international's Donatella Rovera account**
5 August, 2012 - 10:04 AM

Syrian army uses civilians as human shields – video
3 August, 2012 - 07:24 PM

JOURNEY INTO MEMORY S Title 1
3 August, 2012 - 05:54 AM

The Womb of Murder
3 August, 2012 - 05:42 AM

Faces from the Syrian Revolution: Anas Al-Sheghri
31 July, 2012 - 10:04 AM

On the fall of the regime and before

Figure 38: Manar Haidar, "On the Fall of the Regime and Before Statement Number 1," Mundassa, July 31, 2012, personal research archive.

Syrian official media had repeatedly bombarded us with the culture of street battles that were used as a tactic by resistance movements in the region, especially the Lebanese Hezbollah. They talked of the advantages that include ease of movement and maneuvers and that would cause great trouble and disadvantage to any regular army fighting in a street battle. Now, the Syrian regime finds itself trapped into a street battle. They had to understand that by chasing the Free Army out of Damascus countryside, they would find themselves facing it in the heart of Damascus.

What is important for us now is to comprehend the moment and realize that the regime is closer to collapse than ever. We have to be true to what we have been saying for the last year and a half. We have to prove that our revolution is for dignity and that the Syrian people are one. Otherwise, we would be taking our country to the unknown.

Most of us have repeated slogans about national unity and contributed in easing the fears and concerns of minorities even to the point of boredom. Now is the time to see that on the ground. Are we prepared? Or were we just repeating empty slogans? Now is the time to test ourselves..

A civilized victory in popular revolutions is usually demonstrated in two points: tolerance and protecting government establishments. The victory of the revolution is not about toppling the regime and causing chaos in the country. Real success is achieved when the revolutionaries present themselves as a reliable alternative that can secure the country and the people. Otherwise, what good can a revolution be if it results in toppling the regime and destroying the country?

Basher Al-Assad may eventually be gone one way or another and it is quite probable that some of his supporters will not accept his fall. We have to tolerate them and try not to provoke them with gloating. He certainly will not take all his loyalists with him and they would be like a wounded lion and it won't be in our interest to provoke them or inflict more pain on them.

We all know since before the revolutions that he has loyalists and people who love him and people who benefit from his regime. We didn't start the revolution to defeat them or fight them. We started it to build a democratic Syria where we can all be equal in our country.

We also have to be alert and not fall victims of agitation against them. People who would provoke such sentiments may not even be Syrian and their interest may lie in a prolonged conflict that would sever all national ties between Syrians.

Another important point has to be raised in the event of capturing official media by the Free Army in either Aleppo or Damascus, which would obviously mean the defeat of the regime. In such an event, the revolutionaries would have to be up to the challenge and present the most capable to handle such a historic moment.

There is so much to say and one article could never be enough to include what we need to know to prepare ourselves for victory. However, it is important to start spreading a tolerant culture of victory that would overwhelm any minority that would promote feeling of hatred and revenge.

What is ahead of us will be even more challenging; Damascus or Aleppo might be heavily bombed. But even if that happens, it shouldn't distract us from our main objective. What matters the most is the final conclusion and the fall of the regime won't be the final conclusion.

Let's always remember martyr Hadi El-Jendi's will: "remember me when you celebrate the victory." He never said: 'avenge me when you celebrate the victory."

By God's will and blessing, the revolution will triumph

Manar Haidar

On the fall of the regime and before statement number 1
31 July, 2012 - 09:38 AM

Robert Fisk – an elementary mistake
29 July, 2012 - 06:07 PM

Omar Sulaiman died in Damascus
29 July, 2012 - 12:01 PM

Welcome Manaf Tlass
26 July, 2012 - 07:31 AM

The Conditionality of some Intellectuals
26 July, 2012 - 04:30 AM

LATEST COMMENTS

- **The Womb of Murder (1)**
  - Shakeeb Al-Jabri: Please note, this content is copyrighted by Al-Ayyam. Kindly stick to... 0 0
- **Robert Fisk – an elementary mistake (5)**
  - wa7ed mundas barra: well, sometimes common facts need not be repeated. It'd be silly to... 0 0
  - Raymondo: trust me someone somewhere will find a way to critisize those as well 1 0
- **Saint Mary Church and ... (2)**
  - wa7ed mundas barra: uh, pictures of destroyed buildings. How does this prove who the... 0 0
  - Alan: Destruction by FSA of Churches in Homs http://www.youtube.com/watch?v =WylY4psccOw 1 0
- **Syria: Autopsy of a Regime (1)**
  - Youzzers - The Social networking Swiss Made: What is it special about Youzzers? It is a... 0 0
- **A Memory of A Year in the Purgatory: A Testimony about Prison and Journalistic Work (2)**
  - read: Can I simply just say what a comfort to discover an individual who truly understands... 0 0
  - Thiago: Option 4 Syria OKed the strikes and is blirsetung about it in public because it... 0 0
- **Traitors? (5)**
  - Anonymous: I dont agree with your article at all My personal opinion, you have never been... 0 0
- **Me All Along (3)**
  - Hakam: Always powerful Rama! Syria's

Figure 38 (continued)

Bibliography

ʿAbdilhādī ʿAtīq, ʿUmar. *ʿIlm al-Balāgha bayn al-ʾAṣāla wa al-Muʿāṣara* [Rhetoric Between Authenticity and the Contemporary] (Amman: Dār Usāma li-l-Nashir, 2012).

Aboul-Ela, Hosam. "Challenging the Embargo: Arabic Literature in the US Market," *Middle East Report* 219 (2001): 42–4.

Altoma, Salih J. *Modern Arabic Literature in Translation: A Companion* (London: Saqi, 2005).

Appiah, Kwame Anthony. "Thick Translation," *Callaloo* 16, no. 4 (1993): 808–19.

Apter, Emily. *The Translation Zone: A New Comparative Literature* (Princeton: Princeton University Press, 2006).

al-ʿAqīlī, Majdī. *al-Samāʿ ʿInda al-ʿArab* [Listening for Music for Arabs] (Damascus: n.p, 1973).

al-ʿAysh, Yūsuf. *Al-Dawla al-Umawiyya* [The Ummayyad State], 2nd ed. (Damascus: Dār al-Fikr, 1985).

Baker, Mona. "Reframing Conflict in Translation," in *Critical Readings in Translation Studies*, ed. Mona Baker (London: Routledge, 2010).

Barthes, Roland. *The Rustle of Language*, trans. Richard Howard (Berkeley: University of California, 1989).

Barthes, Roland. "Rhetoric of the Image," in *Image, Music, Text*, essays sel. and trans. Stephen Heath (London: Fontana Press, 1977).

al-Bārūt, Jamāl. "Al-ʿAqd al-Akhīr fī-Tārīkh Sūryya: Jadaliyyat al-Jumūd w- al-Iṣlāḥ [Syria in the Last Decade: The Dialectic of Stagnation and Reform] (Doha: Arab Center for Research and Policy Studies, 2012).

Batatu, Hanna. *Syria's Peasantry, the Descendants of Its Lesser Rural Notables, and Their Politics* (Princeton: Princeton University, 1999).

Bishāra, ʿAzmī. *Sūriyya Darb al-Ālām naḥwa al-Ḥurryah: Muḥāwala fī-l-Tārikh al-Rāhin* [Syria: The Path of Suffering towards Freedom (2011–3)] (Doha: Arab Center for Research & Policy Studies, 2013).

Bishop, Russell. "Freeing Ourselves from Neocolonial Domination in Research: A Kaupapa Māori Approach to Creating Knowledge," in *The Landscape of Qualitative Research*, 3rd edition, eds. Norman K. Denzin and Yvonna S. Lincoln (California: Sage, 2008).

Bothwell, Beau. "Minnḥbbuk (ya Baṭa): Musical Rhetoric and Bashar al-Assad on Syrian Radio During the Civil War," in *Tyranny and Music*, eds. Joseph E. Morgan and Gregory N. Reish (Lanham: Lexington Books, 2018).

Bourdieu, Pierre. *Language and Symbolic Power*, trans. Gino Raymond and Mathew Adamson (Cambridge: Polity Press, 1991).

Bourdieu, Pierre. *The Logic of Practice*, trans. Richard Nice (California: Standard University Press, 1990).

Bourdieu, Pierre. *Outline of a Theory of Practice*, trans. Richard Nice (Cambridge: Cambridge University Press, 1977).

Bourdieu, Pierre. *Distinction: A Social Critique of the Judgement of Taste*, trans. Richard Nice (Cambridge: Harvard University Press, 1984).

Büchler, Alexandra, and Abdel-Wahab Khalifa, "Translation of Arabic Literature into English in the United Kingdom and Ireland 2010–2020," *Literature Across Frontiers*, September 2011.

Bukharin, Nikolai, and Evgenii Preobrazhensky, *The ABC of Communism* (Bookyards, 1922).

Butler, Judith. *Bodies that Matter: On the Discursive Limits of "Sex"* (New York: Routledge, 1993).

Casanova, Pascale. *The World Republic of Letters* (Cambridge: Harvard University Press, 2004).

Cheung, Martha Pui You. "On Thick Translation as a Mode of Cultural Representation," in *Across Boundaries: International Perspectives on Translation Studies*, eds. Dorothy Kenny and Kyongjoo Ryou (Newcastle: Cambridge Scholars Publishing, 2007), 22–37.

https://doi.org/10.1515/9783110767698-014

Clark, Peter. *Arabic Literature Unveiled: Challenges of Translation* (Durham: University of Durham, Centre for Middle Eastern and Islamic Studies, 2000).

Coffey, Amanda. "Ethnography and Self: Reflections and Representations," in *Qualitative Research in Action*, ed. Tim May (London: Sage, 2002), 314–31.

Comendador, Maria Luz, Luis Miguel Canada, and Miguel Hernando de Larramendi, "The Translation of Contemporary Arabic Literature into Spanish," *Yearbook of Comparative and General Literature* 48, (2000): 115–25.

cooke, miriam. *Dissident Syria: Making the Oppositional Arts Official* (Durham: Duke University Press, 2007).

Cordesmann, Anthony H. *Israel and Syria: The Military Balance and Prospects of War* (Westport: Praeger Security International 2008).

Court, Deborah, and Randa Abbas. "Whose Interview Is It, Anyway? Methodological and Ethical Challenges of Insider–Outsider Research, Multiple Languages, and Dual-Researcher Cooperation," *Qualitative Inquiry* 19, no. 6 (2013), 480–88.

Creative Memory of the Syrian Revolution (CMSR), *The Story of a Place, The Story of a People: The Beginnings of the Syrian Revolution (2011–2015)*, ed. Sana Yazaji, trans. Rana Mitri (Beirut: Friedrich-Ebert-Stiftung, 2017).

Creative Memory of the Syrian Revolution (CMSR), *Qiṣṣat Makān Qiṣṣat Insān: Bidāyāt al-Thawra al-Sūriyya 2011–2015* [A Story of a Place, A Story of a Human], ed. Sana Yazaji (Beirut: Friederich-Ebert-Stiftung, 2017).

DeGhett, Torie Rose. "'Record! I am Arab': Paranoid Arab Boys, Global Ciphers, and Hip Hop Nationalism," in *The Hip Hop & Obama Reader*, eds. Travis L. Gosa and Erik Nielson (Oxford: Oxford University Press, 2015).

Dente Ross, Susan. "Unequal Combatants on an Uneven Media Battlefield: Palestine and Israel," in *Images that Injure: Pictorial Stereotypes in the Media*, eds. Paul Martin Lester and Susan Dente Ross (Westport: Praeger, 2003).

Dentith, Simon. *Parody: The New Critical Idiom* (London: Routledge, 2000).

Dharīl, ʿAdnān Ibn. *al-Mūsīqā fī Sūriyya: al-Baḥth al-Mūsīqī wa al-Funūm al-Musīqyya Munthu Mi'a ʿĀm* [Music in Syria: The Musical Research and Arts of the Last Hundred Years] (Damascus: Maṭābiʿ Alif Bāʾ – al-Adīb, 1969).

Dib, Kamal. *Tārīkh Suriyya al-Muʿāṣir: min al-Intidāb al-Faransī ilā Ṣayf* [Contemporary History of Syria: from the French Mandate to the Summer of 2011] (Beirut: Dar al-Nahar, 2011).

Dubois, Simon. "Négocier son identité artistique dans l'exil. Les recompositions d'un paysage créatif syrien à Berlin," in *Migrations Société* 174, no. 4 (2018): 45–57.

Dubois, Simon. "Les Chants se Révoltent," in *Pas de Printemps pour la Syrie: Les Clés pour Comprendre les Acteurs et les Défis de la Crise (2011–2013)*, eds. François Burgat and Bruno Paoli (Paris: La Découverte, 2013).

al-Dumayrī, Kamāl al-Dīn Muhammad Ibn Mūsā. *Ḥayāt al-Ḥaywān al-Kubrā* [The Great Life of the Animal] (Damascus: Tlas, 1992).

Fändrich, Hartmut. "Viewing 'the Orient' and Translating its Literature in the Shadow of the *Arabian Nights*," *Yearbook of Comparative and General Literature* 48 (2000): 95–106.

Faraj al-ʿAṣbahānī, Abū. *Adab al-Ghurabāʾ* [The Literature of Strangers] (Beirut: Dār al-Kitāb al-Jadīd, 1972).

Fayyāḍ, Laylā Maliḥa. *Mawsūʿat Aʿlām al-Mūsīqā al-ʿArab wa al-Ajānib* [The Encyclopedia of Famous Arab and Foreigner Musicians] (Beirut: Dār al-Kutub al-ʿIlmiyya, 1992).

Finch, Emily. "Issues of Confidentiality in Research into Criminal Activity: The Legal and Ethical Dilemma," *Mountbatten Journal of Legal Studies* 1, no. 2 (2001): 34–50.

Flick, Uwe. *An Introduction to Qualitative Research*, 4th ed. (London: Sage, 2009).
Foucault, Michel. *The History of Sexuality* vol. 1, trans. Robert Hurley (New York: Pantheon Books, 1978).
Foucault, Michel. *The Archeology of Knowledge*, trans. A. M. Sheridan Smith (New York: Pantheon Books, 1972).
Geertz, Clifford. *The Interpretation of Cultures: Selected Essays* (New York: Basic Books, 1973).
Göksel, Asli, and Celia Kerslake. *Turkish: A Comprehensive Grammar* (London: Routledge, 2005).
Göksun, Yenal. "Cyberactivism in Syria's War: How Syrian Bloggers Use the Internet for Political Activism," in *New Media Politics: Rethinking Activism and National Security in Cyberspace*, ed. Banu Baybars-Hawks (Newcastle: Cambridge Scholars Publishing, 2015).
Halasa, Malu, Zaher Omareen, and Nawara Mahfoud, eds., *Syria Speaks: Art and Culture from the Frontline* (London: Saqi Books, 2014).
Heilbron, Johan, and Gisèle Sapiro. "Outline for a Sociology of Translation: Current Issues and Future Prospects," in *Constructing a Sociology of Translation*, eds. Michaela Wolf and Alexandra Fukari (Amsterdam: John Benjamins, 2007), 93–107.
Heilbron, Johan. "Translation as a Cultural World System," *Perspectives: Studies in Translatology* 14 (2000): 9–26.
Herman, Edward S. *Beyond Hypocrisy: Decoding the News in the Age of Propaganda* (Boston: South End Press, 1992).
Hetzl, Sandra. "Translation of Arabic Literature in German-Speaking Countries 2010–2020," ed. Alexandra Büchler, 5.
Ḥusayn, Ṭāha. *al-Majmūʿa al-Kāmilah al-jiz' 4* [The Complete Collection: Vol. 4] (Beirut: The International Company for Books, 1982).
Ibn Māja. "Bāb al-Nikāḥ" [The Book of Marriage], in *Sunan Ibn Māja* (Riyadh: Maktabat al-Maʿārif, N.D).
ʿIdwān, Mamdūḥ. *Ḥaywanat insān* [Human's Animalization] (Damascus: Mamdūḥ ʿIdwān Publishing House, 2007).
Ismail, Salwa. *The Rule of Violence: Subjectivity, Memory and Government in Syria* (Cambridge: Cambridge University Press, 2018).
al-Jabūrī, Kāmil Salmān. *Muʿjam al-Udabā' min- al_ʿAṣr al-Jāhilī ḥatā 2003* [The Dictionary of Writers from the Pre-Islamic Era to 2002] (Beirut: Dar al-Kotob, 2003).
Jacquemond, Richard. "al-Adab al-Filisṭīnī Mutarjaman ilā al-Firansiyya: Tārīkhuhu wa-Atharuhu" [Palestinian Literature Translated into French: History and Impact], *Alif: Journal of Comparative Poetics*, no. 38 (2018): 94–119.
Jacquemond, Richard. "Translation and Cultural Hegemony: The Case of French-Arabic Translation," in *Rethinking Translation: Discourse, Subjectivity, Ideology*, ed. Lawrence Venuti (London: Routledge, 1992), 139–58.
Jurkiewicz, Sarah. *Blogging in Beirut: An Ethnography of a Digital Media Practice* (Bielefel: Transcript Verlag, 2018).
al-Karīm al-ʿAllāf, ʿAbd. *al-Ṭarab ʿInda al-ʿArab* [Tarab for Arabs] (Baghdad: al-Maktaba al-Ahliyya, 1963).
al-Khuḍarī Bek, Muhammad. *Muḥaḍārāt fī-Tārīkh al-Umam al-Islāmiyya: al-Dawla al-ʿAbbāsiyya*, ed. Najwā ʿAbbās [Lectures on the Muslim Nations: The Abbasid State] (Cairo: Al-Mukhtār, 2003).
Kvale, Steinar, and Svend Brinkmann. *InterViews: Learning the Craft of Qualitative Research Interviewing* (Los Angeles: Sage, 2008).
Lazzarato, Maurizio. "La forme politique de la coordination," *Multitudes* 3, no. 17 (2004): 105–14, https://doi.org/10.3917/mult.017.0105.
Liamputtong, Pranee. *Researching the Vulnerable: A Guide to Sensitive Research Methods* (London: Sage, 2007).

Mehrez, Samia. "Introduction: Translating Revolution: An Open Text," in *Translating Egypt's Revolution: The Language of Tahrir*, 1–2, ed. Samia Mehrez (Cairo: AUC Press, 2012).
Merton, Robert K. "Insiders and Outsiders: A Chapter In The Sociology Of Knowledge," *American Journal of Sociology*, 78, no. 1 (July 1972): 9–47.
al-Muqarrī al-Tilmisānī, Aḥmad Ibn Muhammad. *Nafḥ al-Ṭīb Min Ghuṣūn al-Andalus* [The Pleasant Fragrance from the Tree Branch of al-Andalus], vol. 1 (Beirut: Dār ṣāder, 1988).
Ngugi wa Thiong'o. *Decolonizing the Mind: The Politics of Language in African Literature* (London: Heinemann, 1986).
Othman, Hashem. *Tārīkh Suriyya al-Ḥadīth: 'Ahd Hafezal-Assad 1971–2000* [The Modern History of Syria: Hafez al-Assad Regime 1971–2000] (Beirut: Riad El-Rayyes, 2014).
Parker, *Qualitative Psychology: Introducing Radical Research* (Buckingham: Open University Press, 2005).
Parrilla, Gonzalo Fernández. "Translating Modern Arabic Literature into Spanish," *Middle Eastern Literature* 16, no. 1 (2013), https://doi.org/10.1080/1475262X.2013.775858.
Rajāʾī, Fūʾād, and Nadīm Darwīsh. *Min Kunūzinā* [From Our Treasures] (n.p, 1970).
al-Razzāq ʿĪd, ʿAbd. *Wa-Yasʾalūnaka ʿan al-Mujtamaʿ al-Madanī: Rabīʿ Dimashq al-Mawʾūd* [And They Ask You About the Civil Society: The Buried Damascus Spring] (Beirut: Dār al-Fārābī, 2004).
Roth-Burnette, Jennifer L. "Syria," in *Hip Hop Around the World: An Encyclopedia*, 2 vols., eds., Melissa Ursula Dawn Goldsmith and Anthony J. Fonseca (Santa Barbara: Greenwood, 2019).
Ruocco, Monica. "A Survey of Translation and Studies on Arabic Literature Published in Italy (1987–1997)," *Arabic & Middle Eastern Literatures* 3 no. 1 (2000): 63–75.
ʿAli al-Ṣābūnī, Muhammad. *Rawāʾiʿ al-Bayān: Tafsīr Ayāt al-Aḥkām* [The Beauty of Eloquence: Interpretation of the Commandment Verses], 3rd ed. (Damascus: Al-Ghazali Publishing House, 1980).
Said, Edward W. "Embargoed Literature," *The Nation* 251 no. 8, September 17, 1990.
Said, Edward W. "The Text, the World, the Critic," *The Bulletin of the Midwest Modern Language Association* 8, no. 2 (1975): 1–23.
Sassoon, Joseph. *Anatomy of Authoritarianism in the Arab Republics* (Cambridge: Cambridge University Press 2016).
Sawah, Wael, and Salam Kawakibi. "Activism in Syria: Between Nonviolence and Armed Resistance," in *Taking to the Streets: The Transformation of Arab Activism*, eds. Lina Khatib and Ellen Lust (Baltimore: Johns Hopkins University Press, 2014), 136–71.
Stagh, Marina. "The Translation of Contemporary Arabic Literature into Swedish," *Yearbook of Comparative and General Literature* 48 (2000): 107–14.
Seale, Patrick. *Asad: The Struggle for the Middle East* (Berkeley: University of California, 1990).
Suh, Dae-Sook, *Korean Communism, 1945–80: A Reference Guide to the Political System* (Honolulu: University of Hawaii, 1981).
Ṭlās, Mustafā, ed. *Hakadhā Qāl al-Assad* [Thus Said Assad]. *N.d., n.p.*
ʿUmarīn, Zāhir , Mālū Hālāsā, and Nawwāra Mahfūḍ, eds., *Sūryā Tataḥaddath: al-Thaqāfa wa-l-Fann min ajl al-Ḥurriyya* [Syria Speaks: Culture and Art for Freedom] (Beirut: Dar El Saqi, 2014).
Venuti, Lawrence. "Translation as a Social Practice: or the Violence of Translation," in *Translation Horizons: Beyond the Boundaries of Translation Spectrum,* ed. Marilyn Gaddis Rose (Binghamton: State University of New York, 1996).
Venuti, Lawrence. *The Translator's Invisibility: A History of Translation* (London: Routledge. 1995).
Vogiazou, Yanna. *Design for Emergence: Collaborative Social Play with Online and Location-Based Media* (Amsterdam: IOS Press, 2007).
Walsh, Michael. *Graffito* (Berkeley: North Atlantic Books, 1996).
Wardhaugh, Ronald. *An Introduction to Sociolinguistics*, 6th ed. (Chichester: Wiley-Blackwell, 2010).

Wedeen, Lisa. *Ambiguities of Domination: Politics, Rhetoric, and Symbols in Contemporary Syria*, 2nd ed. (Chicago: University of Chicago Press, 2015).
Wischenbart, Rüdiger, et al., *Diversity Report 2018: Trends in Literary Translation in Europe* (Vienna: CulturalTransfers.org, 2019).
Yazbek, Samar. *Taqāṭu' al-Nīrān: Min Yawmiyyāt al-Intifāḍa al-Sūriyya* [Crossfire: The Diary of the Syrian Intifada] (Beirut: Dār al-Ādāb, 2012).
Yazbek, Samar. *A Woman in the Crossfire: Diaries of the Syrian Revolution*, trans. Max Weiss (London: Haus Publishing 2012).
Yule, George. *The Study of Language*, 6th ed. (Cambridge: Cambridge University Press, 2016).

List of Figures and Tables

Figures

https://doi.org/10.1515/9783110767698-015

Tables

Index

https://doi.org/10.1515/9783110767698-016

www.ingramcontent.com/pod-product-compliance
Lightning Source LLC
Chambersburg PA
CBHW070923300326
R18048100001BA/R180481PG41927CBX00009BA/1
* 9 7 8 3 1 1 2 2 1 4 0 2 2 *